"The authors have written a fascinating and necessary book, tackling the urgency of defining critical steps and actions needed to protect our nature and ecosystems. They do so by proposing practical nature-based solutions and inviting a sense of collective urgency. This is a book that every nature lover and anyone committed to a sustainable future should read."

Iván Duque Márquez, Former President of Colombia

"Never before have the consequences of nature's decline been more obvious and the need for change more urgent; never again will there be a better time to protect and restore what remains of the living systems that make our existence possible. Bravo to the authors for clearly articulating in *Becoming Nature Positive* how everyone, everywhere, can be a part of the transition from steadily losing the natural systems that underpin our existence, and tip toward treating nature as if our lives depend on it – because they do."

Sylvia Earle, Oceanographer and Founder of Mission Blue

W0018557

Becoming Nature Positive

As humanity sits at an existential crossroads, this book introduces the need to build a nature-positive future to secure the functioning and stability of Earth systems essential to the survival and wellbeing of present and future human generations as well as the rest of Earth's amazing diversity of life.

Alongside the change in climate, a more silent but equally terrifying crisis is unfolding: the loss of nature and biodiversity. These twin crises are in fact interconnected. After decades of ignoring our impacts on the natural world, we are beginning to realise that nature conservation is a security issue for humanity, and an imperative for intersectional and intergenerational justice. For these reasons, we must embrace a transition from a nature-negative to a nature-positive society, one that ensures human development and addresses today's inequality, while conserving, restoring and sustainably benefiting from nature's resources and services. A Nature Positive future is one with more nature than today: more forests, more fish, more pollinators, more soil biodiversity, with benefits for the Planet and for us. In this book we define what becoming Nature Positive means from a variety of perspectives, what it takes to deliver it and why it is possible and, most importantly, necessary.

This book is essential reading for those concerned with conserving nature and securing a safe future for humanity in the face of climate breakdown, biodiversity loss, and ecological collapse. The future can be bright. The choice is ours.

Routledge Studies in Conservation and the Environment

This series includes a wide range of inter-disciplinary approaches to conservation and the environment, integrating perspectives from both social and natural sciences. Topics include, but are not limited to, development, environmental policy and politics, ecosystem change, natural resources (including land, water, oceans and forests), security, wildlife, protected areas, tourism, human-wildlife conflict, agriculture, economics, law and climate change.

Positive Psychology and Biodiversity Conservation
Health, Wellbeing and Pro-Environmental Action
Jolanta Burke, Sean Corrigan, Jimmy O'Keeffe, and Darren Clark

Conservation in the Anthropocene
Reshaping Interaction with Nature
Fred Van Dyke

The Maya Forest Waterlands
Shared Conservation, Entangled Politics, and Fluid Borders
Hanna Laako and Edith Kauffer

Saving Biodiversity
Threats, Strategies, and Big Ideas
Matt W. Hayward

Rewilding and Ecological Justice
The Ethics and Politics of Wildlife Regeneration
Cristian Moyano-Fernández

Becoming Nature Positive
Transitioning to a Safe and Just Future
Marco Lambertini, Joseph W. Bull, Leroy Little Bear, Harvey Locke, Eva Zabey, Dorothy Maseke, and Carlos Manuel Rodríguez.

Becoming Nature Positive
Transitioning to a Safe and Just Future

Written by

Marco Lambertini

Joseph W. Bull

Harvey Locke and Leroy Little Bear with Éliane Ubalijoro, Brigitte Baptiste and Fuwen Wei

Eva Zabey

Dorothy Maseke

Carlos Manuel Rodríguez and Sonja Sabita Teelucksingh

Edited by

Peter Vanham, Ross Chainey and Gemma Parkes

Cover design by

Lou Clements

 Routledge
Taylor & Francis Group

LONDON AND NEW YORK

 earthscan
from Routledge

Designed cover image: © Jasmina007/iStockphotoLP

First published 2025
by Routledge
4 Park Square, Milton Park, Abingdon, Oxon OX14 4RN

and by Routledge
605 Third Avenue, New York, NY 10158

Routledge is an imprint of the Taylor & Francis Group, an informa business

© 2025 Marco Lambertini, Joseph W. Bull, Leroy Little Bear, Harvey Locke, Eva Zabey, Dorothy Maseke and Carlos Manuel Rodríguez

The right of Marco Lambertini, Joseph W. Bull, Leroy Little Bear, Harvey Locke, Eva Zabey, Dorothy Maseke and Carlos Manuel Rodríguez to be identified as authors of this work has been asserted in accordance with sections 77 and 78 of the Copyright, Designs and Patents Act 1988.

British Library Cataloguing-in-Publication Data
A catalogue record for this book is available from the British Library

Library of Congress Cataloging-in-Publication Data
Names: Lambertini, M. (Marco), author.
Title: Becoming Nature Positive : transitioning to a safe and just future / Marco Lambertini [and 6 others]
Other titles: Routledge studies in conservation and the environment.
Description: New York, NY : Routledge, 2025. | Series: Routledge studies in conservation and the environment | Includes bibliographical references and index.
Identifiers: LCCN 2025001469 (print) | LCCN 2025001470 (ebook) | ISBN 9781032754543 (hardback) | ISBN 9781032754536 (paperback) | ISBN 9781003474043 (ebook)
Subjects: LCSH: Green movement. | Political ecology. | Biodiversity conservation. | Sustainable development.
Classification: LCC GE195.7 .L415 2025 (print) | LCC GE195.7 (ebook) | DDC 304.2/8—dc23/eng/20250225
LC record available at https://lccn.loc.gov/2025001469
LC ebook record available at https://lccn.loc.gov/2025001470

ISBN: 978-1-032-75454-3 (hbk)
ISBN: 978-1-032-75453-6 (pbk)
ISBN: 978-1-003-47404-3 (ebk)

DOI: 10.4324/9781003474043

Typeset in Times New Roman
by Apex CoVantage, LLC

Access the Support Material: https://www.naturepositive.org/

Contents

Foreword

by André Hoffmann

What is 'Nature Positive', and why does it matter so much that you should read a book about it?

'Nature Positive' is a recently agreed upon global goal to "halt and reverse nature loss by 2030 on a 2020 baseline and achieve full recovery by 2050" [1]. The goal was first put forth by NGOs, scientists, sustainable business associations, and advocacy groups in early 2020, and was later adopted by governments, including those of the G7 [2]. Their idea was that having a clear and simple goal could galvanize the global community for nature in the same way that 'net zero' did for climate. Today, 'Nature Positive' is supported by several United Nations agencies [3–6], and its central goal – to halt and reverse nature loss by 2030 – is part of the Kunming-Montreal Global Biodiversity Framework, agreed upon by 196 UN member governments [7].

As one of the early supporters of the 'Nature Positive Initiative' (the group that put together this book), I hope that in the years ahead, 'becoming nature positive' will become a goal that all of us – individuals, societies, and companies, will fully make our own. Nature is the life-support system of Earth and, therefore, that of all of humanity and the living world. It is not (only) something 'nice to have' or to enjoy in one's spare time – though I do recommend that. Taking care of nature means taking care of our global commons, of ourselves, and of future generations – humans, flora, and fauna alike.

Like so many other people, I started caring about nature at an early age. I spent most of my childhood in the Camargue in Southern France, where my father, an ornithologist, had moved our family to be close to the migratory birds he loved. In the Rhone delta, I saw flamingoes and other migrant birds use the shallow salty water as a yearly refuge. But at the same location, a salt mining company also extracted salt and sold it to customers all over the world. This business process required careful management of water and salinity levels, thereby creating optimal conditions for the birds to nest and reproduce. So nature and humanity were working hand in hand! It made me realize both the immense value of nature and its enormous fragility.

Over the past few decades, finding a balance between human development and maintaining our life-support system has become more difficult. As the global human population exploded from 3 to 8 billion, societies sought to improve their prosperity in any way they could, and businesses scrambled to deliver on the needs and dreams of their customers. In doing so, they often regarded nature as a seemingly endless source of resources to exploit, and they barely accounted for the true cost of using natural resources, nor did they make many efforts to restore what they destroyed.

Sadly, the way we went about counterbalancing this encroachment hasn't been effective either. Nature conservation groups have often focused single-mindedly on trying to defend, protect, and preserve nature in the face of such destructive over-exploitation. But that also prevented them from putting forth a viable alternative model for human development. The result has been a rapid deterioration of our global commons and biodiversity.

In a way, I observed this evolution from both sides. During the past 40 years, I have been both an environmentalist and active in business. As Vice Chairman of the World Wide Fund for Nature, known as WWF (1998–2017), and steward of various nature conservation initiatives (such as the Fondation Tour du Valat), I supported many efforts to protect nature from encroaching human activities. As Vice Chairman of Roche (1996–present), meanwhile, I saw how human ingenuity, combined with the use of ingredients that nature provides, can save millions of lives. But I also had to reckon with the fact that even the best-intentioned company can struggle to make its operations truly sustainable.

Over the years, I realized that the solution to our collective societal schizophrenia lies neither in trying to eliminate or limit the role of business nor in prioritizing nature conservation over all other societal goals. The solution lies in integrating them. That, to me, is what Nature Positive is truly about: finding a balanced and harmonious way to interact with nature, where humans can continue to prosper, but where they do so while allowing nature to regenerate and recover.

Today, we have a chance to strive for this outcome together. The overarching goal of becoming Nature Positive is one that, I hope, everyone can get behind. But if we want to achieve this goal, and regenerate nature, then every stakeholder in society – including business, finance, government, society, and academia – also needs to know *what* they can do about it and *how* they can integrate this goal into their own activities.

That is what this book is about. In the first part, Marco Lambertini, the convener of the Nature Positive Initiative and former director general of the WWF, explains how Nature Positive came about and sets the context for why it is an existential opportunity for humanity and which levers and alliances exist to make it a reality. In the second part, experts from a variety of backgrounds lay out how science, civil society, business, finance, and governments can play their part in achieving this common goal. Throughout, more than two dozen contributors offer their perspectives and show how the organizations they represent are doing their part in the nature-positive journey.

This book is not the be-all and end-all of the quest to become Nature Positive. In fact, at the time of writing, developments to identify nature-positive metrics are in full swing, and they will be a crucial building block in achieving our goal. For our economy and society to become Nature Positive, it is important that we know how to value nature, how to account for it, and how to make the trade-offs that are necessary to achieve both human and sustainable development. That is an exercise which will become fully actionable only after this book is published.

But in the interim, this book does provide a panoramic view of what Nature Positive is, how it came about, and how a broad coalition of stakeholders is going about achieving it. I hope you find this an inspiring read and invite you to join us in our quest to make Nature Positive a reality.

André Hoffmann
Vice-Chairman, Roche; Member, Club of Rome;
and former Vice President, WWF International
Switzerland, 1 December 2024

References

1 Nature Positive Initiative, 2024. The Definition of Nature Positive. Available at: www.naturepositive.org/app/uploads/2024/02/The-Definition-of-Nature-Positive.pdf
2 UK Government, 2021. G7 2030 Nature Compact, 12 July. Available at: www.gov.uk/government/publications/g7-2030-nature-compact/g7-2030-nature-compact
3 UNDP, SCBD and UNEP-WCMC, 2021. Creating a Nature-Positive Future: The Contribution of Protected Areas and Other Effective Area-Based Conservation Measures. UNDP: New York, NY. Available at: www.cbd.int/pa/doc/creating-a-nature-positive-future-executive-summary-en.pdf
4 UN Environment Programme, 2021., Nature-Positive Finance Guidance, July. Available at: www.unepfi.org/publications/new-nature-positive-finance-guidance/
5 UNDevelopmentProgramme,2021.CreatingaNaturePositiveFuture,11November.Available at:www.undp.org/publications/creating-nature-positive-future-contribution-protected-areas-and-other-effective-area-based-conservation-measures
6 UNICEF, 2023. Environmental Sustainability at UNICEF – Our Wwn Footprint. Available at: https://knowledge.unicef.org/CEED/environmental-sustainability-unicef
7 European Commission, 2022. Global Biodiversity: The EU is Committed to Protecting and Restoring Biodiversity. Available at: https://environment.ec.europa.eu/topics/nature-and-biodiversity/global-biodiversity_en

Becoming Nature Positive: a short introduction

by Marco Lambertini

As the title suggests, *Becoming Nature Positive* is a book about the imperative for our society and economy to transition from nature-negative to nature-positive. The Nature Positive goal is about halting and reversing global biodiversity loss by 2030 and achieving a full recovery by 2050. The book reflects the collaboration among members of the Nature Positive Initiative (NPI) [1], a coalition that drives alignment around the use of the term 'nature positive' and supports efforts to deliver nature-positive outcomes. In two parts, nine chapters, and more than two dozen contributions, the book shows how the goal of Nature Positive came about, why it matters, and what it will take to deliver it.

Part I of the book, written by myself, describes a series of macro trends that led to the decline of nature, and the shifts that are needed to embrace a nature-positive future. The various chapters of Part II are written by NPI members and other thought leaders, including Joseph Bull, Harvey Locke, Leroy Little Bear, Eva Zabey, Dorothy Maseke, Carlos Manuel Rodriguez, and Sonja Sabita Teelucksingh. They zoom in on the role that science, society, business, finance, government, multilateral organizations, and NGOs can play in making the nature-positive transition happen. Throughout the book, more than two dozen more contributors also provide their perspectives on various dimensions of Nature Positive.

Below is a shorter description of each chapter.

Part I: Nature Positive: the journey so far

In Chapter 1, 'The genesis of Nature Positive', I set the context and explain how the Nature Positive goal was developed and the collaborative effort that led to the definition of the goal in 2020 and its adoption by the world in 2022.

In Chapter 2, The "Great Rise" (*of humanity*) and the "Great Decline" (*of nature*)', I describe the great acceleration of humanity's impact on nature, and the uncomfortable distant origins as well as the recent drivers of our unsustainable and exploitative relationship with nature. I also discuss how the deteriorating state of the natural world is beginning to represent 'The Great Threat' to humanity's present and future, including the unprecedented, terrifying danger of Earth system tipping points.

In Chapter 3, 'The "Great Awakening"', I discuss what makes our generation the most environmentally aware ever, and examine how the 're-discovery' of our existential dependencies on nature for climate, food, and water security and also health and wellbeing represents the basis and the motivation that could trigger and inform a nature-positive transition.

In Chapter 4 I lay out the direction and the characteristics of 'The Great Transition' that is needed to get us back within safe and just planetary boundaries. It includes a detailed definition of what Nature Positive means, what the goal involves, and how it could be measured; describes the role business and finance can play in the transition to a nature-positive economy; and explores how society can overcome the powerful obstacles to embrace a nature-positive transition and unlock the positive socio-economic tipping points that will lead to nature-positive outcomes. I close this chapter by sharing my reflections on what it takes for society – and for us all as individuals – to build a nature-positive future.

Part II: Transitioning to a safe and just future

In Chapter 5, 'The science of Nature Positive', Joseph Bull summarizes the sources of scientific literature specific to the discussion on Nature Positive, from concepts such as the tragedy of the commons, to the role of biodiversity in supporting life on Earth, to the planetary boundaries framework.

In Chapter 6, 'Building a Nature-Positive society', Harvey Locke and Leroy Little Bear with Eliane Ubalijoro, Brigitte Baptiste, and Fuwen Wei describe the various ways in which societies are trying to live in harmony with nature, taking the reader on a journey from Canada to China, and from Bhutan to South Africa.

In Chapter 7, 'The role of business in a Nature-Positive economy', Eva Zabey looks at the journey business has been on: from being a net-negative force for nature to pioneers undertaking efforts to turn the tide, and supporting frameworks, targets, and disclosures to becoming nature-positive going forward.

In Chapter 8, 'Nature-Positive finance', Dorothy Maseke looks at the critical role that finance can and should play in going from a nature-negative to a nature-positive economy. She also provides an overview of the current gaps and provides descriptions of solution mechanisms such as biodiversity credit markets, payments for ecosystem services, and many others.

In Chapter 9, 'Nature-Positive governance', Carlos Manuel Rodríguez and Sonja Sabita Teelucksingh discuss the role of government, regulation, and international organizations in supporting a nature-positive transition, and the crucial requirement to have domestic, international, and systemic policy coherence.

Marco Lambertini
Convener, Nature Positive Initiative

Reference

1 Nature Positive Initiative, n.d., About Us. Available at: www.naturepositive.org/about/ (Accessed: 2 December 2024).

About the authors, contributors, and editors

Authors

Marco Lambertini is Convener of the Nature Positive Initiative, and former Director General of WWF and CEO of BirdLife International.

Joseph "Joe" W. Bull is Associate Professor in the Department of Biology at the University of Oxford, UK, and Director of Wild Business Ltd.

Harvey Locke is Co-founder of the Yellowstone to Yukon Conservation Initiative and Nature Needs Half movement, and IUCN World Commission on Protected Areas Vice Chair for Nature Positive.

Leroy Little Bear is Member of the Blood Tribe/Kainai Nation of the Blackfoot Confederacy, scholar and Vice-Provost at the University of Lethbridge, and Officer of the Order of Canada.

Éliane Ubalijoro is CEO of the Center for International Forestry Research and World Agroforestry (CIFOR-ICRAF) and Director General of ICRAF.

Brigitte Baptiste is Chancellor of Universidad Ean, Colombia, and former Director of the Alexander Von Humboldt Biological Resources Research Institute.

Fuwen Wei is President of Jiangxi Agricultural University in Jiangxi Province, China, an Academician of the Chinese Academy of Sciences (CAS), and a Fellow of The World Academy of Sciences (TWAS).

Eva Zabey is CEO of Business for Nature, and former Director of Redefining Value and Natural Capital at the World Business Council for Sustainable Development (WBCSD).

Dorothy Maseke is Head of Secretariat at the African Natural Capital Alliance (ANCA), and Africa Lead, Nature Finance and TNFD at Financial Sector Deepening (FSD) Africa.

Carlos Manuel Rodríguez is CEO and Chair of the Global Environment Facility, and former Costa Rican Environment and Energy Minister.

Sonja Sabita Teelucksingh is the Advisor to the CEO of the Global Environment Facility (GEF), and Lead of its Data and Analytics Team.

Contributors

The following contributors provided a 'perspective' on 'Becoming Nature Positive':

Chapter 5: The science of Nature Positive

Perspective 5.1: Nature Positive and Planetary Boundaries
Johan Rockström, Director of the Potsdam Institute for Climate Impact Research

Perspective 5.2: The Ecosystems We Cannot Afford to Lose
M. Sanjayan, CEO, Conservation International

Perspective 5.3: The Role of Earth Observations in Supporting Nature-Positive Outcomes: The Global Ecosystems Atlas
Yana Gevorgyan, Director of Secretariat, Group on Earth Observations

Chapter 6: Building a Nature-Positive society

Perspective 6.1: The Right to a Healthy Environment
David Boyd, Associate Professor, University of British Columbia, and former UN Special Rapporteur on human rights and environment.

Perspective 6.2: A 'Nature-Positive' World Is Possible
Gavin Edwards. Executive Director, Nature Positive Initiative Secretariat

Chapter 7: The role of business in a Nature-Positive economy

Perspective 7.1: The Nature-Positive Journey of the WBCSD and its Member
  Companies
Peter Bakker, CEO, World Business Council for Sustainable Development (WBCSD)

Perspective 7.2: The Natural Capital Protocol and the All-Important 'Plumbing' of Capitals Accounting
Mark Gough, CEO, Capitals Coalition

Perspective 7.3: Pushing Nature up the Mainstream Economic Agenda
Akanksha Khatri, Head, Nature and Biodiversity, World Economic Forum

Perspective 7.4: The Corporate Goalposts for Nature Action
Erin Billman, Executive Director, Science Based Targets Network (SBTN)

Perspective 9.4: Translating Global Nature Commitments into Transformative National Action
Kirsten Schuijt, Director General, WWF International

The Nature Positive Initiative wishes to thank all these contributors for their valuable input.

Acknowledgements

Writing, editing, and producing this book has been a massive collective effort, largely undertaken in 12 months from December 2023 to December 2024.

The book's authors are credited under their respective chapters. The book's commissioning editor, Marco Lambertini, is also the author of Part I of the book. Peter Vanham acted as development editor, and Ross Chainey and Gemma Parkes as copy editors.

The authors and editors wish to thank Gavin Edwards, Executive Director of the Nature Positive Initiative, Lou Clements, book cover designer, and the entire Nature Positive Initiative staff for their support in the process of writing and editing this book.

A note from the Nature Positive Initiative

This book is supported by the Nature Positive Initiative (NPI) as a tool to divulge the Nature Positive goal, explain the social and business case for it, and promote a discussion about the challenges and opportunities around its implementation.

While all the references to the Nature Positive goal definition are consistent with the official NPI definition, due to the wide range of themes covered, the views presented by the authors do not necessarily represent the official position of the Nature Positive Initiative as a collective.

Part I

The journey so far

1 The genesis of Nature Positive

Marco Lambertini

When on 12 December 2015, the President of the UN Climate Change Conference of the Parties (UNFCCC COP15) brought down the gavel, it didn't only signify the approval of the historic Paris Agreement. It also more profoundly signalled that, against the odds and despite powerful pressures against climate ambition, the world was able to come together around addressing the existential challenge posed by climate change. One hundred and ninety-six member governments agreed to an ambitious global goal that was going to be very disruptive to the existing business-as-usual model. After many years of frustration, it felt like the wind was finally in the sail of the climate agenda and the world was about to embrace a new energy and development future.

The crux of what the parties agreed on that day was to limit global warming to well below 2°C, and to do so by balancing carbon sources and sinks [1, 2], later clarified as achieving 'net-zero' anthropogenic greenhouse gas (GHG) emissions by mid-century.

Very quickly, the world came to see this globally agreed goal as the reference point for the future climate agenda. In the months that followed, the 'nature community' interrogated itself on the meaning and the opportunities linked to the Paris Agreement. We all felt that what had just happened for climate could and should be replicated in the fight against the rapid loss of biodiversity, the other interconnected, equally existential, but more silent crisis of our times. Paris gave many of us in the nature community the impetus and the confidence that we could – finally – agree on a transformative global goal akin to 'carbon neutrality' for nature. We all knew that the nature crisis required an unprecedented higher ambition, and that it too needed to be in the form of a global agreement that would set a common ambition and direction on nature conservation and restoration.

How could we push nature higher up the global political, corporate, and economic agenda, and bring the world to agree on a 'Paris-style' accord for nature? What should such an agreement contain, so that it would give the world a 'north star' or 'southern cross' for nature? And what should the specific global goal for nature be, so that it could inspire a common plan to address the accelerating destruction of nature worldwide?

DOI: 10.4324/9781003474043-2

In 2018, while I was Director General of WWF International, I launched the idea of a 'New Deal for Nature and People' to reverse the decline of biodiversity and to protect and restore nature by 2030 for the benefit of both people and the planet [3, 4].

The idea of a New Deal for Nature was resonating throughout the academic and political world and similar ideas were already independently emerging. A group of scientists called for a 'Global Deal for Nature' to protect half of the terrestrial realm [5]. In June 2019, the European Union started working on its ambitious and comprehensive 'Green Deal' package [6]. In the United States, progressive law-makers also proposed the concept of a 'Green New Deal' [7]. In China, President Xi Jinping articulated the need for an '*ecological civilisation*' from a holistic and strategic perspective, integrating ecological wisdom in the development of human civilisation under the slogan of 'Green is Gold' to symbolise the green transition to climate and also biodiversity [8].

At the 2018 UN Convention on Biological Diversity (CBD) meeting in Egypt (COP14), the aspiration and the need for a New Deal for Nature was often mentioned. Nation parties of the convention agreed to develop a post-2020 global biodiversity framework "to safeguard nature and biodiversity for decades to come." They anticipated it would be agreed upon at the next Conference of Parties, COP15, which was scheduled to take place in China in 2020. There was a distinct feeling that after and in conjunction with the Paris Agreement, nature had to be – and was going to be – the next global focus.

To make COP15 a success, those of us advocating for the New Deal still needed to answer the last open question: what exactly should the global goal for nature at the centre of the New Deal be that we should rally behind and that could galvanise change in the same way that 'net zero' had done for climate?

On a personal level, a coffee break conversation at a World Business Council for Sustainable Development (WBCSD) meeting in Morges, Switzerland, in the spring of 2018 gave me a first inkling of what the global goal for nature could look like. A corporate executive present at the meeting challenged me on what I thought the elusive 'apex goal*'* mentioned in my speech should be. It had to be as simple and communicable, but also as transformative as 'carbon-neutral', we agreed. 'Carbon neutral' and 'net zero' were the two buzzwords on everyone's mind on climate, but it surely could not be 'nature neutral'. It had to be *more* than that. So much nature had been lost, and its outlook was ever worsening. So while halting the *loss* of nature would be key to preserve the stability of Earth's systems, a *recovery* element was also necessary, considering that living nature, contrary to climate, can recover in spectacular fashion and do it much faster, if given a chance.

Nature has displayed spectacular comebacks, including of forests, wetlands, rivers, marine ecosystems and wildlife in many localities around the world. Now the ambition had to be to drive nature's recovery at the global level. But how should a new global goal for nature be defined?

Meanwhile, signs of a transition towards a low-carbon economy were starting to emerge in many parts of the world. Renewable energy sources such as solar and wind were becoming ever more ubiquitous and affordable. Energy efficiency became a primary focus area in the built environment, and electric mobility solutions started to appear everywhere. Those developments were inspired and driven by progress in science and innovation. But their most important catalysts were the globally agreed goal of becoming a carbon-neutral society, and the clear and measurable pathway of getting to net-zero emissions by 2050 to deliver the overall ambition of limiting global warming to around 1.5°C. Even though the global transition to a carbon-neutral economy did not yet proceed as fast as needed (and it is still the case today), it was clear that without the agreed climate Global Goal, and the metrics to measure against it, progress surely would have been slower or non-existent. It hardened our belief that we urgently needed to identify a measurable new *global goal for nature*.

Another sign, if needed, was that compared to the growing awareness and the multiplying commitments on the net-zero emissions target, the nature agenda was lagging far behind. Ten years earlier, the United Nations had agreed to make the 2010s the 'Decade on Biodiversity', and to pursue 20 so-called 'Aichi Biodiversity Targets' adopted at the Convention on Biological Diversity COP10, named after the location in Japan where the agreement was reached. But as the end of the decade approached, it became apparent that not even one of the targets was going to be fully achieved [9]. Much of the world remained unaware of what they were and that they even existed. The need to accelerate action on the nature agenda, however, was obvious. A concerted effort to move nature higher up the societal, political and corporate agendas was urgently needed.

As the business and finance sectors bear the responsibility of most of humanity's impact on nature, we decided to bring our nature challenge straight into the 'lion's den', the World Economic Forum Annual Meeting in Davos, scheduled for January 2019. This was where the most influential economic and financial elite would gather. This was where most of the impacts on nature came from, and the main issues regarding the economy of the future were discussed. And this was where nature had been mostly absent from mainstream conversation. We were determined to change that, and we quickly found in the World Economic Forum (WEF) itself a crucial partner.

The Davos gathering turned out to be a key milestone for the new nature agenda. We managed to place nature at the heart of the meeting. Nature images were projected across the main hallway for every delegate and leader making their way to the main hall to see. Influential business and political leaders present at the meeting, such as Roche Vice-Chair André Hoffmann, then Unilever CEO Paul Polman and former Vice-President of the United States Al Gore, backed the idea for a new and transformative global agreement on nature at the first New Deal for Nature dinner, co-organised by the World Economic Forum and the World Wide Fund for Nature (WWF). And, amazingly, Sir David Attenborough brought nature to a Davos plenary session – the first time that had ever happened. The iconic naturalist and documentary maker was interviewed by the United Kingdom's Prince William on stage, followed by a screening from his new series, *Our Planet*, a collaboration between Netflix and WWF [10]. The unprecedented, shocking images of the tragedy of wildlife decline and ecosystem collapse brought the allegedly hard-nosed Forum business audience literally to tears.

Davos 2019 proved to be a pivotal moment. The ultimate proof of this was that biodiversity loss, for the first time ever, made it into the top 10 risks in the Forum's *Global Risks Report 2020*, and it has remained there ever since [11, 12]. The chance to achieve a transformative moment for nature became more real because it now had the attention of its nemesis: big business.

Amid these exciting developments, however, the crucial 'coffee break' question of the 2018 WBCSD meeting was still unanswered. What should the overall science-based goal for nature look like? Two elements were increasingly becoming clear to us. First, we needed to focus on *halting the decline of nature* by protecting more of the intact nature left on the planet and *restoring* what we had degraded. And second, we needed to also see *transformation* in our production and consumption model, the source of human impacts on nature. We needed a global goal that was able to capture all these dimensions, embrace ambition and, most importantly, drive transformative action at scale.

As we feverishly searched for our all-encompassing global goal for nature, our gaze was quickly turning to 2020, the year that was billed as the upcoming 'Super Year for Nature' [13]. Several major nature-related global events were scheduled to take place in 2020, culminating with COP15, due to take place in October in Kunming, China. It was there that the global nature and sustainability community had to come together, capitalise on the extraordinary progress made in elevating the importance of addressing the nature crisis and aim for a higher global ambition. But with a clear articulation of a global goal for nature still lacking with one year to go, time was of the essence.

In December 2019, I decided to candidly write to my leadership peers at global environmental and business sustainability organisations, convening a meeting to explore how we could align on a definition for a global goal for nature and come together around a joint advocacy plan to get it adopted. The replies were instant and in unison: "We're on, let's talk."

We established the 'Global Goal for Nature Group'[14][1] and jumped on a call just before Christmas. We quickly homed in on the contours of the global goal for nature. It needed to be simultaneously *science-based* and something that *inspired action* on a whole new level. And, most importantly, it had to be *measurable*, to increase accountability in delivering it. Two options emerged for the global goal: it could either be to achieve *net-zero loss* of nature, or a *net-positive gain*.

The latter ultimately won everyone's backing. We needed to embrace what was scientifically necessary to avoid ecosystem collapse, we agreed, not simply what was politically palatable. In that regard, a net-positive global goal was the only viable option: a goal that led to the halting and the reversing of nature loss. '*Nature Positive*' was then coined as the less technical, more communicable and inspiring

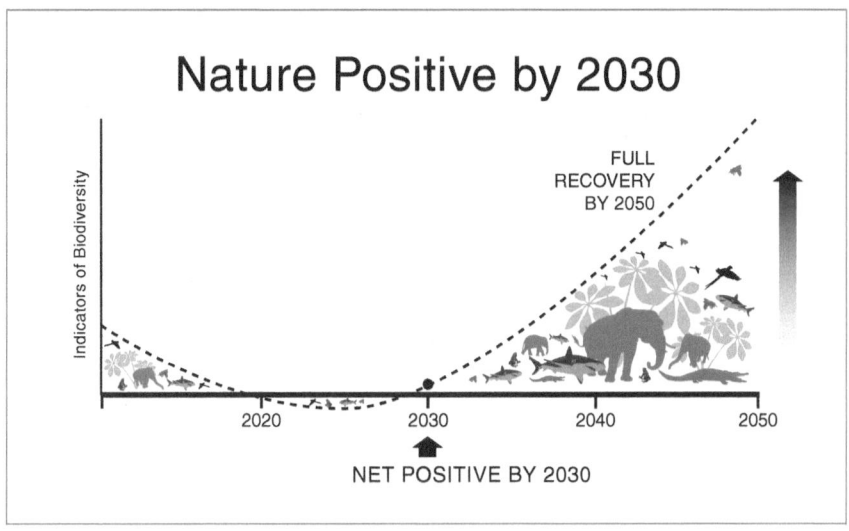

Figure 1.1 Nature-positive graph. The Nature Positive Global Goal: Halt and reverse nature
loss by 2030 on a 2020 baseline and achieve full recovery by 2050.

Source: Nature Positive Initiative

version of net positive. Nature Positive, nature's response to and an ally of carbon
neutral.

At the end of our discussions, we agreed on the following definition of the
Nature Positive Global Goal: *"Halt and reverse nature loss by 2030 on a 2020
baseline, and achieve full recovery by 2050"* [15, 16].

The goal, as we defined it, had two main components: on the one hand, the
2030 deadline captured the scientific imperative and the ambition, while the 2020
baseline indicated the start of a journey, increased its realism and made it measurable. In simple terms, this meant that we needed to have *more* nature at the end of
the decade than we did at the start of it. The ambition to achieve 'full recovery by
2050' signalled that the journey of reversing nature loss would not stop at 2030 but
would have to continue in restoring the huge amount of nature lost, particularly
in the last 100 years. The definition of the Nature Positive Global Goal was first
publicly presented at the New Deal for Nature and People dinner at Davos 2020.

Next, having agreed on what the global goal for nature was, we needed to get it
adopted by an international policy instrument. We identified the unmissable opportunity to codify the Nature Positive definition in the forthcoming UN Convention
on Biological Diversity 10-year plan, which would be agreed in Kunming in October. But if we wanted to succeed in that objective, there was no time to lose: it was
already January, and the negotiations on the next ten-year plan were approaching
their second meeting. We had to hit the ground running, and we did. In the weeks
and months that followed, we embarked on an ambitious concerted advocacy and

communication effort to embed the global goal in the post-2020 Global Biodiversity Framework, more specifically in its mission, and to secure the targets needed to deliver it.

The stakes couldn't be higher. If we succeeded, it would turn the Global Biodiversity Framework's mission into a measurable global goal for nature, defining the necessary level of global ambition for nature, as Paris did for climate. Only in this way would we be able to mobilise and align governments, businesses, financial institutions and consumers towards contributing to the global goal, inspiring a whole society approach and at the same time injecting the necessary degree of accountability. Just as the global goal of achieving a carbon-neutral society and economy by 2050 was disrupting the energy sector, propelling investments away from fossil fuels towards renewables, the Nature Positive Global Goal was designed to disrupt the industries that are drivers of nature loss, with agriculture, fishing, forestry, infrastructure and extractives – the so-called 'big-five' – top of the list accompanied by energy, manufacturing, tourism and others.

In those early months of 2020, as we approached the final stretch of preparations for the Global Biodiversity Framework, the feeling that we were on the cusp of a major transformational agreement understandably kept all of us highly focused and, more importantly, kept us together. Brands and egos set aside, we all focused on the prize ahead, a Global Biodiversity Framework and a Nature Positive Global Goal delivered by a set of ambitious and measurable targets. For all of us, being part of this effort was incredibly exciting and motivating, aware that only by building a broad and diverse coalition of governments, business, finance and civil society would we stand a chance of succeeding. It has been one of the most fulfilling, exhausting and motivating times in my whole nature conservation journey.

But then, of course, the pandemic happened. During the second UN CBD's negotiating meeting in Rome in February 2020, I vividly remember the morning when we were all greeted by the strange sight of Italian police personnel, wearing masks and standing at the entrance with gun-shaped thermometers, ready to take everyone's temperature. By early March, much of the world was in lockdown.

It soon became clear that the COP15 meeting on biodiversity would not take place as planned, at least not in person. The final negotiating meeting, where the Global Biodiversity Framework would be adopted, was first postponed to 2021, then to December 2022, and the venue was moved from Kunming, China, to Montreal, Canada. In the interim, a COP15 virtual meeting was organised in 2021 in Kunming to keep up momentum and to signal political commitment [17].

Meanwhile, the coalition around a Nature Positive Global Goal was growing much bigger and more diverse. Several leading science-based environmental organisations who were part of the Global Commons Alliance got behind the creation of the Science Based Targets Network initiative, aiming to define targets for the private sector that would contribute to the Nature Positive Global Goal [18]. This was followed by the launch of another crucial initiative, the Task Force on Nature-related Disclosures (TFND) which aimed to develop recommendations and guidance to enable business and finance to assess, report and act on their nature-related dependencies, impacts, risks and opportunities [19].

On World Environment Day on 5 June 2021, 64 heads of state from all regions of the world and the president of the European Commission signed the Leaders' Pledge for Nature, calling for a reversal in nature loss by 2030 and a commitment to live in harmony with nature by 2050 [20]. On the same day, businesses that were part of the World Economic Forum's Nature Action Agenda, Business for Nature, WBCSD and other business sustainability platforms, as well as over 300 environmental, humanitarian, human rights, Indigenous, women and faith civil society organisations and the countries who were signatories of the Leaders' Pledge for Nature, united around the global campaign #TheRaceIsOn . . . to a nature-positive future. The campaign focused on driving transformative action to reverse biodiversity loss and secure a nature-positive world by 2030 [21]. A week later, the G7 leaders meeting in the UK issued the G7 2030 Nature Compact agreeing to "halt and reverse biodiversity loss by 2030", stating that "our world must not only become net-zero, but also nature-positive" [22]. By September 2022, two months before the much-anticipated and much-delayed COP15 meeting on biodiversity in Montreal, the number of heads of state that signed the Leaders Pledge for Nature had risen to 91.

When the COP15 meeting finally convened in Montreal in December 2022, the support for the global goal for nature was so large that it finally seemed possible. On 19 December 2022, at 3:34 am local time, after a frantic and dramatic final rush of negotiations to overcome the last attempts to water down ambition, seven years after Paris, the President of the COP hit the gavel once again. This time it was the Chinese presidency of COP15 that sanctioned the agreement of the Kunming-Montreal Global Biodiversity Framework (GBF), and its mission to "Halt and reverse biodiversity loss by 2030", a mission fully consistent with the definition of the Nature Positive Global Goal supported by four overarching long-term goals and 23 action-oriented targets. It was the 'Paris moment for nature', which so many had been waiting for and which was celebrated with tears of happiness and exhaustion. We had a global goal for nature, and the most ambitious, comprehensive and measurable biodiversity plan ever agreed internationally. Now the focus shifted to how we would deliver it!

As you read this book, the original global goal for nature coalition has grown larger, more diverse and has evolved into today's Nature Positive Initiative, bringing together a diverse group of leading global organisations from all corners of society to preserve the integrity of the Nature Positive goal and to align on guidance to support its implementation [23]. The global goal for nature was developed both to provide clarity on the need to halt and reverse nature loss as well as inspire an ambitious nature-positive transition.

Back in January 2020, young people were handing out free copies of *The Wall Street Journal* along the icy central promenade of the Alpine town of Davos during the World Economic Forum's Annual Meeting. The sponsored front covers chosen for the 'Davos editions' are always interesting, as they tend to capture the burning topics of the time. But on this occasion, the newspaper's front page, sponsored by a Japanese multinational, truly stood out to me. It was titled, "A carbon neutral world by 2050? It can happen. Here's how". It was an undeniable sign that climate change had made it to the top of the corporate agenda, a few years after the 2015 Paris Agreement. Nature was missing. I thought. Would we, one day, see similar exposure for the nature goal in a global business paper? That day happened almost five years later, in October 2024, coinciding with the opening of the UN CBD COP16 in Colombia. The *Financial Times* published a full-page article in its 'The Big Read' section with the title "The call of 'nature positive'", republished in the online version under the headline, "The new corporate green goal: being 'nature positive'" [24]. The nature-positive agenda had come a long way.

Carbon neutral and Nature Positive. Net-zero emissions by 2050 and net-positive biodiversity by 2030. These are the two interdependent and mutually supportive global goals for climate and nature. Together, they can guide us towards a safe and sustainable future for humanity and help us address the most pressing challenges of our time, as reflected in the UN's 17 Sustainable Development Goals (SDGs). For as exciting and inspiring they may be, goals are only as good as their delivery. This is the focus of the Nature Positive Initiative. Nature Positive is something exciting to aspire to. Despite the multifaceted crisis we face today, the future can still be a bright one. It is down to choices. This generation, not the next one, has the chance to make it happen, the chance of our lifetime.

But, what led us to become a nature-negative society, and how can we become a nature-positive one? This is what we'll dive into next.

Note

1 The Global Goal for Nature Group included 4 Sustainable Development, BirdLife International, Business for Nature, Capitals Coalition, Conservation International, The Global Environmental Facility, Imagine, Luc Hoffmann Institute, Potsdam Institute for Climate Impact Research, The Nature Conservancy, Wildlife Conservation Society, World Business Council for Sustainable Development, World Resource Institute, World Wide Fund for Nature, and Yellowstone to Yukon Conservation Initiative.

References

1 UNFCCC, 2015. The Paris Agreement. Available at: https://unfccc.int/process-and-meetings/the-paris-agreement
2 IPCC, 2018. 'Summary for Policymakers', In V. Masson-Delmotte, P. Zhai, H.-O. Pörtner, D. Roberts, J. Skea, P.R. Shukla, A. Pirani, W. Moufouma-Okia, C. Péan, R. Pidcock, S. Connors, J.B.R. Matthews, Y. Chen, X. Zhou, M.I. Gomis, E. Lonnoy, T. Maycock, M. Tignor and T. Waterfield (eds.), Global warming of 1.5°C: An IPCC Special Report on the Impacts of Global Warming of 1.5°C above Pre-Industrial Levels and Related Global Greenhouse Gas Emission Pathways, in the Context of Strengthening the Global Response to the Threat of Climate Change, Sustainable Development, and Efforts to Eradicate Poverty. Cambridge University Press, pp. 3–24. Available at: https://doi.org/10.1017/9781009157940.001
3 WWF, 2018. Annual Review, 2018. Available at: http://europe.nxtbook.com/nxteu/wwfintl/annualreview2018/index.php#/p/20
4 WWF, 2019. Annual Review, 2019. Available at: http://europe.nxtbook.com/nxteu/wwfintl/annualreview2019/index.php#/p/Cover
5 Dinerstein, E., Olson, D., Joshi, A., Vynne, C., Burgess, N.D., Wikramanayake, E., Hahn, N., Palminteri, S., Hedao, P., Noss, R., Hansen, M., Locke, H., Ellis, E.C., Jones, B.,

Barber, C.V., Hayes, R., Kormos, C., Martin, V., Crist, E. and Saleem, M., 2017. 'An ecoregion-based approach to protecting half the terrestrial realm', BioScience, 67(6), pp. 534–545. Available at: https://doi.org/10.1093/biosci/bix014

6 Council of the European Union, 2024. Timeline – European Green Deal and Fit for 55. Available at: www.consilium.europa.eu/en/policies/green-deal/timeline-european-green-deal-and-fit-for-55

7 U.S. Congress, 2019. H.Res.109 – Recognizing the Duty of the Federal Government to Create a Green New Deal. 7 February. Available at: www.congress.gov/bill/116th-congress/house-resolution/109/text

8 UNEP, 2016. Green Is Gold: The Strategy and Actions of China's Ecological Civilization. Available at: https://reliefweb.int/report/china/green-gold-strategy-and-actions-chinas-ecological-civilization

9 Secretariat of the Convention on Biological Diversity, 2020. Global Biodiversity Outlook 5. Available at: www.cbd.int/gbo/gbo5/publication/gbo-5-en.pdf

10 World Economic Forum, 2019. Davos 2019 – a Conversation with Sir David Attenborough and HRH the Duke of Cambridge. Available at: www.youtube.com/watch?v=oam6Ca-mkg

11 World Economic Forum, 2020. The Global Risks Report 2020. Available at: www.weforum.org/publications/reader-the-global-risks-report-2020/

12 World Economic Forum, 2024. The Global Risks Report 2024. Available at: www.weforum.org/publications/global-risks-report-2024/

13 President of the 74th Session of the UN General Assembly, 2020. World Environment Day. 5 June. Available at: www.un.org/pga/74/2020/06/05/world-environment-day/

14 Global Goal for Nature Group, 2020. 'A global goal for nature: Nature positive by 2030', Establishing a Goal for Nature Positive Societies. Available at: www.natureposi-tive.org/app/uploads/2024/02/Global-Goal-Nature-Positive-2030-v11092020.pdf

15 Locke, H., Rockström, J., Bakker, P., Bapna, M., Gough, M., Hilty, J., Lambertini, M., Morris, J., Rodriguez, C.M., Samper, C., Sanjayan, M., Zabey, E. and Zurita, P., 2021. A Nature Positive World: The Global Goal for Nature. Available at: www.nature.org/content/dam/tnc/nature/en/documents/NaturePositive_GlobalGoalCEO.pdf

16 Nature Positive Initiative, 2023. The definition of Nature Positive. Available at : www.naturepositive.org/app/uploads/2024/02/The-Definition-of-Nature-Positive.pdf

17 Convention on Biological Diversity, 2021. Kunming Declaration: Ecological Civilization: Building a Shared Future for All Life on Earth. 13 October. Available at: www.cbd.int/doc/c/c2db/972a/fb32e0a277bf1ccfff742be5/cop-15-05-add1-en.pdf

18 Science Based Targets Network, n.d. About. Available at: https://sciencebasedtargetsnetwork.org/about/

19 Taskforce on Nature-Related Financial Disclosure, 2021. Our History, June. Available at: https://tnfd.global/about/history/#:~:text=In%20July%202020%2C%20an%20initiative%20to%20bring%20together,of%20Co-Chairs%20David%20Craig%20and%20Elizabeth%20Maruma%20Mrema

20 Leaders' Pledge for Nature, 2020. Leaders' Pledge for Nature, 27 September. Available at: www.leaderspledgefornature.org/wp-content/uploads/2021/06/Leaders_Pledge_for_Nature_27.09.20-ENGLISH.pdf

21 Leaders' Pledge for Nature, n.d. #TheRaceIsOn: Communicators' Explainer. Available at: www.leaderspledgefornature.org/wp-content/uploads/2021/06/TheRaceIsOn_communicatorsexplainer.pdf

22 G7, 2021. 2030 Nature Compact. Available at: www.consilium.europa.eu/media/50363/g7-2030-nature-compact-pdf-120kb-4-pages-1.pdf

23 The Nature Positive Initiative, n.d. The Nature Positive Initiative. Available at: www.naturepositive.org/about/the-initiative

24 Savage, S., 2024. 'The new corporate green goal: Being "nature positive" FT Big Read', Financial Times, 21 October. Available at: www.ft.com/content/4d12f8d1-c0df-4ab6-b374-741e9517448

2 The 'Great Rise' (*of humanity*) and the 'Great Decline' (*of nature*)

Marco Lambertini

"The fact is that no species has ever had such wholesale control over everything on Earth, living or dead, as we now have. That lays upon us, whether we like it or not, an awesome responsibility." This quote from British naturalist and broadcaster David Attenborough [1] epitomises the unprecedented reality life on Earth is facing today, where one species has developed an unparalleled ability to alter entire Earth systems and affect the Planet's diversity of life. This is what has generated an extraordinary level of impact on the natural world in a surprisingly short period of time in both human and geological history terms. Welcome to the Anthropocene, 'the Age of Humans'.

I still remember the excitement I felt when my mother, Fosca, after picking me up from school at the end of the morning, would take me to the fish market in my hometown on the coast of Tuscany. I was totally fascinated by the diversity, colours and shapes of fish, crabs and shells on display, the only way to admire amazing life forms that were usually hidden beneath the surface of the sea. I imagined them crawling among rocks, swimming through the vast open blue or lurking in the ocean's dark depths. Each visit felt like exploring an oceanographic museum, filled with daily surprises. Often, my mother would humour my insistent requests to buy a live octopus or scorpion fish from the counter, which we would release back into the sea on our way home. I was in primary school then, and nearly all the fish were wild and locally caught, as intensive and large-scale fish aquaculture hadn't yet begun. Today, the same fish market is mostly stocked with farmed fish or imports from distant seas. Across the European Union, over 60% of fish are now imported [2].

I also have vivid memories of flying over Sumatra in Indonesia in the early 1980s on one of my first tropical field trips. Below me lay a vast, uninterrupted blanket of tropical forest. Now and again red lines would appear, like veins running through the green landscape. These were either new logging roads penetrating the heart of largely untouched forests, or old roads being re-conquered by the recovering jungle. The rusted colour was due to the tropical lateritic soil otherwise covered by the closed forest canopy, a sign of its intactness. However, things were about to change on an unimaginable scale. The days of selective logging were about to

DOI: 10.4324/9781003474043-3

give way to large-scale forest conversion schemes, for large and small plantations, mining, settlements: the great forest burning was about to begin.

Today, flying over the same region, you see a very different picture. Seeing a large enough tract of lowland forest, even if it is criss-crossed by a network of logging roads, is a reason for hope, as most of the lowland has been permanently converted to industrial as well as small-scale commercial and subsistence plantations. Today, less than 20% of Sumatra's original lowland forests remain [3].

These are just two of the many displays that I have directly experienced in my lifetime of the *'Great Decline'* of nature which has particularly accelerated across the world in the past few decades, highlighting the universal root cause of nature's degradation: the unsustainable exploitation of natural resources, which has become the foundation of our economic development.

As you may have guessed from the title and the opening paragraphs, this chapter is not going to be particularly cheerful. This is unusual for me, as I am known, and sometimes even criticised, for being a 'glass-half-full' kind of person. However, the reality is that the 'Great Decline' of the natural world at humanity's hand, axe, tractor, nets and bulldozer is so serious that it cannot be described as anything less than an unprecedented existential threat to the diversity of life as we know it, and to humanity itself.

But it is not all darkness. There are plenty examples of successful efforts to conserve and restore nature. However, the net result is undeniably a massively net-negative one. We cannot afford to lose sight of the gravity of the nature loss crisis and its profound consequences. This must remain top of mind in any conversation about human development and nature conservation. There is no room for complacency. It is this accelerating existential threat that forms the rationale for embracing a nature-positive future.

2.1   The 'Great Acceleration'

"Everything in excess is opposed to nature." This quote from ancient Greek physician and philosopher Hippocrates referred to the need to embrace balance and moderation and to the fact that exceeding the limits of what the body can cope with contradicts the order of nature and can produce highly detrimental effects. Today I'd like to re-interpret it in the light of our relationship with nature and as a warning about the consequences of our excessive and unsustainable use of natural resources. Something that has accelerated exponentially in a very short period of time and with such devastating impacts on the natural world. And something that we will need to course correct with equal speed and urgency.

We are living in extraordinary times. Times of multiple crises and paradoxes. And times of consequences. Every generation thinks its time is somehow special. For my grandmother Ester, it was living through the horrors of two World Wars. But I would argue that our time is more profoundly unique than any other in the history of our civilisation, even of our species. The uniqueness of our time is reflected in something that is so deep and foundational, yet has been largely ignored, until now.

Something that has never happened before in human history, let alone the four billion years of the history of life on Earth. Something that has happened so recently and so quickly that it has passed unnoticed. And something that has to do with the most existential relationship of all, our relationship with our 'mother', our home, our lifeline. Nature.

Forget space exploration, the internet or artificial intelligence. When taking a planetary perspective, what's fundamentally unique to our time is something much darker. It is the way the natural world succumbs to the impacts of our actions, our unprecedented ability to alter systems on Earth that have evolved over billions of years and the disastrous consequences we are beginning to see around the world. Humans interact with the natural world in different ways and in response to different spatial and temporal contexts. Today, the extent and depth of our footprint on the planet threatens its stability and our own future – this is what is truly and sadly extraordinary about 'our time'.

This is not 'green alarmism' or 'doom and gloomism'. The planetary crisis cannot be exaggerated. Today the science has never been clearer and more precise in assessing the depth and extent of humanity's impact on the natural world, the effects of which are shocking both in terms of speed and scale.

Life on Earth has endured difficult times before, for example five major 'mass extinctions', which is defined as a period when more than 75% of animal and plant species become extinct in a relatively short period of time (geologically speaking, typically around less than 2 million years) [4]. The first of these events happened 440 million years ago, the heaviest 250 million years ago (when over 90% of all species were wiped out) and the most recent – and famous – 65 million years ago, when all non-avian dinosaurs went bust. Has Earth's sixth mass extinction already arrived?

In all cases, these mass extinctions were caused by dramatic changes in climate and ocean chemistry, driven by catastrophic volcanic activities, glaciations or meteor strikes. As a result, scientists estimate that more than 99% of all species that have ever lived on Earth have gone extinct, and less than 1% are alive today [5].

Nevertheless, today's biosphere hosts an amazing diversity of life, estimated at between five to ten million different species, most of which have not yet been identified by science. In her book *The Sixth Extinction: an Unnatural History* [6], Elizabeth Kolbert suggests that we are entering a new mass extinction event, driven by human activity.

Kolbert is not the first nor the only expert to reach this conclusion. Whether we have already entered the sixth mass extinction, or are about to, the evidence is clear: extinctions are already happening 100 to 1,000 times faster than pre-human background rates [7]. The difference is that this time, there is no 'mega-volcano' or 'meteorite' – the cause is us.

Due to intensifying human impact, one million species are threatened with extinction by the end of the century, many destined to disappear much sooner [8]. These are not just lions, tigers and pandas. Half of these at-risk species are invertebrates, and it is possible that up to 10% of the estimated 5 million species of insect were already wiped out during the Industrial Revolution [9].

In his book *The Insect Will Avenge the Bird*, nineteenth-century Italian naturalist Giuseppe Sancasciani warned that the unbalance created by the hunting of birds would lead to an explosion of insect populations, greatly damaging agriculture [10]. What Dr Sancasciani could not predict was the advent of Nobel Prize–winning 'wonder chemical' dichlorodiphenyltrichloroethane, or DDT, the subsequent generations of pesticides and herbicides, as well as agricultural mechanisation and intensification, habitat conversion and light pollution which led to the '*insect Armageddon*' which has devastated populations of common species and in turn affected insectivorous bird populations.

In an unexpected turn of events, the bird was not avenged by an insect explosion, but instead further suffered due to the collapse of insect populations. The decline of insects also disrupts their vital role in maintaining ecological balance and performing pollination for two-thirds of human crops – contributing to over half a trillion dollars annually in food production [11, 12].

Species extinction is the ultimate result of our impact on the planet's species and ecosystems, driven by five key drivers of biodiversity loss: habitat loss due to conversion or degradation (such as deforestation, wetland reclamation, river damming, agriculture intensification); unsustainable exploitation (hunting, fishing, logging); climate change; pollution (chemicals, plastic, eutrophication); and invasive species (on land, fresh water and at sea) [13].

Population decline is the precursor to extinction. Though less recognised and dramatic than extinction, the global decline in species abundance is one of the most alarming consequences of human impact on the planet. Abundant species play essential ecological roles, and their decline significantly undermines these functions. Viable and functional populations that perform key ecological roles such as predation, pollination, seed dispersal, grazing, decomposition and filtration, or form the foundation of the trophic chain are critical for ecosystem stability and productivity, and ultimately for the functioning of Earth's systems.

A recent study found that nearly half of more than 70,000 animal species evaluated are in decline, while less than half are globally stable, and only 3% are increasing. Most population losses today are concentrated in the tropics, where biodiversity is higher and large-scale impacts are more recent [14].

The *Living Planet Index* [15] tracks the abundance of over 30,000 populations of 5,000 species of mammals, birds, fish, reptiles and amphibians around the world. Since the 1970s when the index started, the average change in population size of terrestrial, freshwater and marine populations of vertebrate species shows a decline of 73% between 1970 and 2020. Although some scientists question the accuracy of the index, the trends are in line with the declines of monitored wildlife populations around the world [16].

The relative abundance of these species is the direct, cumulative result of these major drivers of nature loss, and act as an early warning system for overall ecosystem health. Population decline reflects a growing risk of extinction, providing opportunities for preventative intervention. Population trends are also often highly responsive to both the negative and positive effects of human behaviour, making them an excellent indicator to use in implementing nature-positive strategies, although often limited by data availability and complex methodologies. Population decline, if left unchecked, will ultimately lead to extinction, and long before that, it will disrupt the functioning of ecosystems, including the services these ecosystems provide to humans.

Severe population decline can make a species 'functionally extinct' (i.e. unable to fulfil its ecological functions) far before it disappears from Earth. Population decline can also often go undetected. For example, a forest may still be there, but it could have been emptied of fauna due to poaching and overhunting, a phenomenon known as '*defaunation*', which ultimately affects the health and resilience of the forest itself. Many Southeast Asian and Central African forests are known to suffer from severe wildlife population decline driven by intense and non-selective hunting methods like trapping or snaring.

The International Union for Conservation of Nature's (IUCN) *Red List of Threatened Species* is the world's most comprehensive and up-to-date assessment of global species extinction risk, with more than 160,000 species assessed, of which over 45,000 (28%) are categorised as threatened [17]. Much like other indicators, the number of threatened species has increased four-fold in the last 20 years.

The 'Great Decline' of nature is driven by a multitude of impacts, all of which contribute to today's ecological crisis. We have significantly 'altered' 75% of Earth's land surface and 66% of its ocean. Over two-thirds of the planet's natural habitat is increasingly fragmented [18], and almost 10% of the world's estimated 5.9 million terrestrial species don't have sufficient habitat for long-term survival [19]. Meanwhile, nearly half of the world's natural forests have been destroyed, as well as half of all warm water coral reefs, and over 85% of wetlands [20]. In 2017, one-third (34%) of assessed global fish stocks were overfished, 60% were 'fully' fished and only 6% underfished, globally and in aggregate. Ninety percent of commercial fish stocks are either fully fished or overfished [21]. For one in 10 species classified as globally threatened, invasive alien species represent a threat [22], and the more we study this as a long-time underestimated driver of biodiversity loss, the more we discover new impacts.

One of the most astonishing figures which shows how we have altered the very composition of the biosphere is that today 96% of the mammalian biomass on land is made up of humans (34%) and livestock (62%). Only 4% is made up of all the other wild mammals. Poultry weighs more than twice all wild birds [23]. Through the overexploitation of 'living resources', the eradication of native unwanted species, the introduction of 'alien' species and through the conversion of natural habitats, we have been increasingly replacing the 'natural' with the 'unnatural', altering and destabilising the natural world. This is a different world, far removed from the delicate and sophisticated ecological balance of the '*living planet*' in which such an amazing diversity of life evolved. We have changed and are continuing to alter the makeup of the biosphere. We have created a '*new unnatural order*' that fulfils our specific needs while undermining the stability, resilience and functionality of Earth's systems. The consequences are there for all to see and are most likely underestimated due to the complexity of reinforcing feedback effects that could accelerate the destabilisation of Earth systems in unpredictable ways.

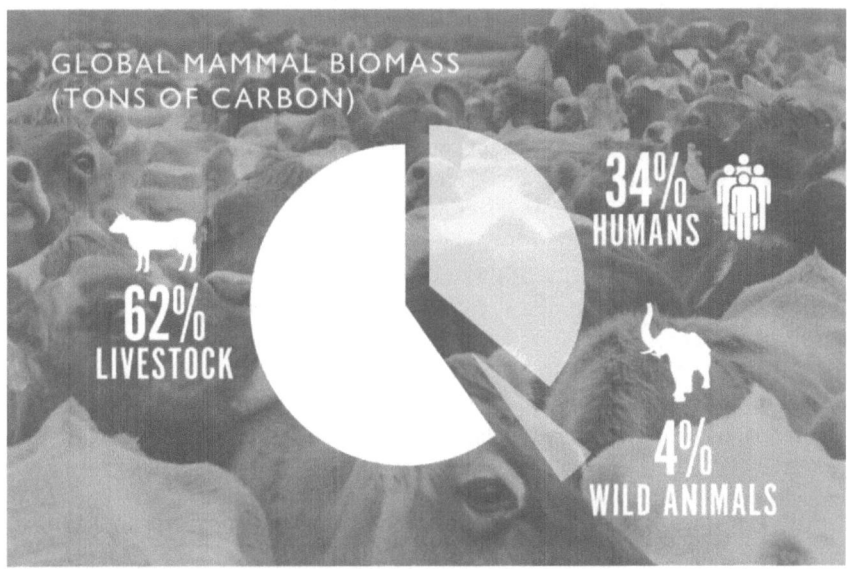

Figure 2.1 Global biomass. Today 96% of the mammalian biomass on land is made up of humans (34%) and livestock (62%). Only 4% is made up of all the other wild mammals.

Source: Adapted from Bar-On, Y.M., Phillips, R. and Milo, R., 2018 [24]

How did we get to this point? Let's go back to when this all began to fall out of sync with the natural world. Around 12,000 years ago, during the so-called Neolithic Revolution, human populations grew rapidly as agriculture began to provide relatively stable and predictable sources of food. It is estimated that only 2.5 million people lived on the entire planet at that time, a quarter of the population of London today. It took around 200–300,000 thousand years (when it is believed that modern humans, Homo sapiens, first evolved) for our species to reach its first billion people at the start of the nineteenth century.

It then took another century to double, and the following half century to triple. Around the time I was born, at the start of the 1960s, the human population stood at about three billion, and since then we have added roughly one billion people every decade. In my own lifetime the human population has almost tripled. Inevitably, alongside this sharp population growth, the overall consumption of natural resources has also been growing exponentially, propelled in more recent times by a new consumption-based economic model.

The spectacular scientific and technological advances of the last 100 years or so have also played a huge role in this demographic explosion, and the rapid rate of consumption. During this period, humanity has experienced some of the most sophisticated and revolutionary technological developments ever, equipping us with more powerful tools to exploit natural resources. In turn, this has allowed us to develop our economies and improve life expectancies and well-being around the world – albeit with huge disparities across geographies.

These three reinforcing elements – population and consumption growth and technological advances – created a perfect storm to ignite what scientists call the '*Great Acceleration*' [25–27]. Since the 1950s, shortly after the end of the Second World War, the consumption of natural resources and commodities – such as water, energy, fossil fuels, minerals, timber, fish, crops, animals, cement, plastics, fertilisers, pesticides and almost anything else you may think of – has increased exponentially. The similarity of these exponential growth curves is striking and marked the beginning of a new era of exponential human population growth and unsustainable consumption.

We have left behind the Holocene with its climatic stability and ecological abundance and entered an era of human domination and overexploitation, Earth systems disruption, nature loss, unpredictability and extremes. The 'Great Acceleration' is characterised by the unprecedented pace of human development and far-reaching impacts on the natural world.

This is the result of deep socio-economic changes which have occurred in the last 70 years. From 1950 to 2024, the global population has more than tripled from 2.5 to over 8 billion, with more than half now living in cities. Global GDP has grown from $9 trillion to over $110 trillion, despite over 700 million people still living in extreme poverty. Annual emissions from burning fossil fuels have increased from just over 10 billion tons to over 37 billion tons. Meanwhile, average CO_2 atmospheric concentrations rose from 350 parts per million (ppm) to over 420 ppm, the highest in 14 million years. Consequently, the ocean's acidity has increased by 30%, and the average global temperature has risen by almost 1.3 °C above the baseline, and counting. Notably, for more than half of 2024, average temperatures were above 1.5 °C, and even the annual average may have exceeded this level for the first time. Plastic production and consumption have grown from 2 million metric tons to over 400 metric tons and meat production four-fold, while the global average consumption of fish and other seafood per person per year reached a record of over 20 kg. This is just a snapshot of the 'Great Acceleration', which is still gaining pace.

Humanity's 'ecological footprint', defined as the demand on nature's resources versus Earth's capacity to provide and regenerate them, has been steadily growing globally and in most countries [28]. This ecological overuse generates an overshoot that depletes natural resources and weakens the planet's regenerative capacity. We cut down more trees than can regrow, catch fish faster than they can reproduce and lose more organic soil than can regenerate. Today, we would need 1.7 Earths to satisfy global consumption, meaning we are using nature 1.7 times faster than the planet's ecosystems can replenish. Some countries use resources so intensively that if everyone on Earth consumed at the same rate, we would need as many as five Earths to keep up. The annual Overshoot Day – the point when consumption surpasses regeneration – has shifted from December in 1971, when we were using one Earth equivalent, to August in 2024. One of the rare inflection points occurred during the COVID-19 pandemic when economic growth slowed and consumption decreased.

2.2 Ancient roots: the reckoning with our past

"I encourage all of us, whatever our beliefs, to question the basic narratives of our world, to connect past developments with present concerns, and not to be afraid of controversial issues." I would like to borrow Israeli historian Yuval Noah Harari's invitation at the start of his thought-provoking book Sapiens [29] to also open this subchapter.

The problems outlined above accelerated but did not start with the 'Great Acceleration' of the 1950s nor with the Neolithic Revolution. It is important to acknowledge the deep and distant origins of our unbalanced relationship with nature. If it is true that our impact on the natural world has increased significantly with the development of agriculture and grown exponentially in just a handful of recent decades, then it is also true that our competitive and exploitative relationship with nature began at the very start of human civilisation.

While it's lovely to embrace the romantic idea that in the old days our distant ancestors' relationship with nature was all about balance and harmony, regrettably this is misguided and misleading. The evidence from fossil records tells a very different story, a story of human impact on other species that dates back tens of thousands of years, far before the advent of agriculture, despite the common perception that human-driven extinction is a recent phenomenon. Humans' adversarial and exploitative relationship with nature and her resources started a long time ago – at the very origin of our species.

As soon as human cognitive ability advanced and we became more creative and organised hunters, we started having huge impacts on many species. Tens of thousands of years later, one of the most compelling indicators of today's ecological crisis is the modern, human-generated mass extinction of species. But megafauna (arguably one of the most valued natural resources for early humans) have suffered at the hands of a growing human presence on every continent since prehistoric times.

Fossils, cave paintings and archaeological remains show how ancient human migrations across the planet were accompanied by a wave of extinctions across all continents. Carbon-14 dating, also known as radiocarbon dating, has confirmed a high number of extinctions of terrestrial megafauna in the prehistoric age, specifically during the Late Pleistocene (129,000–11,700 years ago) and Early to Middle Holocene (11,700–4,200 years ago), indicating human hunting as the likely primary cause [30–32].

In the last 50,000 years, larger animals suffered a wave of extinctions during what is known as the *Quaternary Megafauna Extinction* (QME). This was not technically a mass extinction, as during the QME almost only large animals, from the size of a large dog to the size of a mammoth, disappeared, but it did nonetheless have a dramatic impact on many species. While climatic factors during the last Ice Age likely contributed to population declines among these species, the fact that extinctions predominantly affected large animals suggests a selective human impact through hunting. The QME was also a first signal of what would happen several thousand years later at a much greater scale and faster rate.

All this occurred despite the relatively small global human population, estimated at five million by the end of the QME, 8,000 years ago. By then, humans had already developed more efficient hunting tools, mastered the use of fire, and improved social hunting techniques. Fossil records reveal that the extinction of various species in different regions closely aligns with the timing of modern human arrivals on each continent. For instance, Homo sapiens reached Australia between 65,000 and 44,000 years ago, and by 50,000 to 40,000 years ago, nearly 90% of the megafauna had gone extinct. When humans entered new territories, local species were unprepared for the arrival of such an efficient predator. We see similar patterns today, with many small islands experiencing species extinctions due to the sudden introduction of exotic predators like rats, snakes and cats.

In a planet mainly void of people, an obvious solution to scarcity of resources, competition and defaunation of an area was to simply move to a new, uninhabited and unexploited one. Migrations from Africa to Europe and Asia to the Americas opened vast frontiers for our species.

Whenever humans arrived and settled in an area, the largest animals consistently and quickly disappeared. The expansion of modern humans in Eurasia led to the extinction of 35% of megafauna. Australia, North America and South America were particularly hard-hit, respectively losing, shortly after humans arrived, 88%, 83% and 72% of their megafauna.

Megafauna species were hunted to extinction either for their value as a food source, given prehistoric humans' reliance on game meat, or because they posed a threat and competed for resources.

This is not so different to how we exterminate predators today, like wolves, lions and tigers. Examples of ancient large species hunted to extinction include the European lion and woolly mammoth in Europe, the mastodon in North America, the giant sloth in South America, several species of giant kangaroo in Australia, the large game bird *Sylviornis* in New Caledonia and, more recently, the giant moa in New Zealand and the elephant bird (*Aepyornis*) in Madagascar [33]. Other late Pleistocene *Homo* species like Neanderthals and Denisovans became extinct partly through hybridisation, but most likely at the hands of *Homo sapiens*, an untold and disturbing chapter of the early history of our species.

Additional research explores the impact human arrival had on smaller species. The first humans to reach the Pacific Islands are believed to have caused the extinction of around 1,300 bird species, long before the second wave of extinctions brought about by European colonisers [34]. Only 11 out of 57 species of megaherbivores have survived to today, accounting for an extinction rate of over 80%. Such a decrease is unique throughout the last 30 million years, with high diversity of megafauna remaining stable during this time – except during the more recent millennia [35].

The QME tells us that humans' exploitative and unsustainable relationship with nature and her resources began a long time ago. The largest animals consistently disappeared shortly following human arrival and settlement. Our ancient impacts did not just affect fauna. It is estimated that half of global forest loss occurred from the advent of agriculture between 8000 BCE and 1900 and the other half in the last century [36].

Eventually, the world became pretty much 'fully occupied' by humans, and as they migrated to new territories in search of land and food, bloody conflicts between groups erupted. Transitioning from a migratory species in an empty world to a territorial one on an increasingly 'crowded' planet, resident hunter-gatherer societies quickly learned that unsustainable harvesting of animals or plants led to starvation, malnutrition and disease.

These groups soon learned the limits of what their territories could sustainably provide, developing rules for hunting, fishing and gathering based on a clear understanding of finite resources. This realisation marked a cultural evolution and adaptation to ensure survival in a world that had become 'smaller' with limited resources. By internalising our dependencies from nature, we also developed a deep relationship with nature. There is a Māori saying, "*Ko au Te Taiao, ko Te Taiao ko au*" – "*I am nature, and nature is me*". This holistic approach remains evident today in many Indigenous cultures around the world.

However, with the domestication of animals and plants and the rise of agriculture, resources that were naturally limited until then seemed to expand endlessly. Our 'dormient' competitive and exploitative approach to nature re-emerged more powerfully than ever. But this time it was not just large animals that would pay the price but entire ecosystems. The impact on the natural world grew broader and deeper through relentless clearing and conversion of forests and savannahs to cultivated crops or pastures for grazing.

Spears, arrows and fishing baskets were progressively replaced by guns, snares and large nets, giving any prey, on land or in the ocean, only a slim chance of escape. Axes and fires were replaced by handsaws and then chainsaws and powerful machinery, driving the ever-faster conversion of forests and other natural ecosystems. But it was shortly after the Second World War, when the combination of a fast-growing global population, a political focus on rapid economic growth supported by the need to rebuild post-war economies and the advancement of groundbreaking technologies, that the 'Great Acceleration' began. This started around 70 years ago, the blink of an eye in evolutionary terms, or the equivalent of half a second to midnight on 31 December if you compressed the history of the planet into 12 months.

During the time of the 'Great Acceleration' of human impacts on the natural world, conservation action did also increase considerably. Nearly 300,000 protected areas have been established, including a growing number of Other Effective Conservation Measures (OECMs), covering over 17% of land and inland waters, as well as a significant increase in marine protection over the last couple of decades which now cover over 8% of the ocean [37].

Although some have labelled these areas as 'paper parks' due to inadequate funding, management and governance, rendering them ineffective at conserving nature and supporting local communities, biodiversity is generally richer and more abundant in protected areas, and research from Shandong University in China and Duke University in the United States [38], which analysed 10,000 protected areas worldwide, found that in more than half, they have stimulated local socio-economic development [39]. There is plenty of room for improvement but also confirmation of the value of protected areas to nature and people.

Counterfactual scenarios in the absence of modern conservation action show that nature loss would have been much more severe. A recent meta-analysis of 186 studies found that in two-thirds of cases, the overall state of biodiversity improved, or the decline in biodiversity was slowed down. Conservation actions targeted at species and ecosystems proved particularly effective [40].

Conservation action prevented up to 32 bird and 16 mammal extinctions between 1993 and 2020 and up to 18 bird and seven mammal extinctions since 2010. Extinction rates would have been up to four times greater without conservation action [41]. Despite a generalised and overall sharp decline in biodiversity, the last few decades have also seen many successful stories of local ecosystem and species recovery. Think of the wetland and waterfowl population recovery in North America; the rebound of almost all the world's whale populations after the end of industrial commercial whaling; the many examples of forest, mangrove, wetland, river, seagrass and coral restoration projects; and the spectacular comeback of predators like the grey wolf in Europe, the most densely populated and industrialised of all continents.

We have seen so many inspiring local examples of successful nature conservation and restoration which indicate that when passionate leadership and inspiring vision bring together science, communities, governments and even companies, we can halt and reverse biodiversity loss.

However, the fact that the 'Great Decline' has happened despite significant nature conservation efforts and the many examples of local conservation and recovery successes indicates the systemic issues behind nature loss: overall societal lack of ambition; low prioritisation of nature protection; inadequate recognition of the values of nature; and a lack of political will and low corporate responsibility and accountability. It also highlights how spatial conservation and species-specific actions are crucial but not sufficient to counter socio-economic drivers of pressure, and that only a combined strategy of nature conservation and the systemic reform of our production and consumption model can truly halt and reverse nature loss and build a nature-positive future. The 'Great Decline' and its root causes provide the foundation of why we need a nature-positive goal and ambition and how we can hope to course correct our future economic development journey.

The culture of Indigenous communities and their approach to nature provide many existential lessons for our modern society, pointing at a safe and sustainable way forward. Our modern society, driven by selfish accumulation of wealth and impulsive domination over nature, and blinded by technological might, treats the planet's resources as infinite and is embracing the convenient but delusional idea that we can exploit nature without limits and continue to deforest or overfish or pollute one area after another, without consequences.

During this 'Great Acceleration', many economies grew very fast and made a great deal of progress in terms of human well-being and development, from increased life expectancy to reduced mortality rates in childbirth. However, a model of consumption and production that prioritised unlimited and competitive economic growth was established, not surprisingly accelerating pressure on the natural world and leading to a widening inequality gap in terms of wealth distribution and living conditions. Economic development is commonly perceived as synonymous with improved human welfare. Our more recent development model, which at first led to positive outcomes for humanity, was destined to become dangerously unsustainable on a planet of finite resources, eventually leading to the 'Great Decline' of nature and today's '*Great Threat*' to humanity and life on Earth.

2.3   The 'Great Threat': planetary boundaries and the biophysical tipping points, code red for the planet (and humanity)

"Man does not weave this web of life. He is merely a strand of it. Whatever he does to the web, he does to himself." The awareness of our connectedness and dependences on nature is captured by this famous quote of Chief Seattle of the Suquamish and Duwamish tribes, reflecting the culture, attitudes and practices towards nature of several of today's indigenous populations deeply integrated in nature but lost in much of the rest of 'modern' society.

Over 50 years ago, a milestone report by the Club of Rome called *Limits to Growth* [42] warned of the dangerous dichotomy of human development and the integrity of natural systems. Now we are right in the middle of it. Today, the dangers of infinite economic growth on a finite planet are obvious, but it is still far from mainstreamed in our culture, behaviours and economic models. The more recent science on planetary boundaries has made clear which red lines are not to be crossed: "Transgressing one or more planetary boundaries may be deleterious or even catastrophic due to the risk of crossing thresholds that will trigger non-linear, abrupt environmental change within continental-scale to planetary-scale systems" [43, 44].

Researchers have identified the boundaries (or thresholds) of key Earth systems and processes which, if crossed, could generate disastrous environmental changes: climate change; rate of biodiversity loss; alterations of the nitrogen and phosphorus cycles; stratospheric ozone depletion; ocean acidification; global freshwater use; change in land use; chemical pollution; and atmospheric aerosol loading. Of the nine *planetary boundaries* identified, six (biosphere integrity, climate change, water, land use, novel chemical entities and biochemical flows) have already been crossed, and we are on track to cross others soon [45].

Today there is a greater recognition of the fact that humanity is entirely dependent on the stability of the biophysical systems across the planet's key realms: the atmosphere, geosphere/lithosphere, hydrosphere/cryosphere and biosphere. A variety of intricate and yet not fully understood feedbacks between these realms maintain the equilibrium, stability and predictability that life on Earth has enjoyed for the last 10,000 years following the last great Ice Age. These have been the main

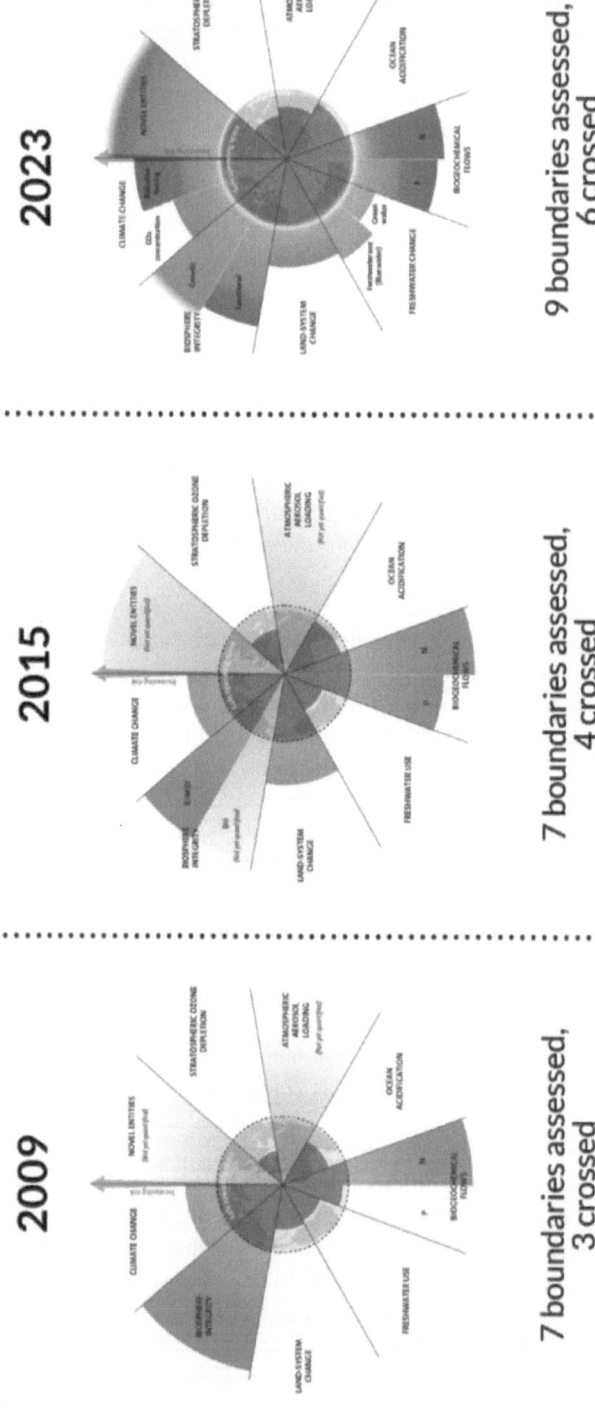

Figure 2.2 The progressive crossing of planetary boundaries.

Source: Adapted from Richardson et al. 2023, Steffen et al. 2015, and Rockström et al. 2009 [46]

characteristics of the Holocene, the geological era which saw the '*Great Rise*' and spectacular development of humanity.

Several scientists have suggested that recent but exponentially increasing human impacts are already significantly destabilising Earth's systems and that we have entered a new era, one that is characterised by high human dominance and disruption of natural systems. The '*Anthropocene*' – the Age of Humans [47–49].

A new era much less favourable to human development and survival and filled with destabilising factors and threats that are all of our own making.

The message is clear, and the warning light is flashing red. We are entering what scientists call a "non-analogue trajectory", characterised by the risk of triggering deep irreversible changes in several of Earth's systems, otherwise known as '*tipping points*' [50]. As scientists who specialise in planetary boundaries have written: "While we have been mainly assuming a linear degradation of natural systems under our impacts, it is today clear that pressure beyond a certain threshold make systems change, rapidly and even abruptly, to a different state" [51].

The state of biological and ecological systems is rarely entirely stable but is rather in a continuously evolving condition determined by various pressures and interactions between their biotic and abiotic components, which causes deviations from a mean condition over time. However, there are situations, when they are forced across a particular threshold, that their state can shift radically, abruptly and irreversibly as the result of cumulative incremental changes. These thresholds are influenced by many variables and can be difficult to predict.

To make it less abstract, when these tipping points are reached, the state of entire biomes like coral reefs, rainforests, ice sheets and glaciers; weather systems like monsoons and ocean currents; or geographic ecoregions like the Amazon or Antarctica will rapidly alter, and it may take centuries, even millennia, for them to recover. And they may never recover.

When this happens, the planet will look and Earth systems will behave very differently from today, and most life on Earth, including our own species, will face huge challenges in adapting and surviving to radically new or extreme environmental conditions and unpredictable weather patterns. We are impacting the very systems that, in the last 10,000 years, have given stability to the biosphere, enabling life and our civilisation to thrive and develop in spectacular fashion.

This represents a fundamental transformation from the stability that has characterised the Holocene, while also representing a new threatening feature of the Anthropocene. Over the course of Earth's history, certain systems have tipped before. These were caused by radical shifts in climate linked to Earth's orbit variations and meteorite impacts but also cumulative changes in atmospheric chemistry (mainly concentrations of carbon dioxide due to volcanic activity) and leading to change in ocean chemistry and, as a result, affecting life across the globe. The unprecedented dimension today is both the speed and the anthropogenic origin of these changes, and their ultimate irreversibility, at least in human time scales.

Ecosystems, ecoregions, biophysical and meteorological systems that are key to the stability, resilience and productivity of System Earth has been recently redefined as the '*Planetary Commons*'. As Johan Rockström and other scientists write:

> The planetary commons are defined by the functions they provide to Earth system stability and resilience and include all critical Earth-regulating biophysical systems and their functions, irrespective of where they are located, because they are essential to sustain all life across the planet [52].

These are the essential Earth systems humanity should preserve and steward, collectively and sustainably govern and equitably benefit from.

Recent research from Exeter University [53] has provided significant insight into the nature of planetary tipping points. It has identified more than 25 elements of the Earth system that can encounter tipping points across the cryosphere (such as the Greenland and Antarctic ice sheets, mountain glaciers and permafrost), the biosphere (ecosystems and biomes like the Amazon rainforest dieback, savannah and dryland desertification, lake eutrophication, coral reef and mangrove die-offs and the collapse of marine life populations) and weather systems dependent on ocean-atmosphere circulations (the Atlantic and Southern oceans overturning currents, as well as monsoon systems).

The main drivers of tipping points are the usual suspects: climate change, habitat loss, species and natural resources overexploitation and pollution. Some tipping points could be very close, such as the state of tropical coral reefs as well as ice sheets or glaciers that could soon 'tip' at current levels of warming. Interactions and negative feedbacks can then generate 'tipping cascades' or domino effects, accelerating the process and affecting other interconnected Earth systems. Up to 15 tipping point elements may be already active [54].

For example, it is estimated that 17% of the Amazon rainforest has already been deforested, and this is predicted to reach 20–25% in the next 20 years. Because over half of rainfall is self-generated by the forest, it is estimated that if deforestation reaches the 20–25% threshold the rainforest simply will not be able to self-generate enough rain to survive in most of the Amazon regions, and it will most likely turn into a drier savannah, possibly in just a few decades [55]. Less forest means less recycled rain, less evaporation to cool the air, less canopy to shield against sunlight. Under drier, hotter conditions the rainforest vegetation will die off. The impacts on local biodiversity will be devastating, and the implications for local rainfall patterns and economic activities in the region will be immense, as well as the knock-on effect on global weather and rainfall systems.

An Amazon-wide human-induced tipping point starts with local deforestation and forest fragmentation impacts. This opens new deforestation frontiers, and more and more adjacent areas undergo the same changes. These impacts accumulate, then begin to affect local climate and microclimates, accelerating the transition over an increasingly large area. The reduction of rainfall intensifies and reaches the threshold beyond which rainforest trees and animals adapted to wetter conditions cannot survive. A change of state into drier and less dense forests occurs. The unprecedented droughts increasingly occurring in the Amazon region are a warning signal of both global climate change but also the approaching tipping point for the Amazon rainforest. A change in state of the Amazon rainforest will impact the so-called 'rivers in the sky', or atmospheric rivers, the water vapour that originates from the Amazon and travels through the atmosphere, influencing precipitation in the region and far beyond.

The Congo Basin rainforest has even higher rates of self-generated rain than the Amazon. The rainforest's survival depends on these rainfall patterns, as – indirectly – do other regions in East Africa, including highly populated and agricultural areas of Ethiopia, which benefit from weather patterns originating in the Congo rainforest.

Ocean acidification is another looming tipping point with huge implications for life in the ocean and the wider global climate. Ocean water has traditionally been alkaline (with a pH above 8). As a result of the human-induced increase of CO_2 concentration in the atmosphere and consequent absorption by the ocean, since the start of the Industrial Revolution, the pH of the ocean has already fallen from 8.15 to 8.05 and is expected to decrease an additional 0.5 points by the end of the century, generating acidic waters [56, 57].

This low level of pH has probably never been experienced in the ocean for hundreds of thousands of years and has developed in such a short time. Two hundred and fifty million years ago, very intense volcanic activity caused an acidification event with a pH decrease of 0.7 over 10,000 years that resulted in the extinction of virtually all marine life [58]. The impact this will have on ocean ecology and the CO_2 sequestration functions of marine living organisms and ecosystems will exacerbate global warming. This will also have a huge impact on food security for the hundreds of millions of people who depend on both fishing and marine aquaculture for their livelihoods and as a source of food. Ocean acidification has been dubbed the '*evil twin*' of global warming [59].

The recent escalation of climate-related events has taken scientists by surprise, and this defiance of conventional predictions can be interpreted as an early warning sign of systemic destabilisation driven by self-reinforcing and external feedbacks (e.g. deforestation, wildfires, etc). In other words, tipping points could be approaching faster than we thought and could be moving from being local to tipping entire Earth systems or even triggering a whole planetary state shift [60]. The role of biodiversity loss in exacerbating climate disruption and accelerating tipping points is firmly recognised as critical [61, 62].

Other signals of approaching tipping points include the unprecedented wildfires in the Amazon, Australia and Canada, and the shocking acceleration of Antarctic, Arctic and mountain glacier ice melt. Due to the complexity of the way Earth systems behave and interact, the timing of eventual tipping points is still uncertain, but this is something we don't want to discover too late.

The Exeter University team concluded:

Harmful tipping points in the natural world pose some of the gravest threats faced by humanity. The effects will cascade through globalised social and economic systems and could exceed the ability of some countries to adapt.

Negative tipping points show that the threat posed by the climate and eco-logical crisis is far more severe than is commonly understood and is of a magnitude never faced by humanity, capable to severely damage our planet's life-support systems and threaten the stability of our societies.

Since the dawn of our civilisation, humans have grown accustomed to a relatively stable, predicable and mostly benevolent Earth system, with abundant resources to sustain our lives and promote our development. The idea that the disruption and degradation of the Earth system may not be linear and progressive, abrupt and practically irreversible, is a new and daunting prospect. In addition, the deep disruption triggered by tipping points is something we are unprepared for – culturally, socially, politically and economically. The 2020 pandemic highlighted the inadequacy and vulnerability of our economically globalised but politically fractured society to a sudden and global acute crisis.

Our biggest challenge perhaps lies in the fact that Earth system tipping points driven by our own impacts are something our civilisation has never experienced on a large or global scale. This makes us unprepared to deal with them or even to fully appreciate that they may occur, and occur with such threatening consequences for our civilisation. Ice ages were caused by climatic tipping points, triggered by changes in the Earth's orbital characteristics that influenced the distribution and intensity of solar radiation reaching the planet. These changes, combined with geophysical processes and interactions within the Earth's climate system, unfolded over thousands of years. The only good news in all this is that because today's ecological and climatic tipping points are essentially self-inflicted, it is theoretically in our power to prevent them. The bad news, however, is that the window of opportunity to prevent them from happening is already very narrow and is narrowing further with each day of inaction.

Human-induced tipping points are, for many people, difficult to comprehend because of their complexity, which is often entrenched in science, their sheer scale and the terrifying consequences they could trigger. Because of this, there is a fascinating and very dangerous psychological dimension to the planetary crisis. Our instinctive reaction to threats that are so big and existential is to perceive them as either impossible or intractable. Either way, our reaction ranges from denial to incredulity and doubt, to anxiety and depression.

The most effective and dangerous of all reactions is perhaps a convenient denial of the problem or the process of readjusting thresholds of tolerance and acceptability to avoid disruption, panic or, worse, to defend self-interests. All these reactions are recognisable today. This is a big challenge in the fractured, nationalistic governance model of our global society. The tribal approach that has driven our evolutionary journey on Earth is perhaps the biggest obstacle to the development of a model of global governance able to co-ordinate the response to global challenges. We require universal rules on how to collectively secure critical Earth systems that provide stable living conditions on Earth for present and future human generations and all non-human life. The nature-positive goal is a key element of this new compact for humanity and the planet.

Yet the notion of irreversible planetary scale destabilisation is complex and disorientating. Denial and dismissal are often an easier and more seductive choice. This psychological element represents perhaps the biggest obstacle in effectively avoiding and mitigating these fast-approaching tipping points. We will explore these aspects in Chapter 4.

As the biophysical limits of what we can do to the planet become apparent, we are timidly beginning to realise that exceeding these limits will impact the very systems that support life, and the entire planet may irreversibly (in human time scale) shift to an unstable and unpredictable state. This is the start of humanity's *'Great Awakening'*.

References

1 Attenborough, D., 1979. Life on Earth. Little, Brown and Company.
2 EUMOFA, 2023. 'The EU fish market', European Union. Available at: https://eumofa.eu/market-analysis?#yearly
3 Uryu, Y., Purastuti, E., Laumonier, Y., Sunarto, Setiabudi, A., Budiman, A., Yulianto, K., Sudibyo, A., Hadian, O., Kosasih, D.A. and Stüwe, M., 2010. Sumatra's Forests, Their Wildlife and the Climate Windows in Time: 1985, 1990, 2000 and 2009. WWF-Indonesia, Jakarta, Indonesia (Accessed: December 2010).
4 Barnosky, A., Matzke, N., Tomiya, S., et al., 2011. 'Has the Earth's sixth mass extinction already arrived?', Nature, 471, pp. 51–57. Available at: www.nature.com/articles/nature09678
5 Stearns, S., et al., 2000. Watching, from the Edge of Extinction. Yale University Press.
6 Kolbert, E., 2014. The Sixth Extinction: An Unnatural History. Henry Holt and Company.
7 Pimm, S.L., et al., 2014. 'The biodiversity of species and their rates of extinction, distribution, and protection', Science, 344(6187). Available at: www.science.org/doi/10.1126/science.1246752
8 IPBES, 2019. Global Assessment Report on Biodiversity and Ecosystem Services. Available at: www.ipbes.net/news/ipbes-global-assessment-summary-policymakers-pdf
9 Achaz, G., Bouchet, P., Cowie, R.H., Lambert, A., Fontaine, B. and Régnier, C., 2015. 'Mass extinction in poorly known taxa', Proceedings of the National Academy of Sciences (PNAS), 112. Available at: https://doi.org/10.1073/pnas.1502350112
10 Sancasciani, G., 1982. 'L'insetto vendicherà l'uccello', In M. Lambertini and E. Meschini (eds.), L'insetto vendicherà l'uccello. La Fortezza Editore.
11 Cardoso, P., et al., 2020. 'Scientists' warning to humanity on insect extinctions', Biological Conservation, 242. Available at: https://doi.org/10.1016/j.biocon.2020.108426
12 IPBES, 2016. 'The assessment report of the Intergovernmental Science-Policy Platform on Biodiversity and Ecosystem Services on pollinators, pollination and food production', In S.G. Potts, V.L. Imperatriz-Fonseca and H.T. Ngo (eds.), Secretariat of the Intergovernmental Science-Policy Platform on Biodiversity and Ecosystem Services. Available at: https://doi.org/10.5281/zenodo.3402856
13 IPBES, 2019. Global Assessment Report on Biodiversity and Ecosystem Services of the Intergovernmental Science-Policy Platform on Biodiversity and Ecosystem Services. E.S. Brondizio, J. Settele, S. Díaz and H.T. Ngo (eds.). IPBES Secretariat. Available at: https://doi.org/10.5281/zenodo.3831673
14 Finn, C., Grattarola, F. and Pincheira-Donoso, D., 2023. 'More losers than winners: Investigating Anthropocene defaunation through the diversity of population trends', Biological Reviews Cambridge, 98(5). Available at: https://onlinelibrary.wiley.com/doi/10.1111/brv.12974

15 WWF and ZSL, 2024. The Living Planet Report. Available at: https://livingplanet.panda.org/

16 Smyčka, J., Storch, D. and Toszogyova, A., 2024. 'Mathematical biases in the calculation of the Living Planet Index lead to overestimation of vertebrate population decline', Nature Communications, 15(5295). Available at: https://doi.org/10.1038/s41467-024-49070-x

17 IUCN, 2024. Red List of Threatened Species. Available at: www.iucnredlist.org/en

18 Haddad, N.M., et al., 2015. 'Habitat fragmentation and its lasting impact on Earth's ecosystems', Science Advances, 1, p. e1500052. Available at: https://doi.org/10.1126/sciadv.1500052

19 IPBES, 2019. Global Assessment Report on Biodiversity and Ecosystem Services of the Intergovernmental Science-Policy Platform on Biodiversity and Ecosystem Services. E.S. Brondizio, J. Settele, S. Díaz and H.T. Ngo (eds.). Available at: www.ipbes.net/global-assessment

20 Convention on Wetlands, 2021. Global Wetland Outlook: Special Edition 2021. Gland, Switzerland: Secretariat of the Convention on Wetlands.

21 FAO, 2020. The State of World Fisheries and Aquaculture 2020: Sustainability in Action. Rome. Available at: https://doi.org/10.4060/ca9229en

22 IUCN, 2022. The IUCN Red List of Threatened Species. Available at: www.iucnredlist.org/

23 Bar-On, Y.M., Phillips, R. and Milo, R., 2018. 'The biomass distribution on Earth', Proceedings of the National Academy of Sciences, 115(25), pp. 6506–6511. Available at: https://doi.org/10.1073/pnas.1711842115

24 Bar-On, Y.M., Phillips, R. and Milo, R., 2018. 'The biomass distribution on Earth', Proceedings of the National Academy of Sciences of the United States of America (PNAS), 115(25), pp. 6506–6511. Available at: https://doi.org/10.1073/pnas.1711842115

25 Steffen, W., Broadgate, W., Deutsch, L., Gaffney, O. and Ludwig, C., 2015. 'The trajectory of the Anthropocene: The Great Acceleration', The Anthropocene Review, 2(1). Available at: https://doi.org/10.1177/2053019614564785

26 Engelke, P. and McNeill, J.R., 2016. The Great Acceleration: An environmental history of the Anthropocene since 1945. Harvard University Press.

27 Steffen, W., 2022. 'The Earth System, the Great Acceleration and the Anthropocene', In S.J. Williams and R. Taylor (eds.), Sustainability and the New Economics. Springer. Available at: https://doi.org/10.1007/978-3-030-78795-0_2

28 Global Footprinting Network. Available at: www.footprintnetwork.org/our-work/ecological-footprint/

29 Harari, Y.N., 2015. Sapiens: A Brief History of Humankind. Vintage Books.

30 Svenning, J-C., Lemoine, R.T., Bergman, J., et al., 2024. 'The late-quaternary megafauna extinctions: Patterns, causes, ecological consequences and implications for ecosystem management in the Anthropocene', Cambridge Prisms: Extinction, 2, p. e5. Available at: https://doi.org/10.1017/ext.2024.4

31 Andermann, T., Antonelli, A., Faurby, S., Turvey, S.T. and Silvestro, D., 2020. 'The past and future human impact on mammalian diversity', Science Advances, 6(36). Available at: https://doi.org/10.1126/sciadv.abb2313

32 Lock, P.L. and Barnosky, A.D., 2006. 'Late quaternary extinctions: State of the debate', Annual Review of Ecology, Evolution, and Systematics, 37, p. 215. Available at: https://doi.org/10.1146/annurev.ecolsys.34.011802.132415

33 Crees, J.J. and Turvey, S.T., 2019. 'Extinction in the Anthropocene', Current Biology, 29(19). Available at: https://doi.org/10.1016/j.cub.2019.07.040

34 Blackburn, T., Boyer, A. and Duncan, R., 2013. 'Magnitude and variation of prehistoric bird extinctions in the Pacific', Proceedings of the National Academy of Sciences [online]. Available at: https://doi.org/10.1073/pnas.1216511110

35 Svenning, J.-C., et al., 2024. 'The late-quaternary megafauna extinctions: Patterns, causes, ecological consequences and implications for ecosystem management in the Anthropocene', Cambridge Prisms: Extinction, 2, p. e5. Available at: https://doi.org/10.1017/ext.2024.4

36 Ritchie, H., 2021. 'Deforestation and forest loss', OurWorldinData.org. Available at: https://ourworldindata.org/deforestation

37 UNEP-'WCMC, 2025. Protected Planet. The World Database on Protected Areas. Available at: https://www.protectedplanet.net/en

38 Gray, C., Hill, S., Newbold, T., et al., 2016. 'Local biodiversity is higher inside than outside terrestrial protected areas worldwide', Nature Communications, 7(12306). Available at: https://doi.org/10.1038/ncomms12306

39 Li, B.V., et al., 2024. 'The synergy between protected area effectiveness and economic growth', Current Biology, 34(13). Available at: https://doi.org/10.1016/j.cub.2024.05.044

40 Langhammer, P.F., et al., 2024. 'The positive impact of conservation action', Science, 384(6694). Available at: https://doi.org/10.1126/science.adj6598

41 Bolam, F.C., et al., 2020. 'How many bird and mammal extinctions has recent conservation action prevented?', Conservation Letters, Society for Conservation Biology. Available at: https://doi.org/10.1111/conl.12762

42 Behrens III, W., Meadows, D.L., Meadows, D.H. and Randers, J., 1972. The Limits to Growth. The Club of Rome.

43 Rockström, J., Steffen, W., Noone, K., et al., 2009. 'A safe operating space for humanity', Nature, 461, pp. 472–475. Available at: https://doi.org/10.1038/461472a

44 Steffen, W., et al., 2015. 'Planetary boundaries: Guiding human development on a changing planet', Science, 347, p. 1259855. Available at: https://doi.org/10.1126/science.1259855

45 Richardson, J., Steffen, W., Lucht, W., Bendtsen, J., Cornell, S.E., et al., 2023. 'Earth beyond six of nine planetary boundaries', Science Advances, 9(37).

46 Richardson, J., Steffen, W., Lucht, W., Bendtsen, J., Cornell, S.E., et al., 2023. 'Earth beyond six of nine planetary boundaries', Science Advances, 9(37); Steffen, W., et al., 2015. 'Planetary boundaries: Guiding human development on a changing planet', Science, 347, p. 1259855. Available at: https://doi.org/10.1126/science.1259855; Rockström, J., Steffen, W., Noone, K., et al., 2009. 'A safe operating space for humanity', Nature, 461, pp. 472–475. Available at: https://doi.org/10.1038/461472a

47 Subramanian, M., 2019. 'Anthropocene now: Influential panel votes to recognize Earth's new epoch', Nature. Available at: https://doi.org/10.1038/d41586-019-01641-5

48 Lewis, S. and Maslin, M., 2015. 'Defining the Anthropocene', Nature, 519, pp. 171–180. Available at: https://doi.org/10.1038/nature14258

49 Crutzen, P.J. and Stoermer, E.F., 2000. 'The "Anthropocene"', In L. Robin, S. Sörlin and P. Warde (eds.), The Future of Nature: Documents of Global Change. New Haven: Yale University Press, 2013, pp. 479–490. Available at: https://doi.org/10.12987/9780300188479-041

50 Lenton, T., 2011. 'Early warning of climate tipping points', Nature Climate Change, 1, pp. 201–209. Available at: https://doi.org/10.1038/nclimate1143

51 Rockström, J., et al., 2024. 'The planetary commons: A new paradigm for safeguarding Earth-regulating systems in the Anthropocene', Proceedings of the National Academy of Sciences (PNAS). Available at: www.pnas.org/doi/10.1073/pnas.2301531121

52 Rockström, J., Kotzé, L., Milutinović, S., Biermann, F., Brovkin, V., Donges, J., Ebbesson, J., French, D., Gupta, J., Kim, R., Lenton, T., Lenzi, D., Nakicenovic, N., Neumann, B., Schuppert, F., Winkelmann, R., Bosselmann, K., Folke, C., Lucht, W., Schlosberg, D., Richardson, K. and Steffen, W., 2024. 'The planetary commons: A new paradigm for safeguarding Earth-regulating systems in the Anthropocene', Proceedings of the

National Academy of Sciences of the United States of America, 121(5), e2301531121. Available at: https://doi.org/10.1073/pnas.2301531121

53 Lenton, T., et al., 2023. Global Tipping Points. University of Exeter. Available at: https://global-tipping-points.org

54 Lenton, T., et al., 2019. 'Climate tipping points – too risky to bet against', Nature, 575, pp. 592–595; Correction April 2020. Available at: www.nature.com/articles/d41586-019-03595-0#correction-0

55 Lovejoy, T.E. and Nobre, C., 2018. 'Amazon tipping point', Science Advances, 4(2). Available at: https://doi.org/10.1126/sciadv.aat2340

56 The Royal Society, 2005. Ocean Acidification Due to Increasing Atmospheric Carbon Dioxide. Available at: https://royalsociety.org/news-resources/publications/2005/ocean-acidification

57 Frölicher, T.L., Joos, F. and Terhaar, J., 2023. 'Ocean acidification in emission-driven temperature stabilization scenarios', Environmental Research Letters, 18(2), p. 024033. Available at: https://doi.org/10.1088/1748-9326/acaf91

58 Clarkson, M.O., et al., 2015. 'Ocean acidification and the Permo-Triassic mass extinction', Science, 348(6231), pp. 229–232. Available at: https://doi.org/10.1126/science.aaa0193

59 Rogers, A., 2013. 'Global warming's evil twin: Ocean acidification', The Conversation. Available at: https://theconversation.com/global-warmings-evil-twin-ocean-acidification-19017

60 Barnosky, A., Bascompte, J., Hadly, E., et al., 2012. 'Approaching a state shift in Earth's biosphere', Nature, 486, pp. 52–58. Available at: https://doi.org/10.1038/nature11018

61 UN Environment, 2019. Global Environment Outlook – GEO-6: Healthy Planet, Healthy People. Nairobi. Available at: https://doi.org/10.1017/9781108627146

62 IPBES-IPCC, 2021. Workshop Report: IPBES-IPCC Cosponsored Report on Biodiversity and Climate Change. Available at: www.ipcc.ch/site/assets/uploads/2021/07/IPBES_IPCC_WR_12_2020.pdf

3   The 'Great Awakening'

Marco Lambertini

Of the many quotes from famous psychologists and philosophers about our tendency to escape hard realities and inconvenient truths, I cannot think of a better one than what my grandmother Ester used to say: "You cannot hide behind your finger". This is today's shameful and dangerous widespread attitude of ignoring or denying the overwhelming evidence of our impact on nature and the consequences that derive from it. But this is finally beginning to change.

There is something simple yet extraordinarily powerful about our time. Today, we know. We are the most environmentally aware generation in human history. We know a great deal more about how natural systems work, and we have the best understanding of how our actions are leading to the destabilisation of Earth systems. Perhaps more importantly, we are also beginning to recognise that this impact is not just restricted to the natural world but affects us in return.

The awareness of the 'Great Decline' and the 'Great Threat' has in turn triggered a *'Great Awakening'* regarding the risks that the degradation of nature poses to all life on Earth – including ourselves. Within this new awareness lies our chance to embrace change. This new awareness is also what makes our time truly unique as it is this generation, not the next one, that has the incredibly exciting opportunity to build a safer, nature-positive future.

3.1   The 'Development Paradox'

"Things have never been so good for humanity, nor so dire for the planet." The title of the article of Canadian journalist Arno Kopecky [1] captures perfectly one of the most striking paradoxes of our time and perhaps the most existential challenge for humanity.

For most of human history, around half of newborns died before reaching puberty. This is a rate common in many other wild mammal species. In 1900, around 40% of newborns didn't survive childhood [2]. By 1950, that figure had declined to around one-quarter globally. Today it has fallen to less than 4%. In 1900, global mean life expectancy was around 30 years, it fell below 50 in 1950

DOI: 10.4324/9781003474043-4

and it is over 70 today [3]. At the start of the nineteenth century, three-quarters of the world lived in poverty, unable to afford enough food to stave off malnutrition. In 1950, the proportion of the world's population living in absolute poverty was nearly 60%, and today it is 10% [4]. Even though these global statistics hide a deep and unacceptable level of inequality across regions, progress in human development terms is undeniable in many parts of the world. One reality is that the nature-negative, exploitative development model which exploded particularly during the 'Great Acceleration' has made most of us live better. Until now.

But while human society was developing, the natural world was declining as a direct result. For every city founded, for every road, railway and port built, and for every piece of land taken for farming or grazing, a piece of the natural world was sacrificed. We simply did not pay any attention to the impact this was having on the natural systems that provide us with a stable and predictable climate, the air that we breathe and the water we drink and use to grow crops and run our immense extractive and manufacturing industries. In our frantic race to dominate the natural world, we had forgotten that, in fact, we depend on nature more than nature depends on us.

How blind and foolish it was of us to forget that we depend on nature, that we are part of nature. But also, how natural. In nature everything is in balance, and the availability of resources, or the lack of them, is a main driver of adaptation and evolution of every life form. Living sustainably, within the boundaries of resource availability, is a basic requirement to enable the survival of every individual of every species. This is effectively the case everywhere in the natural world, but more by evolutionary pressure than by choice.

As Richard Dawkins writes in his book *Blind Watchmaker*: "Nature's complex web of life, its living forms and their every adaptation are not the result of . . . conscious design, but natural selection of mechanisms, adaptations and behaviours with high efficiency and survival significance" [5]. This is the essence of Darwinian evolutionary theory.

As a result, sustainable living is something ubiquitous and intrinsic to the natural world and everything that lives in it. It is, however, the result of interdependencies, interactions and feedback between living and non-living elements of nature and not necessarily of conscious decisions.

For example, northern red deer live in balance with the forest as their population is kept in check by both the availability of vegetation and their primary predator, the wolf. If the wolves are eradicated, the deer population grows out of control, and the impact on the forest increases until food scarcity rebalances the population size through diseases and starvation. The lack of food naturally reduces embryo and offspring survival rates but, in many species, also induces a decrease in ovulation, naturally driving lower birth rates and population decline. Quite simply, with population growth, competition increases and resources diminish until the population gets down to more sustainable levels.

Examples of the lack of conscious sustainable behaviour are common in nature. Highly intelligent creatures like elephants pull trees down to the ground to reach the greenest leaves of the canopy, apparently oblivious to the damage they cause to local vegetation. As their natural habitat has now been dramatically reduced by human encroachment and most elephant populations are confined to protected areas, this impact can affect their survival and that of other herbivore species.

Even more extreme is the extraordinary recent behaviour discovered in southern orcas – who kill great white sharks for their liver and young whales just for their tongues, discarding the rest of the body, echoing the wasteful and unsustainable human behaviours we blame for the state of the planet.

Of course, wild species don't have the means of 'mass destruction' that humans do, so the impact they can have on nature is mostly limited and recoverable. Yet in these stories lies one important reflection, and a stark reminder. The reflection is that sustainable behaviours don't necessarily come naturally, or instinctively. In the natural world sustainability is secured through feedback mechanisms, mainly related to resource availability, leading to ecological balance. Sustainability is an anthropogenic concept that evolved in response to the realisation of the impacts that our behaviour was having on the natural world, the destabilisation of Earth systems, the depletion of natural resources and the negative consequences this causes to our own survival. This recognition, new in 'modern' societies, has been present for millennia in many Indigenous populations who directly rely on natural resources for their survival, and it is recently being re-embraced by a growing share of the global population.

Being sustainable requires a deliberate effort and comes at the end of a cognitive process of internalising the need to be sustainable and the benefits that derive from it. This is also a stark reminder that the more imbalance we create in the natural world, the more drastic the rebalancing act will be.

The notion of sustainability, unlike instinctive traits like survival and self-interest, does not seem to have been built into our species, or other species for that matter. Usually, it is triggered when we can forecast negative consequences and, once again, act in our self-interest and the interest of our family and community.

This is evident in some coastal communities that apply seasonal fishing bans of certain species, rotate fishing areas and set fishing quotas to avoid depletion of stocks. In the Middle East, the traditional concept of 'hima' means to set aside natural features such as a forest [6], a spring or a river and to protect it from destructive activities. This is enforced through community agreements to preserve the services nature provides.

There are also examples of how these traditional resource governance systems have been rapidly disrupted by the arrival of technologies such as firearms and snares for hunting, gillnets and explosives for fishing and tractors and bulldozers for conversion to agriculture. Blast or dynamite fishing was widespread and is still a practice present in many coastal communities in Southeast Asia, East Africa and other regions of the world, leading to huge damage to coastal habitats and fish populations [7].

In many local Indigenous communities in Central Africa and Southeast Asia, subsistence hunting with nets and arrows has given way to metal snares and guns, where the wildlife killed is often not consumed directly but sold to local markets or traders, sadly often in exchange for alcohol or drugs [8]. The so-called 'bushmeat' is also traded domestically or internationally, mainly illegally, to diasporas in cities around the world and is no longer confined to subsistence hunting [9].

In modern society, dominated by a physical and cultural separation of people from nature, mass-production, advanced exploitative technologies and hyper consumerism, the notion of sustainability is struggling to hit the mainstream. It requires long-term thinking, an understanding of the negative, often indirect consequences of our actions and a belief in equitable and shared prosperity. During our modern development journey, we have been increasingly discarding natural realities like trees, rivers, the ocean, fish, bees and earthworms and caring more for artificial constructs like cities, economies, technologies, digital tools, manufactured goods and processed food. And at the same time, we have ignored the fact that ultimately our 'artificial world' still depends on the 'natural' one.

To transition to a society in balance with the natural world that operates within planetary boundaries, adaptation, collaboration and natural resource governance are key [10]. Sustainability also implies an empathic level of consciousness that certain behaviours contribute to a healthier society, a better and safer future and leaves no one behind. The Brundtland Commission of the United Nations [11] defines sustainable development as: "Development that meets the needs of the present without compromising the ability of future generations to meet their own needs." Will a better understanding of the natural world, its fragilities, our impacts, the damage we cause and the consequences we will ultimately suffer change the way our brain processes information, develops perceptions and drives behaviours? Will we be able to internalise the need for sustainability and apply it to the way we think and live our lives?

The ultimate driver of a transition to a nature-positive future resides primarily in our mind. The emergence of a new stream of neuroscience, enhanced by increasingly sophisticated brain imaging technology, analyses the response of our brain to sustainability challenges. Neurosustainability is a new, fascinating approach that studies the relationship between neuroscience and the field of sustainability. As a result of our cognitive ability to internalise threats and select 'safe behaviours', will our brain 're-wire' to 'automatically' prioritise sustainable instead of unsustainable behaviours?

The fundamental failure of our mainstream and globalised development model is the assumption that we can endlessly develop our economies at the expense of a limited natural capital. In *The Economics of Biodiversity* [12], Professor Partha Dasgupta powerfully describes and visualises how, over the last few decades, the world has continued to see a steep increase in *produced capital* (human-made goods and infrastructures), a much flatter increase in *human capital* (education, health) and a steep decline in *natural capital* (natural infrastructure, resources and services). Between 1992 and 2014 produced capital per capita doubled while natural capital decreased 40% and human capital increased a mere 13%. This highlights both the inequality gap within and between societies, and the nature loss crisis driven by a consumptive rather than regenerative economic model. These three essential 'capitals' that together are the foundation of human well-being, equity and sustainability as well as planetary health, in our modern development model have been in fact cannibalising each other.

Until now, financial capital has been largely generated by the conversion of natural capital, and its growth has not equitably enhanced human capital. The world's biggest companies destroy $7.3 trillion worth of natural capital every year at no cost to themselves, and this is the reason why they are profitable today [13]. The cost of the externalities is paid by society while the profits remain private. The costs of inaction are staggering. From 1997 to 2011 the OECD estimates that the world lost $4–20 trillion per year in ecosystem services on account of land cover change only [14]. For as shocking and persuasive as they seem, these figures remain part of a rather academic conversation as they are not yet accounted for in our economic assessments.

We are approaching the point where the scarcity of resources and the weakening of natural processes are not only beginning to hinder socio-economic development but are also beginning to reverse the development gains made in previous decades. Extreme poverty, disease, malnutrition and human displacement have already started to increase in some regions of the world because of the impacts of climate change. In 2022, the average human life expectancy of men and women was two and three years lower respectively because of the COVID-19 pandemic, a zoonotic disease that almost certainly originated from wild animals that had been traded or farmed. Clearly, we can't continue to unsustainably exploit the natural world without facing significant consequences. The consequences are happening now and will only worsen if we don't correct course. We ignore this at our peril.

This is the first paradox of our times, the *'Development Paradox'*. The paradox of a 'suicidal', 'dead-end' development model that destroys and erodes its own foundations and whose ultimate collapse is therefore inevitable. As the environmentalist Edward Abbey once put it: "*Growth for the sake of growth is the ideology of the cancer cell.*" Except cancer is not a cell but an externally induced 'malfunction' of the cell that ends up destroying itself. Does this analogy ring any bells?

The neoclassic economic growth model goes back to the 1950s and the work of American economist and Nobel laureate Robert Solow and Australian economist Trevor Swan. The model looks at capital accumulation as the result of three main factors: population growth (boosting consumption), labour and technology (both to boost production). It is fascinating to notice that the model reflects exactly the drivers of the 'Great Acceleration' of natural resource use that started exactly at the same time.

Solow most likely took inspiration and confidence from the population explosion and technological advancements which occurred after the start of the Industrial Revolution and accelerated during the post–Second World War economic boom. The model, unsurprisingly but foolishly, also implicitly assumes an unlimited availability of natural resources. We had to wait over a decade for the publication of the visionary landmark report *The Limits of Growth* [15], which was the first to highlight the unsustainability of our economic model vis-à-vis the limits of the planet's resources. However, even the Club of Rome report focussed primarily on raw materials and not on the whole picture, including nature's services. Two

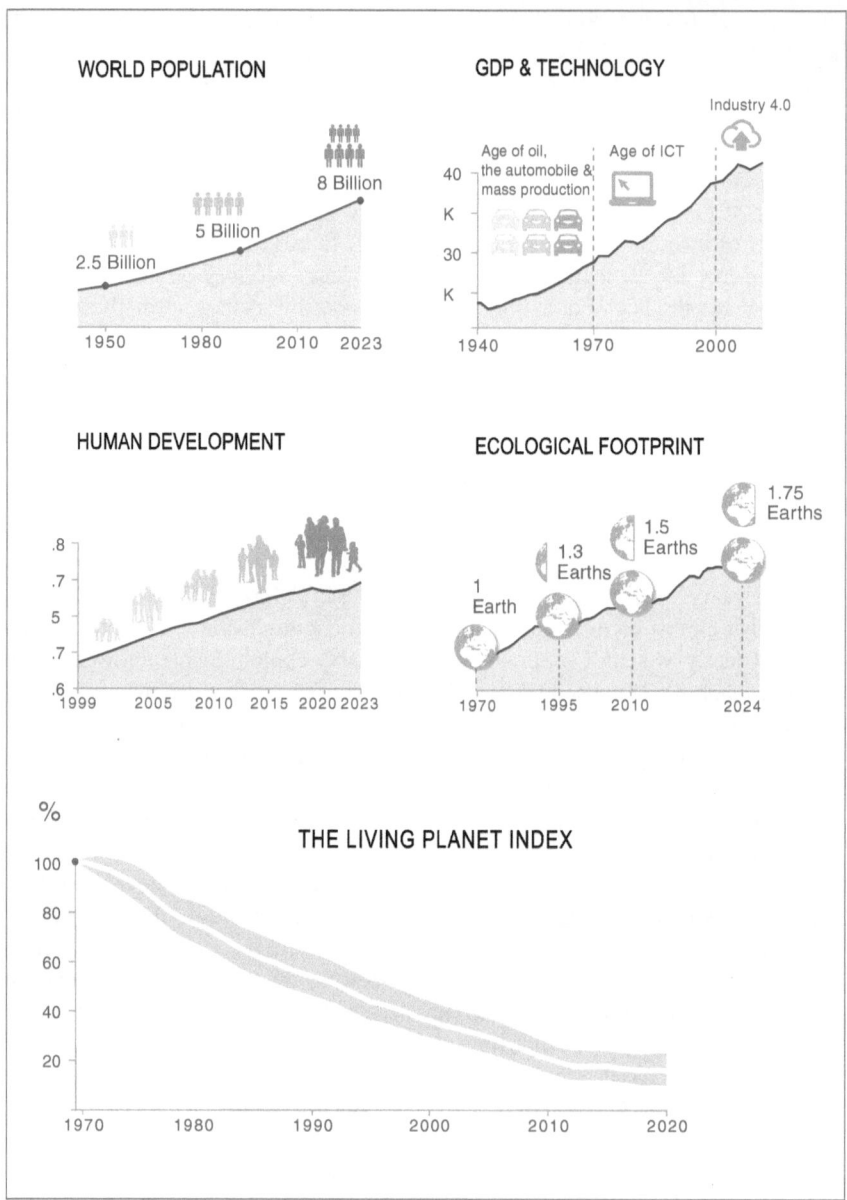

Figure 3.1 The 'Development Paradox': a) Human population (source: Our World in Data.); b) GDP and technology (source: UNCTAD.); c) Human development index and inequality (source: World Bank.); d) Ecological footprint (source: Global Footprint Network.); e) Living Planet Index. (source: Zoological Society of London and WWF.)

decades later, the Earth Summit definitively launched the concept and imperative of sustainable development, calling on governments and businesses to rally around the newborn global conventions on climate, biodiversity and desertification. But it was the UN Environment Programme's *The Economics of Ecosystems and Biodiversity* (TEEB) report [16], which was the first to assess the value to our economy and society not only of natural resources but also of ecosystem services.

Extensive evidence suggests that neoclassical economic (NCE) theory, which underpins much of today's economic systems, has been a driving force behind rapid global economic growth. However, it is also implicated in exacerbating environmental degradation and socio-economic inequality. A recent analysis critiques the foundational principles of individualism, instrumentalism and market equilibration that inform neoliberal economic practices, such as minimal government intervention and market-driven decision-making. These principles are shown to be limited in their sensitivity analysis and predictive accuracy. The study advocates for reforming neoclassical micro and macroeconomics, proposing a shift toward a more integrative and interdisciplinary framework of social-ecological economics to achieve ecological sustainability and social equity [17].

And here we arrive at the second paradox of our age, what we could call the '*Awareness Paradox*'. On the one hand, our impact on the planet is continuing, in fact increasing. The science has never been clearer. The natural world is in crisis. On the other hand, public awareness has never been greater, not only of the impact we are having on nature and biodiversity but – more importantly – of the consequences this will lead to.

This is beginning to challenge our delusional idea that we can continue to act irresponsibly towards the natural world, taking nature for granted, exploiting its resources and distributing them unequally and without consequences. Today, we know that there are consequences. Some of them are here already, including death and economic losses from extreme weather; aggravated poverty and food insecurity from droughts and floods; social unrest and increased conflicts and forced migration; and devastating zoonotic pandemics that can bring the whole world to its knees.

3.2 Why Nature Positive? From moral duty to human security

Climate change and nature loss are increasingly perceived not only as moral or isolated ecological issues but as issues that affect our economy, social stability, our individual health and wellbeing and our happiness.

Conserving nature and stabilising the climate are not just linked to a moral duty to coexist with the amazing diversity of non-human life on Earth. This is a powerful argument and is strongly felt by an increasing number of people (and the reason why I started to work in nature conservation), but on its own it has not been sufficient to persuade enough of us to course correct from our destructive path.

The *eco-centric* moral argument for conserving nature is still there and is in fact more powerful than ever because today it includes us, alongside the rest of life on Earth. After all, most people care most about . . . people. It is not a coincidence that over 80% of philanthropic donations are for human health, poverty, education or disaster relief. Alongside moral motivations, a material argument has more recently and powerfully emerged where climate change and biodiversity loss equate to huge risks and costs to our society, economy and individual quality of life. Our impact on the natural world has become a security issue for humanity. It is also a justice issue, as the most vulnerable populations are already the first and most affected, and an inter-generational justice issue as by losing nature and her services we are leaving a terrible legacy to our children, their children and future generations to come.

This new awareness is driving the 'Great Awakening' of our time, a recognition of our impacts on nature, and the consequences the loss and degradation of nature is having and will increasingly have on humanity. This is a huge cultural shift and is perhaps what's most needed to embrace the deep systemic change we need to see in our societies and economies. The opposite of our traditional approach of taking nature for granted, treating it like it is something that will always be there for us, something that is plentiful and limitless.

This represents an historic shift in perception about nature loss, a true cultural revolution. Extinctions and deforestation are not just reasons to be sad, but to be concerned, fearful even. Concern and fear have joined sadness and guilt as the emotions we experience for failing to coexist with nature and Earth's diversity of life and putting the future of our children at risk. We are living perhaps the inevitable paradox of a society at the start of a deep transition where the growing impact on the natural world is accompanied by an unprecedented rise in awareness and concern. A paradox in which resides the foundation for change, and the reason for hope.

Nature touches all aspects of our lives, simply because we are a part of it. For most of our history as a species we have been directly dependent and fully integrated in nature. With the agricultural revolution, the sense of dependence gave way to an illusion of control over nature. Every technological advance has strengthened that perception and grown the (delusional) idea of our independence and separation from nature. The culture of thinking of ourselves as separate from nature is finally being challenged by the reality of the impacts of climate change and nature loss on our lives and well-being. Today, the evidence is clear about our dependencies on nature, and therefore the links between a healthy and productive natural world and a safe and just future for humanity. We have progressed from an early stage of total dependence, followed by a stage of growing false perception of independence, to today's new recognition of a deep interdependence.

As a pacifist, I feel uncomfortable using the words of the former British Prime Minister Winston Churchill, from the pre-World War Two war period, but nevertheless the sentiment rings true.

Owing to past neglect, in the face of the plainest warnings, we have entered upon a period of danger. The era of procrastination, of half measures, of soothing and baffling expedients of delays, is coming to a close. In its place we are entering a period of consequences, he said in 1936.

These words resonate today vis-à-vis the current climate and biodiversity crises. The "plainest warnings" come from the destabilisation of Earth systems, and "the past neglect" is the impact of our unsustainable development model and attitude of ignoring the environmental externalities of our actions. The difference, however, is that defending ourselves from the collapse of the natural world will not require war, but in fact peace. Peace with nature.

Indeed, we have entered times of consequences. Times where risks are turning into threats and impacts. Economic dependencies into instability and costs. It is not about optimism or pessimism, as is often said. It is not about our capacity to respond. We have all it takes to avert the worst consequences: the intellectual, financial, technological ability. It is simply about choices and resolve. And in this lies, absurdly and tragically, the greatest threat: to think that someone or something else will save us from the destabilisation of the Earth systems.

3.3 Nature Positive is People Positive: the basis for social equitability and justice

It is intuitive to think that preserving nature benefits both present and future generations of humans. Success in conserving and restoring nature and securing nature's contributions to society is a key element of the Nature Positive Global Goal, and to delivering a nature-positive, equitable and carbon-neutral future.

On the other hand, any strategy or action to deliver a nature-positive outcome must ensure that conserving nature also enables the economic and social development needed to increase equity, end poverty and generally improve the quality of life. Equity is both an outcome of and the condition for a nature-positive future, as social inequity is the product of our unsustainable economic model as well as a barrier to a sustainable future.

The evidence that justice, equity and sustainability are mutually supportive is well-established [18]. The concept of a just world on a safe planet highlights the inextricable connection between ensuring the stability of the Earth's biophysical systems and justice and equity principles that guarantee fair access and distribution of resources to enhance human well-being [19].

The effectiveness and perceived fairness of environmental policies is key in driving social acceptability, participation, inclusion, compliance and permanence. At the same time, unsustainable practices, such as the overexploitation of local resources, can undermine equity, development and therefore also justice.

The rise of environmental human rights is a testimony to the growing awareness of the strong dependencies between a healthy environment and the fulfilment of fundamental human rights at a time when the United Nations General Assembly has recently recognised the human right to a clean and healthy environment [20–22].

A growing section of today's youth population feels that climate change and biodiversity decline, driven by today's development model, compromises their future and represents a breach of intergenerational justice. The issue is, however, full of complexity. The perception of justice is often connected to rights and, ultimately, needs and aspirations. But rights to nature's services and resources are perceived differently by different stakeholders. Poor subsistence agriculture populations living on forested hills may claim the right to food security, jobs and a decent income by expanding their crops and cutting down forest. Farmers and urban populations downstream, meanwhile, may have already converted the forest on their land but claim their right to access water, and would rather see the upstream forests protected as the source of the water supply they depend on. This is where environmental protection, social equity and justice meet and must be integrated while at the same time trade-offs must be carefully considered and managed.

Furthermore, individual and collective rights often collide, and managing this with fairness and ensuring shared benefits is key, but not easy. People around the world claim their right to a stable climate and oppose deforestation and forest fires, but the people living in forest-rich regions see clearing forests as their right to local economic development. At the end of the day, it comes down to power dynamics and asymmetries. In other words, who has decision-making powers, and who is marginalised, be it young people, women, Indigenous people or other groups.

Delivering the Nature Positive Global Goal in complex social systems, while ensuring equitable access to nature's contributions for all, involves navigating numerous trade-offs and challenges. This task is further complicated by an urgent question: As humanity faces unprecedented levels of environmental degradation – threatening our collective survival and demanding radical shifts in governance, regulation, social norms and value systems – should we reconsider the relationship between human rights and environmental security in the Anthropocene? In this era of climate and ecosystem crises, which impact everyone but disproportionately affect the most vulnerable, does this new reality necessitate a fundamental rethinking of our approach to rights? [23] Hugo Slim suggests that humanitarianism needs updating and that the new humanitarianism "is not only for humans but for nature too". [24]

With rights comes *responsibilities*. The extremely dangerous path we are currently following highlights the need for combining rights and responsibilities in a much more stringent way. Delivering a nature-positive society in the new dimension of the Anthropocene certainly reinforces the need to redefine the boundaries of rights and responsibilities and for the integration of these two dimensions.

A nature-positive future also enhances nature's contributions to humanity. 'Nature-based solutions', defined as

> actions to protect, conserve, restore, sustainably use and manage natural or modified terrestrial, freshwater, coastal and marine ecosystems, which address social, economic and environmental challenges effectively and adaptively, while simultaneously providing human well-being, ecosystem services and resilience and biodiversity benefits [25],

capture some of the most direct contributions that nature conservation and restoration can offer to address some of the most pressing socio-economic and welfare issues people face today.

Examples of this include improving water security by conserving and restoring the regulation and purifying functions of ecosystems like forests and wetlands; fighting malnutrition, ensuring food security; supporting economic livelihoods via sustainable fisheries and resilient and regenerative agriculture; creating job opportunities for local communities through ecotourism; improving people's quality of life and mitigating extreme heatwaves in the urban environment through green spaces; enhancing protection from extreme weather events like floods and storms through 'natural infrastructure' like wetlands, coastal forests or reefs; and conserving natural carbon storage or actively absorbing CO_2 from the atmosphere through ecosystem conservation and restoration.

These are only some of the potential socio-economic benefits of nature-positive outcomes translated into nature-based solutions, highlighting the interconnectedness between nature and society. They also demonstrate how nature-based solutions selected or designed to address one specific issue can in fact contribute to addressing multiple developmental goals all at once.

Nature Positive is a global goal that will benefit people through nature, and it supports society to address most, if not all, of its most pressing challenges: from climate change and water and food security to poverty and health. In fact, a nature-positive future is the only chance for a safe, prosperous and equitable world.

Environmental and social objectives, from being perceived as separate and even conflicting, have converged into a reinforcing and interdependent agenda. There is no doubt that both carbon-neutral and nature-positive goals are required to transition away from today's environmentally unsustainable development model, but they are also necessary to sustain human development in a way that manages trade-offs with positive outcomes for both people and nature.

This can be achieved by ensuring that nature-positive transitions in society and key economic sectors are just and adequately supported through the right policies and fiscal and financial incentives, such as the redirection of public subsidies. We must also ensure that countries or communities where biodiversity is higher but economic capability to invest is lower receive the financial means needed to preserve it, and that we adequately invest in preserving and restoring ecosystem services crucial to people's well-being in the areas where they are most needed. Socio-economically just transitions and benefits are another critical aspect of the intersection between nature-positive outcomes and social equality and justice.

As with the energy transition, transitions in key sectors that impact biodiversity such as agriculture, fishing, forestry, extractives, infrastructure, manufacturing and tourism must tailor their plans to the cultural, social and economic local realities. While a nature-positive transition requires moving away from an anthropocentric approach that undermines the importance of nature, it will also involve recognising that people should be at the centre of such a transition and should benefit from it equitably.

In essence, this is why Nature Positive is therefore also *'People Positive'*, as a healthy environment is the foundation for healthy individuals and societies.

3.4   Acknowledging nature's value(s)

"*Green is gold*" said China's President Xi Jinping [26], to signal the need to value nature for her immense contributions to society and economy. We all instinctively know that nature is the foundation of our existence. It purifies and regulates the water that we use for drinking, growing crops and enabling industrial production. It provides us with wonder materials like wood and fibres. Many revolutionary and essential medicines are derived from compounds found naturally in wild plants and animals, while innovations in engineering are often inspired by 'biomimicry', the practice of imitating structures and functions designed by nature.

Two-thirds of our crops are somehow dependent on animal pollinators such as bees, wasps, flies, butterflies and moths but also birds and bats; while animals, plants and algae harvested from freshwater and the ocean supply proteins and essential nutrients to over three billion people [27, 28]. In traditional economic terms, it has been estimated that about half of global GDP (as currently calculated) is moderately or highly dependent on healthy ecosystems providing a variety of goods and services [29]. The reality, however, is that our entire economy depends on productive and stable Earth systems and biodiversity, just as our society does, and just as we do as individuals.

In addition to these tangible and material contributions, there are many other equally crucial but less tangible ones: from the inspiration, beauty and happiness nature transmits to us, to the spiritual and cultural identity natural features and events have given to a multitude of people and groups since the dawn of civilisation. And of course, for over three billion years, photosynthetic plants, algae, phytoplankton and bacteria on land, in freshwater and in the ocean have been clearing the atmosphere of high concentrations of primeval carbon dioxide, filling it with oxygen and transforming it from toxic to breathable, leading to the birth of oxygen-based life and aerobic respiration to which we, and all complex life as we know it, belong. They continue to do so today by providing the oxygen we breathe – roughly half coming from the world's land vegetation and the other half from the ocean – while absorbing over half of the CO_2 emissions we emit into the atmosphere. In essence, nature's contributions are existential and invaluable.

Yet, breathable air and water, essential to our biological functions, survival and health, are economically valued at nothing, or at the very least much less than most other non-essential commodities we produce and consume.

The intuitive but disregarded essential value of nature suggests that the first and perhaps biggest challenge to a nature-positive transition is a cultural one: acknowledging the multitude of values of nature and properly accounting for them when we make decisions that can affect nature. The inconvenient truth is that we have always and are continuing to take nature for granted. This is what I call the '*Great Neglect*'. Nature has always been dominant and plentiful, surrounding us and providing for us since the birth of our species. At the same time, nature is regulated by unforgiving laws that are the source of continuous challenges and threats. Nature has been both our greatest ally and a terrifying force.

Life fully integrated in wild nature is very tough. We evolved primarily to adapt and survive in nature, but then strived in any way possible to protect ourselves from the dangers of the wild by isolating ourselves from nature, in ever more secure homes, in ever bigger villages and, more recently, cocooned in our modern, urbanised lifestyles supported by social norms, welfare and judicial systems and surrounded by increasingly sophisticated technologies. This has created an unprecedented level of security and welfare – 'modernity'. After all, we have been deploying our growing cultural, cognitive and social abilities to fulfil the same old universal aspiration of well-being and survival. We became more and more efficient at extracting the natural resources we wanted and relied on nature's ability to replenish herself, which she has been remarkably able to do for most of human history. Most of us have never questioned this fundamental assumption or considered that the stability and productivity of nature could, one day, no longer apply. Until recently, we have broadly been able to live within the limits of nature's regenerative capacity. Today that is no longer the case.

The more recent capitalist economic model, based on exponential production and consumption, still relies on the assumption that the planet's natural resources are infinite. We are finally fully understanding how utterly wrong this assumption is, and how it has led to unsustainable practices and a dangerous degradation of natural systems.

When Adam Smith, in the eighteenth century, formulated the principles of a capitalistic economy driven by self-regulating market forces, the global human population was less than one billion. Today, with eight billion people on the planet (projected to grow to 10 billion by mid-century), unparalleled technological power and a huge inequality gap in distribution and consumption of resources (which is exacerbated by the impacts of climate change and nature loss), we have developed a global ecological footprint that is totally unsustainable socially, economically and environmentally. And it is growing larger every year.

Most humans today contribute to a society that operates outside planetary boundaries, while many live in communities that are increasingly socially unstable. The cultural imperative of our time is to recognise that the current global economic and governance model is failing to steward the '*global commons*' we all depend on, and to understand the consequences that a degrading natural world has on the lives of present and future generations. In essence, we must stop taking nature for granted. We must recognise nature's material, emotional and spiritual contributions to humanity and shift to sustainable practices of production and consumption, not only to fulfil our moral responsibility to recognise the intrinsic value of nature and to co-exist with all other life on the planet but also to act in our own best interest.

Most people, perhaps all people, love nature, and conserving nature is still largely perceived and presented as a positive, altruistic purpose. But while awareness of the material threat of biodiversity loss to our economy and society is growing, nature conservation is not yet a top priority for politicians, companies, CEOs and for us individuals in our daily lives. The 'Great Acceleration' of humanity and the 'Great Decline' of nature have happened too recently and too quickly compared with the interminable time we have spent in a dominant and rich natural world.

I personally feel very strongly about nature's intrinsic value and our moral duty to coexist with non-human forms of life, and I wish this was enough to make more people respect nature and lead us to a more sustainable society and economy. Sadly, there is much evidence that moral arguments alone have not been the most powerful drivers of behavioural change towards sustainability. Even though they can play a significant role in shaping societal norms and individual values over time, culture and behaviour change often require a combination of factors, such as personal experiences, social influences and practical incentives. Perhaps an unlikely ally here can be Richards Dawkins' *Selfish Gene* – the powerful biological driver of our instinct for self-preservation, which normally extends to our family, relatives and friends closest to us in our social vicinity [30]. If only we were able to internalise the fact that living in harmony with nature and in equity with other humans is not only an act of moral altruism but also self-interest in ensuring that we and our children are able to live safe, secure, happy lives, perhaps our dominant 'selfish gene' will help us live more sustainably and equitably.

Our Common Future defines sustainability as "meeting the needs of the current generation, without sacrificing the needs of future generations" [31]. This reflects a moral imperative that underpins sustainability and the intergenerational justice dimension of human-induced nature loss, which is exacerbated even further by the approaching phenomenon of ecosystemic tipping points.

Today, we use as many natural resources as possible to meet our growing expectations without assessing the real need for doing so and without considering the impact this is having on present and future generations. This is the absolute opposite of the aspirations that every mother and father have for their children, for whom building a safe future is naturally their priority. Intergenerational sustainability implies that people alive today are prepared to make choices that will also benefit future generations. If we can, we invest in the education of our children and save money for their future. Why are we not doing the same in the way we treat nature, the most important insurance policy for our children? Because nature has always been there, and we assume that it always will be, unchanged.

– The painful, difficult debate around pricing living nature

Our value system must change. The way we value nature must change. We must view nature as precious, finite and indispensable. We must see nature loss as a threat to our survival, and our children's future, not just as a reason to feel guilty about the many other life forms we are wiping out.

The perceptions of nature's values are many and diverse, and they respond to different contexts, cultures and even individual experiences and sensitivities. The *Assessment Report on Diverse Values and Valuation of Nature* characterises the different ways in which people relate to and value nature as living 'from', 'in', 'with' and 'as' nature [32]. This reflects different worldviews, philosophies, social structures, religious beliefs and knowledge systems. People who see themselves as living *from* nature emphasise nature's ability to provide resources and livelihoods and sustain our needs. A river, a forest, the sea are valued for the water, the medicinal plants, the fish they provide for people's use. This is a dominant view.

Meanwhile, people who see themselves as living *with* nature value nature's goods and services more holistically and in support of 'non-human' life forms that are seen as part of a whole, with identities and rights to exist. Living *in* nature refers to the importance of natural features in people's lives, practices and cultures. A river, a forest, the ocean are perceived as a place that is part of their identity. Finally, people who see themselves *as* nature perceive it as a physical, mental and/or spiritual part of themselves. In this case, a river, a forest or the ocean are valued based on perceived kinship, belonging and interdependence.

The report emphasises that these interpretations of nature are not mutually exclusive and manifest themselves in various combinations depending on context, but also finds that the dominant focus is on short-term profits, often excluding multiple other values. How we see ourselves *from*, *in*, *with* and *as* nature influences our ability to live up to a shared vision of prosperity for both people and the planet, aligning "the mutually supportive goals of justice and sustainability and their intertwined economic, social and environmental dimensions."

The 'super driver' of global biodiversity decline is linked to our inability to adequately value nature and the way that this is reflected in individual, political and economic decisions. How we apply value to certain things is the foundation of today's economics. In our current economic paradigm, "value determines price and only things that have a price have a value", as the economist and academic Mariana Mazzucato puts it [33]. The economic value of an item is determined by the perceived benefits it provides, often influenced by demand and an individual's willingness to pay, typically measured in monetary terms.

Understanding the true value of nature's multiple dimensions and services is critical to decision-making and to managing trade-offs around development policies, investments and land use planning and to avoid unintended negative consequences.

The problem is that production and consumption remain key metrics of economic success and the foundation for GDP calculations, while both the services derived from nature and the loss of those services are not accounted for. The result of this is that clean air and the rainmaking function of a forest, although having huge positive effects on our lives, have no value in financial markets, as opposed to many traded commodities which ironically affect the air quality or destroy forests.

Even more perversely, destroying a forest for agriculture or reclaiming a wetland for urban development increase GDP and are counted as 'growth' because they involve financial inputs and outputs, while preserving a forest counts for nothing because it has 'always' been there and its crucial services come free of charge. This is a fundamental structural issue in our current economic valuation system which leads to the expendability and destruction of the natural world. The loss of wetlands or the decline of pollinators are not captured by traditional economic frameworks such as GDP even though crop failures caused by water shortages and decrease in pollination have a financial cost.

The inevitable consequence is that converting a wild forest for agriculture or mining will grow the country's GDP without considering the loss of services that the forest provides. This is, in essence, an unrecognised *market failure*, a central concept of environmental economics, in this case the failure to generate environmental and, consequently, social welfare [34], generating negative externalities that affect the environment and society.

The Economics of Ecosystems and Biodiversity was arguably the first comprehensive and structured scientific assessment of the economic benefits derived from nature's goods and services [35]. Its aim was to incorporate the value of biodiversity and ecosystem services into decision-making and accounting, including the non-market economic value of nature through the Total Economic Value (TEV) framework. TEV combines and measures the value of monetary and non-monetary environmental assets. The 'total economic valuation' categorises the benefits derived from ecosystems into '*use values*' related to tangible benefits from direct uses (such as water use from lakes and rivers, timber from natural forests, wild fish stocks) or indirect uses (storm defence from mangroves and coral reefs, flood mitigation and water purification from wetlands, the cooling effect of trees in urban environments) and '*non-use values*' where benefits are of a different nature (cultural, spiritual, emotional or mental). '*Option values*' are based on a potential future use of the resource, and '*existence values*' are based on preserving a natural feature for its intrinsic value and without the prospect of future use.

In our current economic model, the value of everything is primarily based on market exchanges driven by demand and supply, which often leads to a profound distortion of the true value and leaves room for speculation for private profit. Economic values have been classically attributed to land, natural resources, products, labour and capital. Today, the price of something is determined uniquely by demand and supply and reflects the cost of production, while the environmental value or damage are unaccounted for. Nature is only valued economically when it becomes a tradeable commodity. Nature's services, which underpin our economy and well-being but are not traded and monetised, are economically invisible.

As a result, despite understanding that our existence, health and happiness rely on the natural environment, we often overlook this simple and undeniable principle when making decisions about how we live, invest our money or develop our economy. Why? Because we continue to take nature for granted and to consider it to be expendable, and we focus on short-term financial gain. Putting an accurate valuation on nature can rectify, at least partially, this market distortion.

When it comes to valuing nature's contributions to humanity, one of the most controversial but central elements in this debate is its monetary valuation under the current economic system. The idea of assigning an economic value to nature has been heavily criticised. Some of the arguments against are ethical and philosophical. It is simply immoral or impossible to put a price on nature while pointing to the impossibility of accurately capturing nature's extraordinary diversity and intrinsic dimensions. Nature is so existential and immensely valuable to our life that it is ultimately priceless.

It could be argued that the monetary value of an ecosystem's services is equivalent to the cost of replacing that service and the benefits it provides. However, most of nature's services are so essential to life on Earth that they are irreplaceable and inestimable. Also, putting a price on nature is unethical, because nature is not our property and is not there to be traded or monetised.

Other arguments, meanwhile, are more practical and point to the difficulty and the risks associated with valuing a public good, particularly one that is as vast and diverse as nature. Placing a value on nature would be an incredibly complex task that would force us to focus on the distinct components of the natural world, ignoring the fact that they are interconnected, interdependent and often synergic. Commodifying nature based on its contribution to the economy is reductive and ignores all the other important values nature carries. Natural features that make a demonstrably higher financial contribution to the economy could, in theory, be prioritised over other less material but not less important features. Finally, should we really recognise and reward nature's contribution to *this* economy, which is the main driver of her destruction? It is difficult to disagree with the values behind these views.

For others, establishing the economic value of nature's services is crucial, as it will help us to recognise that losing nature comes at a cost not in abstract terms but with a quantifiable monetary value attached to it. If 'who pollutes should pay' is true, then 'who destroys nature should pay' should also apply. In turn, preserved and well-managed goods and services from nature would provide material benefits to the real economy and its stakeholders. And in doing so, the economic value of living nature will appear in the GDP of economies, investors' portfolios and on company balance sheets. It will finally bring many of the hidden costs of environmental degradation out into the open, putting pressure on governments, investors and companies to address them and helping to quantify the financial impact of externalities that until today have never been accounted for.

If nature is not priced (or is priced too cheaply), as is the case today, then converting a tropical forest to grazing land for livestock will continue to be an economic no-brainer. Furthermore, ending the *under-pricing* of natural ecosystems, species or services will help address today's *subsidies paradox* where environmentally harmful subsidies will come at a double cost, firstly as a direct spending cost for the state and taxpayers, and secondly for driving the loss of a financially valuable natural asset. By addressing this issue of under-pricing, we would drive a truly systemic shift inside our economic system.

Natural capital is a concept that looks at nature through a financial and economic lens, recognising that natural infrastructure, resources and processes contribute to the stability and prosperity of our economy and society. Natural capital is the "world's stock of natural resources from geology, soil, air, water and all living organisms" [36, 37]. Put simply, the notion of natural capital reflects a way of looking at the natural world in economic terms, where nature is regarded as a combination of goods and services that benefit society and the economy. The choice of the world 'capital' comes from the fact that like financial capital, natural systems support our economy. Natural capital is made of renewable or non-renewable *assets* like natural abiotic (e.g. water, minerals) or biotic resources (e.g. forests, wetlands, sea grass meadows) and *ecosystem services*, products or processes generated by ecosystems that benefit society (e.g. water regulation and filtration, pollination).

Ecosystem services can be raw materials or products extracted directly from eco-system assets or the outcome of ecological processes that generate a resource or service: *supporting* services (e.g. carbon and nutrient cycles) make ecosystems function; *regulating* services (e.g. pollination) regulate processes and functions that support biodiversity; *provisioning* services (e.g. timber, water filtration) are producing material natural resources; *cultural* services are nature's non-material benefits of cultural, emotional and spiritual value.

This leads to attaching a monetary value to natural assets or services that today have no direct market valuation. Elements of natural capital can be monetised rela-tively accurately, such as carbon storage or flood prevention and mitigation. Others less so [38]. By considering nature's resources and services as capital, we recog-nise their economic importance, but this also forces us to acknowledge the limits of the stocks and the need for their sustainable use. Ensuring the permanence or growth of these stocks will generate value, while their degradation will determine a loss. Nature loss will become bad practice, while its conservation and restoration will become an economic and business no-brainer.

As an extension of the economic concept of capital, natural capital has attracted criticism from an ideological and political perspective from those who believe in a deeper, more radical reform of the current liberal market-based model, while oth-ers recognise that the application of this concept could help us to transition from a uniquely profit-oriented 'shareholder capitalism' to a value-based 'stakeholder capitalism', which is able to address issues of environmental sustainability and social equity.

This has also led to the concept of *natural capital accounting* – calculating the total stocks and flows of natural resources and services in physical or monetary terms. In 2016, the Natural Capital Coalition (now known as Capitals Coalition) released the Natural Capital Protocol [39], a standardised framework to identify, measure and value the direct and indirect impacts and dependencies on and from natural capital. This approach was approved in 2021 by the UN Statistical Com-mission and marked a major step towards moving beyond the evaluation model that has dominated economic reporting since the end of World War Two, ignoring nature's contributions [40].

This measure will lead to natural capital ecosystems and their services being included in economic reporting. The revolutionary aspect of natural capital accounting is that it expresses, in monetary values, the contribution that ecosys-tems make to society and the economy compared to more commonly priced traded goods and services. This allows us to compare values, make informed decisions and assess trade-offs. More generally, it will help change society's perception of nature's economic relevance and its contributions to our lives.

The reality today is that services provided by nature, like insects that pollinate crops and wetlands and forests that regulate and purify water, have always been taken for granted despite their direct and indirect relevance to our economy. This is reflected in the way we measure the performance and development of the economy.

As mentioned earlier, GDP is the most used macroeconomic indicator to measure an economy's inputs and outputs in pure monetary terms but fails to consider nature's critical contributions to society and the economy. Most crucially, it also fails to recognise the economic costs incurred because of nature loss or the unsustainable management of natural capital. In line with the concepts of natural capital and natural capital accounting, and to account for the economic value of nature's goods and services, new ideas have been put forward such as '*Green GDP*' [41], which subtracts the cost of nature loss and degradation from GDP.

More recently, the concept of '*Gross Ecosystem Product*' (GEP) has emerged. GEP not only accounts for the economic costs of nature loss but also for nature's contributions to the economy. GEP has five key areas of focus: nature's contributions to people; the measurement of ecosystem assets as stocks and ecosystem services as flows; the quantification of ecosystem service use; an understanding of ecosystem service contributions along the supply chains; and the disaggregation of benefits across groups [42, 43]. The desired result will be a GDP that measures both financial and environmental stocks and flows and gives a holistic picture of the value and resilience of the economy and nature's goods and services that underpin it.

In 2013, the Economics of Ecosystems & Biodiversity business coalition published a report that concluded that [44], of the world's top industrial sectors ranked by environmental impacts, none would be profitable if environmental costs were fully accounted for. It estimated that the world's primary production and processing sectors are responsible for 'environmental externality' costs totalling a staggering $7.3 trillion annually, 13% of global annual GDP at the time. This came from greenhouse gas emissions (38%), water use (25%), natural habitat conversion (24%), air pollution (7%), land and water pollution (5%) and waste (1%). A year later, another study estimated that the total value of the world's ecosystem services amounted to twice as much as global aggregate GDP at the time – a staggering $124.8 trillion per year [45].

Whether by design, convenience or omission, nature today is – economically speaking – mostly invisible. This has led humanity to use, destroy and waste nature's resources with impunity and without giving it any economic or social consideration. While we tend to preserve our financial capital or to invest it to generate returns, to do our best to avoid overspending, debt and bankruptcy, natural capital has, until now, been simply viewed as expendable. Nature's journey towards 'bankruptcy' has gone unnoticed, and her stakeholders – all of us – have not been and are still not sufficiently aware of the trends and consequences.

Financial capital drives most of today's decision-making. The lack of understanding and accounting for the impacts and dependencies on nature makes corporate executives and shareholders unaware of the risks and costs associated with nature loss and, on the flipside, of the opportunities that arise from the sustainable use, conservation and restoration of nature. This is leading to uninformed decisions that can be harmful to their own businesses, let alone society. In the last few decades, however, it has become increasingly clear that the destruction and misuse of natural capital does indeed lead to economic losses and negative social impacts, such as the impact of water scarcity on agriculture and industrial production.

Giving biodiversity a financial value will help internalise and recognise the cost of its loss, the value of its preservation and drive decisions accordingly.

No number will ever capture the holistic value of nature, let alone its intrinsic value. However, with an economic value on living nature currently at zero dollars – or negative dollars if we consider the public subsidies that incentivise nature's destruction – then valuing nature, imperfect as this is, may nevertheless help reduce the rate at which it is being destroyed.

The ideological argument for not putting a price on nature ignores the reality that a price on nature in fact already exists – but only if it's dead. Trees gain a market value when they are cut for lumber, and not for the oxygen they produce, the soil erosion they prevent, the water regulation and purification function they provide while alive. Fish, meanwhile, are valued when caught and sold to feed humans – or our farmed animals – not for contributing to the health of rivers and seas which in turn generate immensely positive services to our economy, society and individual well-being. And water gains some value after it has been extracted and used for agriculture, industries and households, most likely being polluted as a result – not for sustaining all life on Earth. Even flowers gain value when they are cut and sold for our anniversaries and other special occasions, or are grown for our gardens and flats, not for their beauty and ecological function when living wild in meadows or forests.

The moral responsibility to respect nature in all her dimensions is strongly felt by an increasing number of people and remains a powerful and fundamental driver of sustainability and coexistence. But sadly, evidence shows that this has not been sufficient to change human behaviour and halt the destruction of the natural world. Whether moral or material, both arguments for nature preservation and sustainable management have a valid rationale and do not exclude each other.

We have failed to account for the value of living nature in economic decision-making. This blindness to the values of nature has led to overexploitation, degradation and destruction of the natural world which, in turn, is beginning to affect our economy and society.

The lack of an economic and market value for living nature has, in practice, meant free license for some to exploit, destroy and create waste with no cost to the private actor, but huge, unaccounted costs for society. Consequently, the conservation of nature has always been seen as an economic cost, a forgone economic opportunity and not as an asset or an investment. If nature was seen as a separate asset class with a value attributed to it, and its loss as a risk and a cost, then most likely we would treat it very differently.

From a short-term perspective, attaching a price to nature loss may be seen as an unwelcome additional cost by companies, but only until the moment when nature's goods and services become so scarce that they become hugely expensive and start to have a major impact on the local-to-global economy. At that point, an adequate price on nature's resources and services that supports their conservation, regeneration and sustainable use will become acceptable and even necessary. However, we know that by the time this moment comes, it may be too late, and natural systems may have reached irreversible tipping points. The transformation needs to happen now.

This quote from Ed Begley Jr. captures the cultural challenge that lies at the root of our current perception and valuation of nature: "*I don't understand why when we destroy something created by man we call it vandalism, but when we destroy something created by nature we call it progress.*"

3.5   The nature of our food

"First we eat, then we do everything else." American food writer M.F.K. Fisher's quote [46], or the Spanish proverb "The belly rules the mind", both emphasize the 'power of food' and how we prioritise food over everything else.

Securing calories, nutrients and water needed to sustain biological functions is the most basic preoccupation for most species. Food is one of the key foundational blocks of our existence, and the quest to find and secure food has shaped an amazing variety of anatomical and behavioural adaptations across all living forms. Procuring food, actively or passively, from shoals of tuna wandering across vast areas of the ocean to the stationary tactics of a spider in its web, accounts for the way most creatures spend the majority of their active time.

For humans, securing food for us and our families has always been a fundamental part of life, deeply embedded in our identity. Its influence is so powerful that it shapes many of our conscious and subconscious decisions about what we eat and how we produce it.

From predators stalking their prey to herbivores enduring harsh winters or dry summers, finding food in the wild is a constant struggle – marked by starvation, malnutrition, disease and the looming threat of ageing, when feeding oneself becomes even more difficult.

About 12,000 years ago, at the dawn of the so-called Neolithic Revolution, humans cracked their food insecurity problem by domesticating animals and plants and giving rise to early forms of farming. The use of fire to cook food broadened the variety of plants we could eat, including fundamental staples like rice, wheat and potatoes that were otherwise indigestible and even toxic when eaten raw.

This had profound ramifications on the evolution of our species. More stable, predictable and abundant sources of food boosted population growth. Farming led to more permanent, larger communities, generating more complex social structures and requiring new social and administrative norms and rules. Soon there was no need for all members of society to devote time to providing food, and a much more sophisticated form of diversification and specialisation within society emerged, leading to specialised roles and technological innovation. Humanity was about to embark on the most extraordinary and unprecedented journey any species that had ever lived on planet Earth had ever experienced.

Fast forward thousands of years and today, our food system is designed to fulfil the same fundamental function of nourishing us and supporting our survival, health and well-being. Over two billion people, a quarter of the world's population, depend on farming for their livelihoods, and agriculture is the biggest employer in low- and middle-income countries, with a combined global workforce of around

900 million people, plus many more working in the informal farming sector, particularly in the world's least-developed economies [47, 48]. In addition, there are also the people employed by fishing and aquaculture, as well as the processing, distribution and retail sectors. Farming and fishing keep hundreds of millions of people in employment and out of poverty [49, 50]. The food sector is of existential value to humanity and deserves great attention and respect.

But it is also true that, 12,000 years on, our current food system, defined as the sum of production, transport, processing, distribution and consumption, requires deep reform. In fact, many would go as far as saying that our food production and consumption model is broken, environmentally, socially and economically.

Environmentally speaking, agriculture generates around one-quarter of all anthropogenic GHG emissions [51], is responsible for 70% of global biodiversity loss and 80% of deforestation, is the largest consumer of water worldwide, accounting for over 70% of all freshwater use, and is a major polluter of water (rivers, lakes and the ocean), mainly through runoff of fertilisers, herbicides and pesticides. Current farming practices erode the soil on which production depends up to 100 times faster than soil formation processes. As a consequence, it is not surprising that an estimated 52% of agricultural land, an area twice the size of China, is, to some extent, degraded [52].

Beef, soy and palm oil are the primary drivers of deforestation [53]. The farming of animals, a highly inefficient way to produce calories and protein, has a huge impact on both the environment and our health. Meat and dairy alone account for 12–14.5% of global GHG emissions, primarily from deforestation and methane emitted by ruminants like cows, goats and sheep [54]. Livestock farming alone occupies 35% of habitable land [55].

Although the expansion of pastures is peaking globally, pastures for cattle alone account for almost half of all tropical deforestation, mainly in Latin America, but increasingly in Africa [56, 57]. The figure is likely higher if we include deforestation driven by cultivation (soy, for example) used to feed farm animals.

As production of animal products increases, more of the cereal production is being directed to animals. The production of meat and dairy requires significantly more land, water and other inputs than plant-based foods, and livestock production is the single largest human use of land. An incredible 75% of all agricultural land (which occupies already nearly half of the world's habitable land) is directly (grazing) or indirectly (growing feed for animals) dedicated to animal farming. Producing 100 g of beef requires an average of 164m² of land, whereas 100 g of tofu requires just 2.2 m² – a land-use footprint 75 times smaller [58].

To meet growing demand, industrial meat and dairy production is increasingly the norm in both developed and developing economies and comes with appalling and unethical levels of animal suffering. Levels of waste are also shocking. Each year, a staggering 18 billion chickens, turkeys, pigs, sheep, goats and cows, equalling 52.4 million tonnes of meat, are killed without reaching our plates [59]. Reducing meat production and consumption would help fight climate change, deforestation and prevent the unnecessary suffering of animals. It is also important

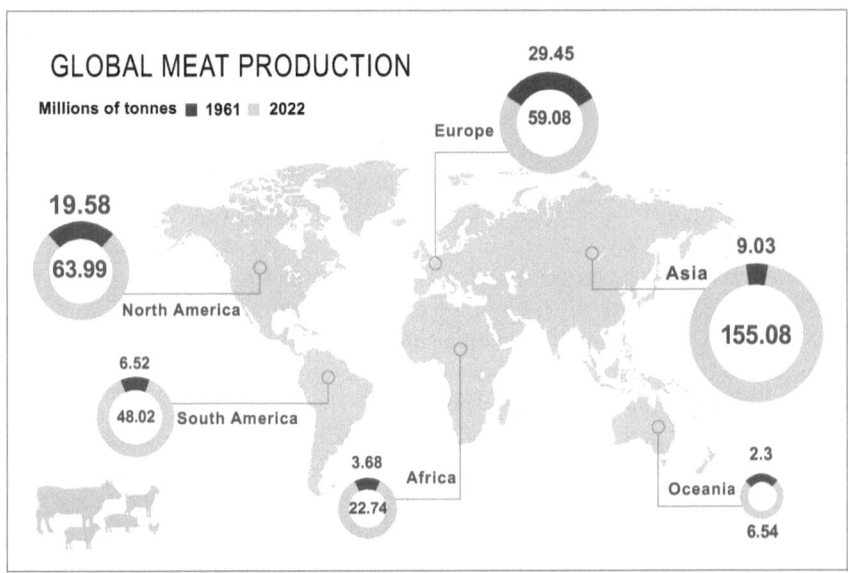

Figure 3.2 Global meat production.

Source: Our World in Data 2023. Ritchie, H., Rosado, P. and Roser, M., 2019. "Meat and Dairy Production" revised in 2023. Published online at https://ourworldindata.org/meat-production

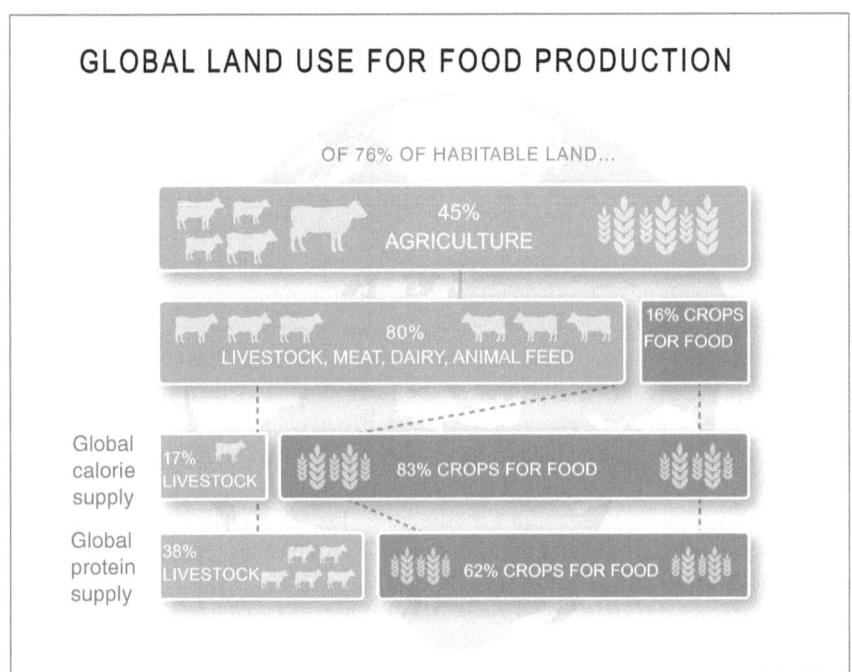

Figure 3.3 Global land use and food production.

Source: Our World in Data, 2024. Ritchie, H., and Roser, M., 2019. "Half of the World's Habitable Land Is Used for Agriculture" revised in 2024. Published online at https://ourworldindata.org/meat-production

to note that today, 68% of deforestation is driven by small-scale farming for crops and livestock, and the rest by industrial farming [60].

The impacts of farming on biodiversity are many and devastating. Even today's strictest food security and agro-environmental standards are utterly insufficient to limit the damage. For example, the *European Farmland Bird Index* shows a staggering 34% decline of farmland bird populations since 1990, while the *Forest Bird Index* remained basically stable with a slight increase of 0.1 %, highlighting the scale of the negative impacts of European agriculture on biodiversity. This is also confirmed by the *European Grassland Butterfly Index*, which shows a decline of 39% in the same period. Agriculture and animal farming, both in Europe and globally, are still far from being ecologically sustainable. Modern farming is not a 'steward' of the environment, as some vested interests would have you believe, but is in fact an environmental slaughterhouse.

The impact of agriculture and animal farming is felt far and wide, and affects freshwater systems and the ocean. Many herbicides used on land are photosynthetic inhibitors and have the same impact on freshwater and marine ecosystems when washed into rivers and subsequently end up in the sea [61]. This impacts the hugely important carbon sequestration and oxygenation function of phytoplankton, but also affects the entire marine food chain which is founded on this minute but abundant organism. As a result, an unabated inflow of herbicides may soon start seriously affecting the ecological stability and productivity of coastal waters and their fisheries.

In the tropics, herbicides are also considered a contributing factor to the bleaching of corals as they affect the symbiotic algae hosted within them. In a fine example of sophisticated interaction between natural realms, the decline of marine phytoplankton also impacts agriculture on land. Phytoplankton is responsible for the production of a thin lipidic surface microlayer (SML) on the ocean's surface. Apart from reducing water evaporation and mitigating global warming (as water vapour is the most potent greenhouse gas), the SML also contributes to the formation of aerosols, which play a key role in cloud formation and precipitation. With fewer aerosols, fewer clouds form, leading to less frequent but more intense precipitation, causing droughts and floods with significant damage to agriculture [62].

Socially, food systems have impressively kept up with decades of population growth, overconsumption and waste, and have contributed to the reduction of malnutrition and poverty and an increase in life expectancy. However, progress has been hugely inequitable around the world, and the evolution of how we produce and consume food is contributing to some of today's greatest societal challenges, from climate change and biodiversity loss to social and health-related crises such as persistent hunger and malnutrition but also obesity, diabetes and cancer.

Today, our food system leaves over 700 million people in the world facing hunger, while nearly one in three (2.4 billion) are suffering food insecurity and lack of access to adequate food, leading to malnutrition and serious health and developmental consequences. In 2019, around three billion people could not afford a healthy diet [63]. At the same time, 2.5 billion people are overweight, of which 890 million are living with obesity. The fact is that, on average, at the global level, we produce more food than we need per capita, but around 30% of all food produced is lost on the farm and post-harvest stages or wasted afterward [64]. In 2022, the world wasted 19% of food available to consumers, 13% of it at the retail level, 27% in food service and 60% at household level. That is in addition to the 13% of the world's food lost in the supply chain, from post-harvest up to and excluding retail [65].

Modern intensive food production still largely depends on high levels of chemical inputs both in agriculture and animal farming, with harmful effects on both environment and human health.

It is simply astonishing that 70% of all antibiotics produced worldwide are administered to farmed animals, including in aquaculture [66]. This has become a routine practice since it started just a few decades ago, alongside the 'Great Acceleration' of meat consumption, to stimulate body growth and to prevent diseases in the intensive animal farming industry where high standards of hygiene are very difficult to maintain, and the high concentrations of animals make it easy for disease to spread. The consumption of meat treated with antibiotics is believed to be one of the main risks of antimicrobial resistance (AMR) in humans that is already responsible for about 1.27 million deaths globally, rising to almost five million if including associated deaths [67].

Administering hormones to animals farmed for the meat and dairy industry in order to promote growth or to increase and extend milk production has also been shown to disrupt the endocrine system in humans, possibly contributing to the decrease in male fertility and rise in cancer, diabetes and obesity. Farmers themselves are not exempt from the impacts of the current food production system either. Economic pressure, labour exploitation and other socio-economic factors often lead to farmers experiencing much higher suicide rates compared to other professions [68], and this seems to be particularly associated with conventional production models and use of pesticides [69].

Economically, the uncounted cost of today's food system has been calculated at $15 trillion (of which $11 trillion is related to health and $3 trillion is in environmental costs), well above what the food system contributes to the global economy [70]. Moreover, much of our food production would not survive without heavy public subsidies. After fossil fuels, agriculture is the most heavily subsidised sector, with more than half a trillion dollars per year. The fishing sector, particularly its oceanic and more industrialised fleet, is also heavily subsidised. These subsidies are often inefficient, inequitable, damaging to human health, and a shocking 87% are harmful to the environment [71].

In the post-war period, the increasingly subsidised, mass use of fertilisers and pesticides, as well as the mechanisation of agriculture, lifted millions out of poverty, hunger and malnutrition. However, the damage to the environment has been accumulating over the decades to such an extent that land has been degraded and yields are falling in many areas [72]. It is the misallocation of public agricultural subsidies, influenced by the lobbying of the industry and farmer associations which has driven this degradation. Without profound reforms and redirection, these public subsidies will continue to turn healthy land into unproductive land.

Apart from adopting increasing mechanisation and chemical inputs, the modern model of food production has not changed its basic exploitative approach to nature since it started to open forests to make space for crops over 10,000 years ago during the Neolithic revolution. It continues to convert nature and destroy biodiversity to make space for highly controlled crop production, due to the illusory assumption that we can replace nature's complex services with simple chemical or mechanical alternatives.

Today, agriculture is still the main driver of deforestation and conversion of natural habitats. In the first two decades of the twenty-first century, cropland increased by almost 10%, primarily due to agricultural expansion in natural areas of Africa and South America [73]. This does not only affect climate stability and ecological services but, paradoxically, also people's health and food security. It also generates huge hidden costs to the environment, society and public health. Externality costs are not reflected in the market prices of food. While they contribute to private profits, they negatively impact society as a whole.

Evidence from the last few decades shows that with economic development comes greater consumption of resource-intensive foods such as meat and processed food. In a demand-supply relationship, logically demand drives supply. But what and how much we consume is greatly incentivised by availability, price and marketing strategies. Essentially, the more we produce, the cheaper food becomes, and the more we are stimulated to consume and waste. Increased availability of cheap food drives its consumption. This has been the vicious consumeristic circle behind the widespread consumption of cheap and processed food of the last seven decades or so.

The intensification of agriculture has made grains cheap enough to be used for animal farming, boosting industrial production of meat and driving a shift in diets in many regions of the world. In these same regions, cheap food has on the one hand led to overconsumption and on the other made waste economically negligible. While affordable food is essential to tackling hunger and malnutrition and a key dimension of social equity, "the 'cheap food paradigm' drives a set of overlapping and often self-reinforcing mechanisms, in which the ratcheting up of production and liberalisation of global markets incentivise economic behaviour that creates negative outcomes for society and the environment" [74].

The consumeristic view that we should produce more and cheaper food to allow consumers to spend their income on other goods and services, and in so doing stimulate further production, consumption and associated neoclassic economic growth, has dominated our recent economic model over delivering equitable development, human well-being and planetary health. The result is growing inequality and social and environmental negative impacts. Government support for agriculture has played a fundamental role in driving, cementing and reinforcing some of the most intensive and unsustainable production practices as well as unhealthy consumption patterns that characterise today's global food system and 'modern diets'. If this support were adequately repurposed through new policy reforms, fiscal and incentive measures, and social and environmental performance-based incentives, it would undoubtedly accelerate the transition of our current food system towards one that is sustainable, healthy, equitable and efficient.

Transitioning to a better model of food production and consumption is possible. Currently, a large percentage of global agriculture is monoculture. In these deeply unnatural conditions, productivity relies on a high level of costly chemical inputs and intensive farming practices and is often highly mechanised [75]. Regenerative agriculture, meanwhile, is based on the concept of growing food by enhancing soil health and fertility and controlling pests through agro-ecological practices that conserve biodiversity while using zero, or at least significantly less, chemical or mechanical inputs.

The concept of regenerative agriculture is not new, and in fact it has ancient origins [76]. It involves producing food profitably, but in a way that preserves and increases soil quality and biodiversity. This includes practices like polycultures, crop rotation and diversification, zero- or low-tillage, avoiding or reducing bare soil exposure, encouraging plant and invertebrate diversity, preserving and restoring natural features such as wild meadows, hedges, forest patches, ponds and streams either by retaining these pockets of natural habitat for wildlife within the farm or as 'set aside' areas for nature in the broader landscape, ultimately leading to a 'nature-positive agriculture' that supports more biodiversity below and above ground.

There are two distinct dimensions to the impacts of agriculture and animal farming on biodiversity. On the one hand, there are the direct effects on natural ecosystems due to expansion and land conversion, including the opening of access to intact natural areas, which leads to human encroachment and other pressures. On the other hand, there are impacts on the biodiversity of working land, resulting from the removal of natural or semi-natural features (such as trees, hedges, ponds and wildflowers) and species (perceived as predators or competitors), as well as the effects of pesticides and herbicides, leading to the decline of insects, birds and soil biodiversity. A nature-positive agriculture strategy must tackle both these dimensions.

The land use approaches to deliver net-positive biodiversity outcomes in agricultural land include 'sparing' and 'sharing'. Sparing means setting aside areas for biodiversity within the farm, preserving natural habitats and natural features to allow the presence and movements of local biodiversity. Sharing is about adopting practices that allow and encourage the presence of biodiversity, from adopting harvesting practices that prevent damage to nests, dens or newborn animals, to maintaining or restoring natural features like hedges, trees, ponds and strips of wildflowers interspersed with crops [77].

Monoculture and pesticides are known to lead to reduced insect diversity and simplified ecological food chains and have also been proven to be an underlying cause of resistance and cyclical 'pest' invasions, with devastating impacts on crops. Higher diversity of invertebrates, birds and other wildlife on farms is directly linked to greater plant diversity and reduced pesticide use. This increased biodiversity serves as a powerful and more resilient pest control mechanism through interspecific competition and predation.

Winter cover crops, crop rotation, field boundaries and strips or patches of natural vegetation are examples of how to reduce the impact of pests while reducing or progressively eliminating chemical intervention. Emerging research on the profitability of regenerative agriculture has produced counterintuitive results. For example, a study in the northern plains of the United States found that agricultural pests were ten times less abundant in pesticide-free regenerative farms than in chemically treated fields [78]. The study also found that although regenerative practices generated one-third (29%) lower production, farm profits were more than three-quarters (78%) higher, driven by lower input costs and higher market premiums [79].

The reduced costs of the social (for example, healthcare) and environmental (floods and droughts) externalities of regenerative agriculture, even on a purely economic basis, would justify the redirection of subsidies towards regenerative rather than intensive agriculture, to increase its competitiveness.

Current industrial crop and animal farming is increasingly impacting the stability, productivity and health of the very natural systems that support the industry itself, mirroring our overall not-fit-for-purpose economic model. Our current industrial food production model is in opposition to nature. Yields are increased through practices which are damaging biodiversity and ecosystem services, both above and below ground. We have reached a paradoxical point where, both in the case of agriculture on land and fishing and aquaculture in the ocean and inland waters, the same food systems that rely on stable and healthy natural systems are at the same time driving deforestation, soil erosion, pollinator decline, water pollution, overfishing and biodiversity loss. The very farming system that relies on a stable climate and nature's services is now the biggest contributor to climate change and the destruction of nature. As nature's services that support food production are progressively weakened and destroyed, our food security and access to healthy, nutritious food is also increasingly under threat.

– Blue food

Of course, food does not just come from land. The 'blue food' which comes from freshwater systems and the ocean is of immense importance.

Fish and other seafood, including algae, provide a crucial source of nutrition to some 3.3 billion people [80], and supplement key elements of our overall health, such as acting as sources of vitamin A, iodine, calcium, iron and omega-3 fatty acids. Globally, marine and freshwater fisheries provide food and/or livelihoods for 560 and 160 million people respectively, often in low-income communities [81, 82].

When it comes to wild catch or farming of aquatic organisms, the challenges resemble those of farming on land, especially in terms of impact on ecosystems (for example, bottom trawling, mangrove deforestation and wetland conversion for aquaculture), pollution (eutrophication and antibiotics from aquaculture, plastic waste and abandoned fishing equipment) and waste (bycatch in wild fisheries). Bycatch, defined as unintended capture of non-target species by unselective fishing methods, represents a significant ecological impact and a waste dimension of 'blue food'. Whilst bycatch in some instances may be sold, in many cases it is unusable for regulatory or economic reasons and thrown back into the sea, most of the time dead.

Beyond fish and marine invertebrates, unintended mortality due to bycatch is a threat to half of all marine mammal, seabird and turtle species, accounting for 300,000 small whales and dolphins, 250,000 endangered loggerhead turtles and critically endangered leatherback turtles and 300,000 seabirds including majestic species like the albatross [83, 84]. It is estimated that between 38 million tonnes of sea creatures are unintentionally caught every year, or between 10% and 40% of the total fish catch worldwide [85, 86]. Almost half of bycatch comes from bottom trawling, and for some species, the proportion of bycatch is even higher: one kilogram of shrimp can lead to 5–20 kilos of bycatch.

Aquaculture depends significantly on feedstock derived from wild-caught fish. 'Reduction fisheries' – industrial fleets targeting small pelagic species such as anchovy, whiting and sardine – account for an estimated one-sixth to one-third of the global marine catch, depending on the year. Approximately 70% of this catch is processed into feed for farmed fish, while the remainder is used for land-based animal feed, dietary supplements, cosmetics and, to a lesser extent, human consumption [87].

Recent research suggests that aquaculture globally relies on much larger quantities of wild-caught fish than previously estimated. The fish-in: fish-out (FIFO) metric, which evaluates the ratio of wild-caught fish biomass used as feed to the farmed fish biomass produced, reveals inefficiencies in the system. For carnivorous farmed species such as salmon and trout, the biomass of wild fish used for fishmeal often exceeds twice the biomass of the farmed fish produced. This raises crucial questions about which fish species can be more sustainably farmed to minimise the impact on wild fish stocks and what alternative, lower-impact aquaculture feeds could be adopted. Despite advancements in sustainable feed options, the rapidly expanding offshore aquaculture industry often adds these alternatives alongside wild-caught fish feed rather than replacing it, perpetuating the pressure on marine ecosystems [88].

The result of all these impacts is that one-third (34%) of assessed global fish stocks are overfished, 60% are maximally fished, while only 6% are underfished (in other words, are in a healthy condition) [89]. A recent study assessing 230 fisheries involving 125 species found that the state of populations of many overfished species is far worse than reported, and the sustainability of fisheries overstated [90].

This disregard for living marine resources is representative of our broader perception of the ocean. We have always looked at the ocean as a limitless resource, primarily for food and, more recently, sand, minerals and metals. The ocean looks so vast, and what happens there is mostly out of sight, and thus out of mind. We never considered, until very recently, that human activities could ever represent a threat to the health, productivity and stability of the world's largest ecosystem, which covers 71% of the planet, so much so that we kept dumping plastic, sewage, industrial waste, dredge from mining, chemical pollutants and even radioactive waste into the sea. Huge amounts of fertilisers (nitrogen and phosphorus-based) used on land ultimately end up in the sea, causing eutrophication and leading to abnormal or toxic algal blooms which in turn cause hypoxia and create so-called marine 'dead zones.' Pesticides and herbicides affect marine life at all levels and, most worryingly, plankton at the base of the ocean food chain.

Another indicator of our lack of care and attention for the ocean and – among the many things it gives us – its contribution to food security is the manner in which Marine Protected Areas (MPAs) are grossly undervalued. Just over 8% of the ocean surface is protected compared to over 17% of land [91, 92]. Of this, less than 3% of the ocean is strictly protected through 'no take zones', where any significant fishing is forbidden, and these are the type of protection regimes that have proven to be more effective in replenishing fish stocks. MPAs have proven to be effective at both protecting what is left of marine life and to help restore what has been depleted. There are many successful examples of fish stock recovery using a well-designed and managed network of MPAs and benefits to fisheries generated from the 'spill over' effect into adjacent or even distant fishing areas [93].

The benefits of MPAs to fisheries include increased fish stocks and catch volumes outside the protected area but also at greater distances through larval dispersal, and larger fish size which in many species is in direct correlation to the number of eggs produced and therefore leads to a more rapid recovery of depleted stocks. In seabass, for example, a female of 40cm produces 250,000 eggs, compared with 1.3 million for a 60 cm fish and 3.3 million for an 80 cm fish [94].

Many successful local examples show that managing a network of MPAs represents one of the best strategies for maintaining the sustainable harvesting of marine resources yet still face opposition and are seen as 'set aside', unproductive areas rather than what they are – an investment in the health of fish However, many MPAs worldwide are not effectively managed or do not cover the most critical habitats and locations. This means that on the one hand they are not delivering the benefits they were designed to deliver, and secondly, they give a false sense that the ocean is being protected. Much more needs to be done to protect more of the ocean and to do it where it matters.

The economic benefits of MPAs go beyond fishing. There is plenty of evidence that the declaration of a MPA attracts tourism with benefits for local communities, businesses and local governments. In the past decade we have seen an acceleration of the establishment of MPAs around the world, and it is encouraging that the UN Kunming-Montreal Global Biodiversity Framework includes a target of protecting and effectively managing 30% of the ocean by 2030.

Then there is a legal issue. The ocean is administratively divided into two broad categories: waters under national jurisdiction where national laws apply (39%), and international waters beyond national jurisdiction (61%) – the 'high seas' or *mare liberum* – with open access. The UN Convention on the Law of the Sea (UNCLOS) is a post–World War Two international treaty that provided access to the high seas (which are 200 nautical miles from the coast) to all states, with no state being able to subject any part of it to its sovereignty. The purpose of this was to facilitate operations like fishing, navigation, seafloor cabling and seabed exploitation in the interest of the global community.

The institution of Regional Fisheries Management Organisations (RFMOs), tasked to ensure sustainable management of fish stocks in a particular region by setting fishing quotas and regulating fishing methods, and the very recent UN Biodiversity Beyond National Jurisdiction or 'High Seas Treaty', have moved us a step closer to a more coordinated and equitable governance of the ocean and to recognising it as one of humanity's most precious 'global commons'. The perception of the ocean is slowly shifting from 'too big to fail', leading to overexploitation, to 'too big to ignore' [95].

– Wild food

All food comes from the 'wild', either directly or through domestication of plants and animals. Beyond the many species of wild fish and other marine organisms which are widely consumed and traded both locally and across the world, there is another, less known type of wild food: so-called 'bushmeat'.

With 800 million individuals worldwide still suffering from hunger, millions of people in rural areas of poor countries are still resorting to wild foods from local ecosystems to fight hunger and malnutrition, often made worse by social unrest, conflicts and environmental stress. Wild meat or 'bushmeat' refers to meat from wildlife hunted for human consumption, and it is a contributor to food security and nutrition for millions of individuals throughout the developing world [96], representing an important or even primary source of protein and bioavailable nutrients in areas where animal farming is difficult or expensive.

Humans have relied on wildlife for food for millennia, employing a variety of traditional hunting methods to harvest diverse species. Many Indigenous communities developed sustainable practices for managing wildlife resources, guided by a combination of factors. These included their relatively small population sizes, the use of simple hunting tools and techniques and a primary focus on securing sufficient food to meet the needs of their families or villages.

In the last few decades, however, the 'Great Acceleration' has reached many of these remote areas and traditional societies. Population growth, habitat destruction, the loss of traditional hunting control measures and the advent of advanced and unselective hunting tools such as guns or metal snares have put unsustainable pressure on wildlife, driving a sharp decline of populations and even local extinctions. Meanwhile, the rapid human migration to cities and growth of urban populations has generated a new demand for bushmeat and triggered an unprecedented bushmeat trade [97]. At the start of the new millennium it was estimated that up to five million tons of mammal bushmeat were extracted every year from the Afrotropical forests [98], equivalent to approximately two-thirds of the bovine meat produced in the European Union at the time [99].

The result is that bushmeat hunting in the tropics has been largely unsustainable across all regions for several decades, and even when it is conducted as genuine subsistence hunting [100]. The cause is a combination of human population growth, also due to relatively recent local migration and the decrease in extent and degradation in condition of the forests, with the addition of the fact that wildlife population densities are naturally low in tropical forests.

Whether to maintain ties with past traditional lifestyles or as a matter of taste preference, delicacy or status symbol, the consumption of bushmeat by urban and expatriate communities has fuelled an unprecedented national and international trade of bushmeat through well-organised networks, mainly illegal [101].

There is also another dark and threatening side to bushmeat consumption and trade. Many of the wildlife species consumed or traded as bushmeat may carry pathogens. The hunting, handling and consumption of bushmeat has been linked to several serious disease outbreaks including Ebola, HIV and SARS [102]. A recent analysis of 58 species of bushmeat found that 48 species were hosts to one or more pathogens. At the same time, the awareness of health risks is low among hunters, traders and consumers [103].

In a disturbing parallel with many other examples of resource overexploitation, the overharvesting of wildlife for food has become one of the greatest threats to biodiversity in many remote regions of Africa, Asia and also Latin America, while at the same time threatening food security of the marginalised populations who still rely on this type of food [104]. Once again, biodiversity conservation, food security and human health are different sides of the same challenge: a sustainable, healthy and safe future for humanity.

– Time for a 'food system transition'

Alongside predicted population and income growth and dietary shifts towards higher consumption of animal-based products in recent decades, global food demand will continue to significantly increase. At the same time, it will also increase the impact on environmental and human health unless we embrace a transition that leads to a more sustainable way to produce and consume.

Knowledge is the key pillar of every transition, and, undoubtedly, today's better understanding of the social and environmental impact of food systems has resulted in a raised awareness of the ecological footprint and health risks attached to our diets. More people and farmers are beginning to come to terms with the fact that decimating the pollinators of our crops or poisoning our own food is a not exactly a smart strategy! However, much more needs to be done to connect the food on our plate to its environmental impacts. Most meat products top the list of foods with the highest environmental footprint, but the global average per capita consumption of meat continues to grow, driven by increasing individual incomes and population growth as well as cultural and psychological factors.

Meat consumption growth rates vary across different regions, with consumption in high-income countries stable or declining and in middle-income countries moderately to strongly increasing [105]. As global incomes increase, diets typically shift to ones with a greater proportion of meat, dairy and eggs. This transition from plant to animal-based diets has been called the 'Livestock Revolution'. It is estimated that 40% of the world's population will undergo this transition to more consumption of animal products by mid-century [106].

Meanwhile, vegan, vegetarian and flexitarian diets are on the rise and vary from a total avoidance to a significant reduction of animal products in favour of plant-based ones. Vegan and vegetarian are, however, still a small proportion of global diets (3% and 5 % respectively), while flexitarian diets account for 14%. Unsurprisingly, there is a great variability between countries, with India accounting for 11% and 25% of the global number of vegan and vegetarians respectively [107, 108].

The rationale for plant-based diets from an environmental viewpoint is that the reduction of animal-based foods would lead to less livestock farming with, as a result, less destruction of biodiversity and emission of GHGs, including methane, a GHG that is 25 times more potent than CO_2, albeit with a shorter permanence in the atmosphere. Reduced beef consumption and production will also free up formerly deforested areas that could be restored to their natural state, with great benefits in terms of carbon absorption, biodiversity recovery and ecosystem services. According to Intergovernmental Panel on Climate Change (IPCC) estimates, as much as 7 giga-tonnes of CO_2 equivalent per year (GtCO2-eq year) could be saved by 2050 (about one-third of total food system emissions!) through sustainable, healthy diets as a consequence of lower methane emissions and the restoration of freed up land [108, 109].

If 54 high-income and high-meat-consuming countries adopted the EAT-Lancet diet of reduced meat consumption, also known as the Planetary Health Diet, they would not only be able to cut their agricultural emissions by 61% but also free up over 400 million hectares of land, a surface larger than the European Union, with opportunities to restore natural habitats and create huge additional benefits for climate and biodiversity [109]. Investing in the consumption and production of plant-based food has an average return on investment in terms of reduction of GHG emissions five times higher than the investment in renewables and four times the investment in electric vehicles. It could release at least 1.6 bn hectares of land which could in turn return to a more natural condition, benefiting biodiversity and unlocking an enormous additional carbon sequestration, equivalent to up to two times the sequestration from the Amazon and Congo forests combined [110].

The relatively recent exponential global shift to animal protein–based diets is driving intensive, industrialised production systems. Under the pressure of increasing demand for meat [111], livestock production has increased fourfold in the last few decades and has shifted from using mostly waste products, post-harvest crop residues and unutilised land to high-energy feed crops grown deliberately [112, 113].

With this comes an increased need for high-energy animal feed to raise more animals and make them grow faster in weight, alongside administering growth hormones and drugs. Today, shockingly, around 75% of soy and maize production worldwide is used to feed mostly chickens and pigs, rather than humans. Growing food exclusively for direct human consumption could feed an additional 4 billion people, almost twice the population growth expected by mid-century. This is because converting plants into meat is a very inefficient and wasteful way to produce calories. Reducing meat consumption will reduce the use of crops to feed farmed animals and could significantly increase global food availability and security [114].

The mass use of synthetic fertilisers, pesticides, antibiotics and hormones, alongside policies and incentive schemes that reward high-input, unsustainable and non-regenerative farming, are at the centre of our modern food production system. To highlight the power of food in our lives and societies, it is interesting to look at modern history and note that both opposing capitalist and socialist political models and ideologies focused on industrialised mass production of food through monocultures with high chemical inputs and heavy mechanisation.

This was and still is synonymous with modern agriculture, progress, food security and economic development. The result of intensive, large-scale and monoculture agriculture also has social implications: 72% of all farms are smaller than one hectare and control only 8% of all agricultural land. The 1% of farms that are larger than 50 hectares control 65% of the land [115]. This system focuses primarily on intense, short-term production instead of resilience and long-term sustainability, leading to high levels of food loss and waste. While this approach has served at least a part of humanity relatively well for several decades, it has now passed its own sustainability boundaries, let alone the planetary ones, reducing the diversity of crops, eroding the soil, decimating pollinators and increasing the risk of crop failure and food insecurity, while also threatening human and planetary health. The model is based on the hypothesis that by producing huge quantities of food, we are supporting food security and farmers' livelihoods, but in reality, it has created a socially, economically and environmentally unsustainable system and an uncertain future for the food sector itself.

The challenge posed by the food system transition is substantial but not insurmountable, and a transition is absolutely necessary for our food security and our health and well-being. It is estimated that by mid-century we will need to feed an additional two billion people as the global population increases to over 10 billion. We will also need to eradicate hunger, which affects about one billion people, and malnutrition, affecting two and a half billion [116].

We know how to transition from today's unsustainable farming model to a nature-positive one. Adopt more sustainable farming methods by resorting to environmentally friendly practices such as crop rotation and natural pest control, or by using innovative, low-footprint, low-input technologies like precision agriculture and alternatives to synthetic fertilisers. Revert to more plant-based diets. Reduce unnecessary consumption and eliminate loss and waste. Sustainably increase the productivity of already converted or degraded land. This is called 'sustainable or ecological intensification' [117]

The anxiety surrounding this model stems from the fact that the intensification of agriculture over the last few decades has been the source of disastrous impacts on climate, biodiversity and human health. Supporters of this approach point to the tension between growing demand for food and the need to reduce the conversion of natural habitats. As dietary shifts happen very slowly, while deforestation and other forms of habitat conversion continue unabated, growing more from the land already used for agriculture seems like a logical and practical approach.

However, the concept of sustainable intensification is rightly met with suspicion and distrust due to its implementation so far, though it should not be dismissed entirely as greenwashing. The key lies in how it is done. Agricultural intensification means growing more on the same plot of land, and sustainable intensification must demonstrate that, alongside increased yields, it also delivers net-positive biodiversity outcomes, net-zero emissions, animal welfare and social equity. Transitioning from traditional to sustainable intensification requires increasing production efficiency while reducing environmental impacts by adopting more sustainable alternative inputs and practices [118].

The current food production model will continue to exacerbate competition for land (what the Land Carbon Lab of the World Resources Institute call the 'land squeeze') and water to even more unsustainable levels [119]. The transition to a future-proof, nature-positive food system will require a deep systemic change that prioritises more sustainable and healthier food and a move away from quantitative production targets of food that is bad for people and bad for the environment. This includes more careful planning for how we use our land, and the redirection of approximately half a trillion dollars of harmful subsidies towards sustainable regenerative practices.

Technology can help reduce many of the environmental and social impacts of our food system. Innovations that help us to produce 'better with less' – less land, less water, less emissions, less chemical inputs, less loss and waste – have a huge role to play. It is about using technologies that enable a reduced and more efficient use of inputs and adopt more sustainable alternatives than current synthetic inputs and unsustainable practices.

Precision agriculture, cultivated proteins and regenerative practices are just some of the approaches with the potential to produce better quality, more sustainable food, reduce waste and keep nutritious food affordable and accessible. However, this requires a cultural and systemic change that is currently opposed by many 'lock-ins' created by today's system, the first of which is agro-subsidies.

To achieve nature-positive agriculture, alongside changing practices and adopting new technologies, we should follow a strategy guided by crucial global systemic shifts, as well as local sociopolitical contexts. The Food System Economics Commission has identified five very clear and pragmatic steps which could have a transformative impact [120]:

1. Shift consumption patterns towards healthy and sustainable diets.
2. Reset incentives: subsidies from governments.
3. Reset incentives: fiscal policies and revenue allocation to support the transformation.
4. Innovate to increase labour productivity and workers' livelihoods, especially the most vulnerable.
5. Scale up safety nets to keep food affordable for the poorest.

The Commission also identifies critical tensions that must be eased to acceler-ate the food transformation, including fears of food price rises, job losses, policy siloes, global inequalities and entrenched vested interests.

In reforming food systems, first and foremost, we must focus on farmers and fishers. Not by supporting the status quo but by ensuring a fair transition to sus-tainable practices, innovative technologies and financial mechanisms that improve livelihoods and reward farmers and fishers for meeting not only quantitative but also qualitative targets for producing quality food while protecting the climate, nature and our health. This means shifting from a monoculture and high-input industrial production towards sustainable and regenerative production practices and new technologies able to improve the resilience, productivity and profitability of the half billion small-scale farms and the families and workers who run them.

Secondly, we must focus on consumers. Transparency about the environmental and health impacts of food can help consumers to make better choices, and increase demand for healthy, affordable, nutritious and environmentally and socially sus-tainable food. Although changes in dietary behaviours are generally difficult and slow, surveys indicate that consumers' awareness of the environmental impact of food is growing, particularly regarding climate, and especially among younger age groups [121]. However, it is also clear that the majority of people are more likely to embrace dietary transitions when associated with health benefits and when they can find alternatives that taste good, satisfy their hunger and are competitively priced. This highlights the need for accurate information and incentive policies to support a dietary shift [122].

3.6   Climate and nature, the two interdependent sides of today's ecological crisis

The tragic 2019–2020 Australian forest mega-fires perhaps marked a turning point in the global public perception of the climate and nature agenda. Apocalyptic images of a fire stretching as far as the eyes could see, immense columns of smoke rising to the sky, vast areas of incinerated forests, iconic Australian wildlife like koalas and kangaroos burned alive and the red night skies like those of an alien planet were broadcast worldwide, reaching billions of people around the world. Those images symbolised most powerfully and clearly than ever before the interconnect-edness between climate change and the destruction of nature. After being treated as separate topics for decades, even dealt with by two separate UN conventions, the interdependence of climate and biodiversity has never been clearer. Today, science is unequivocal about the contribution of nature destruction to climate change and the fact that the goal of net-zero anthropogenic emissions alone won't be enough to prevent global temperatures rising above the agreed threshold of 1.5°C. We also need to preserve and restore nature's ability to do what it has done for the last three billion years – sequester and store carbon from the atmosphere.

It is crystal clear that preserving biodiversity is essential to also preserving cli-mate stability, ensuring that existing carbon stocks remain locked up in natural systems away from the atmosphere, and that the carbon sequestration functions of

natural ecosystems and species continues. This, in parallel with reducing anthropogenic greenhouse gas emissions. Without the carbon sink function of natural
ecosystems on land, in freshwater systems and in the ocean, the planet's global
temperature would be much higher than it is today because of our emissions, and
with much more disastrous impacts on life on Earth and human society, particularly
the most vulnerable groups and those less well-equipped to adapt.

The figures of bio-carbon are stunning. Terrestrial vegetation absorbs 110–150
gigatonnes of carbon each year from the atmosphere through photosynthesis [123].
The world's forests store approximately 900 gigatonnes of carbon, of which 44%
is stored in soil, 42% in living biomass and 13% in dead wood and organic lit-
ter. Tropical forests store most of their carbon in above-ground vegetation, while
boreal forests store most of it in soil [124].

In total, this is equivalent to nearly a century's worth of current annual levels
of anthropogenic CO_2 emissions. Many of the natural carbon sinks, such as peatlands and old growth forests, have stored carbon for millennia, making their carbon
stocks effectively irrecoverable once released [125]. These areas with the highest
carbon storage, which are the most difficult to restore once degraded, account for
30% of tropical/sub-tropical forests and a similar proportion of boreal/temperate
and tropical peatlands.

The more we study the complex interactions between species and natural processes, the more we realise how little we know about Mother Nature. A recent
discovery has demonstrated that trees absorb not only CO_2 from the atmosphere
but also methane. Methane is a much more powerful heat trapping gas than CO_2,
although it persists in the atmosphere for a shorter period. It is estimated that methane has contributed to almost one-third of global temperature rise, and it is increasingly released in the atmosphere by human activities.

While CO_2 is absorbed through photosynthesis and stored in plant biomass as
carbon, methane is broken down by microorganisms hosted within trees. Many
trees, especially those in swampy areas, absorb methane from the soil and release
it into the atmosphere while the upper part of the tree trunk houses methane-
eating bacteria, or methanovores. Trees remove more methane than they emit, and
research estimates this could amount to as much as 50 million tonnes per year. This
makes forests, particularly tropical and temperate ones, even more important in
mitigating climate change and global heating [126].

The crucial carbon storage function of trees has been constantly undermined
by human activity for millennia, particularly in the last century as humans have
increasingly cleared forests, cut down trees and drained peatlands. The global number of trees has fallen by approximately 46% since the start of human civilisation. Today around three trillion trees survive on Earth, of which approximately
1.3 trillion are in tropical and subtropical forests, 0.7 trillion in boreal forests and
0.7 trillion in temperate ones. However, over 15 billion trees are cut down each
year [127]. During the last decade, in addition to tree-felling and forest conversion
for farming, we have seen the rise of forest fires of a direct and indirect (climate
change) anthropogenic nature. Recent data show that wildfires have become more
intense and widespread, destroying nearly twice as much forest today than they

did just 20 years ago. This increase affects every forest typology and geography, accounting for almost one-third of annual forest loss globally [128].

The increase of forest fires could be a symptom of one approaching tipping point where self-reinforcing events will continue to drive deforestation even without direct human intervention. The 2023 record-breaking Canadian forest fires destroyed over 15 million hectares of forest, releasing one billion tons of CO_2, equivalent to the entire annual output of Japan, the world's sixth largest emitter. Just three years earlier, following record-breaking temperatures in December, devastating wildfires in Australia released a similar amount of CO_2 and destroyed 13 million hectares of forests and nearly 3,000 homes, killing 33 people and likely killing three billion wild mammals, birds, reptiles and amphibians. Fires in the vast forests of Siberia have seen similar devastation and emissions levels during the same period. The 2024 extreme drought affecting the Amazon forest has caused a record high number of fires, rising to a record 145,000. Fires are continuing to ravage many regions of the world and are directly triggered by humans, mainly for agricultural expansion and exacerbated by prolonged droughts.

It is not just about violent events like wildfires. Terrestrial and marine ecosystems are losing their ability to absorb and store carbon. A global study found that the overall ability of biological systems to act as carbon sinks is weakening. In 2023, the increase of CO_2 concentration in the atmosphere was 3.37 ppm, reaching a record high since observations began in 1958. However, global fossil fuel CO_2 emissions only increased by 0.6%. This implies an unprecedented weakening of land and ocean sinks such as an abnormal carbon loss in the drought-affected Amazon, extreme fire emissions in Canada and a loss in Southeast Asia, concluding that record warming in 2023 had a strong negative impact on the capacity of terrestrial ecosystems to function as carbon sinks [129].

These findings demonstrate the fact that, under continued and growing human pressure, the carbon sink function of ecosystems is beginning to weaken with huge implications for the climate, and that the worsening of climate change further affects the functions of these sinks. By eliminating human-induced wildfires and allowing degraded ecosystems to recover, we would also recover their ability to absorb carbon, but the continued and growing impact of climate change could make it impossible for some regions and ecosystems to recover once climatic or ecological tipping points are reached.

Another stark warning comes from Finland's forests. For decades, the country's forests and peatlands had reliably removed more carbon from the atmosphere than they released, absorbing anthropogenic emissions. But from 2010, the amount of carbon absorbed started to decline. A decade later, Finland's natural carbon sink disappeared. By 2022, increased emissions from overexploited forests and the impact of high temperatures and droughts on trees, soil and peat made them a net emitter of GHG emissions, contributing to rather than mitigating the climate crisis [130]. The case of Finland highlights the crucial importance of nature in achieving a net-zero target. Despite cutting anthropogenic emissions by around 40% in the last 30 years, Finland's net emissions are back to the levels of 1990 due to failing natural carbon sinks. More than 100 countries are relying on natural carbon sinks to meet their net-zero climate targets, but the combination of direct human impacts and the impacts of the climate crisis itself, such as forests dieback because of extreme droughts, nature is turning into a source of GHG rather than a sink, reversing a function that has been active for hundreds of millions of years.

It is not just forests that are powerful carbon sinks. Wetlands, both inland and coastal, are also known to perform a powerful carbon sequestration function. Wetlands are one of the planet's largest stores of carbon, and when drained or degraded they can release the three main heat-trapping GHGs: carbon dioxide, methane and nitrous oxide. Since 1700, we have lost 85% of wetlands, and since 1970, natural inland and coastal wetlands have declined by approximately 35% [131]. If this has released huge quantities of GHGs, there is huge potential for restoration and to absorb back large amounts of carbon, especially considering how fast wetland habitats can recover [132, 133].

Peatlands, both boreal and tropical, are defined by having a layer of dense and partially decomposed organic material that is 30 cm to several meters thick preserved under waterlogged and anoxic conditions. They represent the ecosystem with the highest carbon concentration and store about 30% of all terrestrial carbon, equivalent to up to 700 gigatonnes [134]. Ignoring their carbon storing function, about 15% of the world's peatlands have been drained for agriculture, forestry plantations and even grazing, triggering the oxidation of the peat and the release of CO_2. Drained or degraded peatlands contribute to 5% to the total global anthropogenic emissions despite covering only 3% of the global land surface [135]. It's estimated that the northern hemisphere's frozen soils and peatlands contain approximately 1,700 gigatonnes of organic carbon, 88% of which locked away in perennially frozen soils [136].

This accounts for approximately 50% of the estimated global below ground organic carbon pool, and it is four times more than humans have emitted since the Industrial Revolution, and twice as much as is currently in the atmosphere [137]. Global heating primarily but also deforestation, forest fires and other pressures are contributing to the thawing of the permafrost and releasing huge quantities of methane and CO_2. Permafrost thawing is one of the most dangerous climatic and ecosystemic tipping points humanity is confronted with.

Grasslands are not intuitively associated with high levels of carbon storage because of their herbaceous appearance and relatively low biomass. In grasslands, most of the carbon is not stored in the above ground vegetation but in the soil. The carbon relevance of grasslands firstly derives from their extent. Grassland ecosystems virtually cover 40% of the Earth's land surface (excluding Greenland and Antarctica) and store 34% of the terrestrial carbon, 90% of which is below ground as root biomass and soil organic carbon [138]. At the same time, grasslands have been the most affected ecosystems by conversion for agriculture and livestock production. Through conservation, restoration, sustainable grazing and regenerative agricultural practices, the carbon sequestration potential of carbon dioxide equivalents per year has been estimated in the order of up to 8 gigatonnes [139].

– Blue carbon

Then there is 'blue' carbon. The ocean plays three critical roles in mitigating global heating: carbon absorption through marine ecosystems and organisms, and the absorption of both atmospheric heat and CO_2.

Mangroves, seagrass meadows, coral reefs, phytoplankton and other marine species and ecosystems absorb CO_2 and store it away in biological matter, structures and sediments, similar to their terrestrial counterparts. This is referred to as the *ocean biological carbon cycle*. Marine and coastal ecosystems are some of the world's most powerful carbon stores and sinks, with mangroves, salt marshes and seagrass topping the list.

In the past 50 years, 50% of mangrove forests have been lost or degraded under the pressure of ports, tourism and aquaculture (mainly shrimp farming, agriculture, fuelwood and timber extraction). The removal and degradation of mangroves also has negative impacts such as reducing coastal protection from erosion and extreme weather events, and the intrusion of salt water to inland coastal areas. However, awareness of the ecological importance of mangroves and their contribution to people is growing, and, globally, their rate of loss is slowing down. Furthermore, around 42% of all remaining mangroves are in protected areas [140].

And then there is the mighty 'floating forest' of phytoplankton, organisms that are miniscule in size but astronomically high in number that absorb carbon through photosynthesis, just like plants do on land. They are responsible for half of global net primary production by photosynthetic activity [141]. When they die, they fall to the ocean floor, carrying carbon with them and locking it away in sediment, a process also known as 'marine snowfall'.

Some scientists estimate that between 20–35% of global annual anthropogenic CO_2 emissions are sequestered by the ocean, mostly by phytoplankton [142, 143]. Phytoplankton is also the source of a thin lipidic surface microlayer (SML) that covers most of the ocean and has been found to play a key role in generating aerosols, influencing cloud formations and regulating heat and gas exchanges between the ocean and the atmosphere. The decline of phytoplankton will reduce the ocean's SML and its regulating functions. Firstly, a thinner SML will most likely produce and increase the evaporation of water. Water vapour has a much greater heat trapping capacity than carbon dioxide and even methane, and it is the primary source of the heating of the Earth's atmosphere.

The reason why we don't focus on water vapour is because it is virtually impossible to control. Instead, we focus our efforts on reducing anthropogenic emissions of GHGs which contribute to global warming and increase evaporation. But the impacts on phytoplankton and the alteration of biophysical mechanisms like the ones related to SML cannot be underestimated. This is yet another powerful example of the direct interaction between biodiversity and the climate.

Secondly, as CO_2 increases in the atmosphere, it dissolves in ocean waters. And thirdly, as the atmosphere warms because of increased anthropogenic emissions, ocean waters absorb that thermal energy. In essence, the ocean acts as a sponge for both heat and atmospheric CO_2, and, as a consequence, over 90% of the heat and over 30% of CO_2 have been absorbed by the ocean [144, 145].

Without this crucial contribution, global temperatures would already be well above 1.5°C compared to pre-industrial levels. Not surprisingly, this relentless absorption of heat and CO_2 isn't without consequences. Alarming figures show rising water temperatures and acidity beginning to have a deep effect on ocean ecology and the survival rates of an ever-increasing number of species.

Coral bleaching, the death of tropical corals due to rising water temperatures, is probably the most well-known impact of a warming ocean, with the first ever mass coral bleaching event recorded back in 1988. Since then, three more destructive events have already occurred, in addition to many local ones. It is conservatively predicted that 70–90% of shallow water coral reefs could disappear with a 1.5°C global temperature rise and 99% at 2°C [146].

The northward shift in migratory and distribution patterns of many marine species, including high-commercial-value species like tuna and deep-sea crabs, is also happening in response to warming waters and food chain disruptions, negatively affecting ocean ecology and local economies. Warmer waters are also less able to store gasses like oxygen, and again this has a damaging impact on marine life. The warming, acidification and deoxygenation of ocean waters are raising additional concerns around the stability of ocean ecosystems, bioclimatic cycles and marine life overall.

In fact, the more we monitor industries within the 'blue economy', the more we discover new impacts on both climate and biodiversity. For example, bottom trawling is well known for destroying marine habitat and marine life. Up to 70% of catches are unselective and unintended bycatch of non-target or non-commercial species, while the practice also contributes to the release of carbon stored in sediments at the bottom of the sea, the planet's largest carbon reservoir. Emissions from bottom trawling have been estimated at 1 gigatonne of carbon per year at current rates, more emissions than the global aviation industry.

Similarly, deep-sea mining of polymetallic nodules scattered on the sea floor can directly release carbon and methane because of disturbance. It can also indirectly undermine carbon sequestration functions by interfering with the biological carbon sinks and pumps, for example affecting the food chain by releasing huge plumes of fine sediments into oceanic waters, spread far and wide by underwater currents. Recently, we have discovered the ability of these nodules to generate oxygen through seawater electrolysis in the total darkness of the ocean depths, hence dubbed 'dark oxygen'. The disruption of this newly discovered oxygenation mechanism will further exacerbate the problem at a time when warming sea temperatures are inducing deoxygenation of the ocean's waters.

The inextricable dependencies and synergies between the climate and biodiversity makes it imperative that we fully integrate these agendas into our ways of thinking and do all we can to highlight the huge benefits from addressing both in an integrated way.

3.7   Water doesn't come from the tap

It is stating the obvious to say that water is one of the building blocks of life. It is also obvious to say that water doesn't come from the tap, but where it actually comes from is too often ignored or forgotten. The answer, of course, is nature. The way in which we manage the world's water resources is another blatant example of how we take the natural world and everything it provides for granted. We overuse, waste, pollute, heat and deoxygenate fresh water all over the world. We degrade freshwater ecosystems crucial to healthy local and global hydrological cycles, which ensure water quality and provide us with a predictable, reliable source of water.

Water doesn't come from the tap or wells or pipes. Both superficial and aquifer water is supplied, regulated and purified by natural ecosystems and the living organisms they host. With the increasing frequency of extreme drought and flooding events in all regions of the world, we are finally beginning to recognise that we face a worsening water crisis which undermines our health and the health of the planet. The threats to water security mirror our overall impacts on the natural world, including over-extraction, climate change, ecosystem destruction or degradation, pollution and the interruption or modification of water flows such as river dams.

Only one-third of the world's largest rivers are still free flowing, with the rest being dammed once or multiple times [147]. We have destroyed over 85% of wetlands, with one-third reclaimed since 1970 [148]. A powerful proxy indicator of freshwater system health comes from the *Living Planet Index*, which shows that freshwater vertebrate populations have, on average, suffered a shocking 85% decline in the last 50 years, much more severe than on land or in the ocean [149].

The conservation of groundwater aquifers and surface water bodies is inextricably linked to stable climate patterns, as well as the proper functioning of natural ecosystems – such as wetlands and forests – which are increasingly recognised as key players in the local and global hydrological cycle. Agriculture is the largest user of water, accounting for almost 70% of the water used globally, even though approximately half of agricultural produce comes from rainfed subsistence cultivation and is even more vulnerable to drought [150].

Freshwater ecosystems like rivers, lakes and wetlands directly support more than one-third of global food production, not just through irrigated farmland but in the often-overlooked wild fisheries and aquaculture, and indirectly by maintaining the fertility of deltas and estuaries and their adjacent lands. The world's deltas are global food baskets, contributing to 4% of global food production, despite their limited size [151]. Meanwhile, land ecosystems like forests and grasslands contribute to the regulation of water flows, the replenishment of aquifers and the overall hydrological process.

The vulnerability of the agriculture sectors to water stress is clear. Yet agriculture is also the main driver of deforestation and wetland reclamation, impacting ecosystems key for water security. This highlights yet another facet of the 'development paradox.' In today's world, especially in the least economically developed regions that rely on rainfed agriculture, frequent droughts severely impact crops. In this context, preserving healthy freshwater ecosystems emerges as a critical solution for adaptation. In other words, it can be a nature-based solution to food insecurity, malnutrition and poverty through securing water resources.

Energy is also a water intensive sector (for cooling processes and hydropower generation), accounting for 10% of global freshwater use [152]. Falling water levels, but also extreme floods, threaten the viability of hydropower generation which accounts for 15% of global energy production. In 2022, due to heavy drought, French nuclear power plants were forced to reduce their output. Every sector of our economy, from textiles to steel, manufacturing and tourism, depends on access to quality water.

Wetlands, natural floodplains and rivers also play a key role in climate mitigation and adaptation by absorbing and storing carbon from the atmosphere and by mitigating the effects of extreme floods. They also provide water reserves in times of drought, replenish aquifers, protect against storms and erosion, slowing the speed of currents and regulating the local climate. Rivers, lakes and wetlands also influence the local microclimate by reducing air temperatures and combatting heatwaves, particularly in urban areas [153].

Mangroves, where land, freshwater and sea meet, are magical coastal forests which have adapted to survive high salinity levels. They rely on the regular flow of sediments and freshwater from land to provide nutrients and mitigate salinity. Mangroves are truly unique ecosystems not just for their extraordinary ecological adaptations and stunning beauty but also in terms of their capacity to store carbon and their function as nurseries for many marine species. Freshwater ecosystems also harbour extraordinary biodiversity. Few people know that 51% of all fish species are found in fresh water [154]. These habitats are critical breeding, wintering and stopover grounds for billions of migratory birds too.

The world urgently needs to start valuing freshwater ecosystems for all the benefits they provide to people and nature. They are our life support system and the best buffer and insurance against rapidly worsening climate impacts. Levels of water stress all over the world, either from drought or flooding, show that it cannot be resolved by technology and grey infrastructure, not just because of high costs but because of the ever-worsening impacts. The most cost-effective, long-term solution is to invest in the natural systems that have evolved to regulate and preserve the water cycle.

A recent report states that the value of water to our society and our economy is inestimable [155]. However, the report, using the average price of water in different countries, estimates that the total value derived from the direct and indirect use of freshwater in 2021 was $58 trillion, equivalent to 60% of global GDP. Meanwhile, the value of direct use for households, agriculture and industry is estimated at a minimum of $7.5 trillion annually, equivalent to 7% of global GDP. Indirect services such as water regulation, purification, as buffers against drought and floods, sediment transport and delivery, are equally critical. Almost every business and industry depend on water supply – from agriculture and manufacturing to energy, mining and technology.

Nature not only secures water supplies but also water quality. This includes drinkable water for human consumption, food processing and beverages, and good quality water for use in many manufacturing sectors.

While we mostly pay attention to the '*blue water*' of rivers and lakes, the '*green water*' stored in plants and released as moisture as part of an extraordinarily efficient recycling system perfected over billions of years is often overlooked. And yet, studies show that the green water that flows from plants into the atmosphere plays a very important role in regulating precipitation levels in many regions of the world. It is calculated that in normal conditions, a large tree in the Amazon can release up to 1,000 litres of water in the atmosphere per day, and the hundreds of billions of trees in the Amazon over 20 billion metric tons of 'green water', more than the 17 billion litres of 'blue water' that the Amazon River pours into the ocean per day [156].

While much atmospheric water originates from the oceans, globally, more than half of terrestrial precipitation originates from water vapour generated by green water moisture recycling [157, 158]. While climate change influences vegetation cover, change in vegetation cover (e.g. deforestation) can change the climate. Countries are interconnected via atmospheric rivers that, like sky waterfalls, turn into rain in downwind areas. This concept of green water strengthens the notion of the water-biodiversity nexus. So, in addition to climate change, land use changes such as deforestation have a disastrous impact on the provision and predictability of precipitation levels, even thousands of miles away from the deforested area.

We have long undervalued and overexploited water while ignoring the role of ecosystems in regulating water supply and quality. We have introduced and supported unsustainable policies and practices in sectors like agriculture, manufacturing, mining, energy and infrastructure that overuse and pollute water and convert and degrade freshwater ecosystems. As a result, we face the prospect of a 40% shortfall in freshwater supply by 2030 [159]. We have breached the planetary boundaries for water – a key dimension of a safe and just future for humanity and all life on the planet.

Our approach to water is another example of our economic paradox. All the industries that depend on a regular supply of water are at the same time responsible for extracting unsustainably, polluting and impacting the natural systems, from wetlands to forests, that provide us with water security. The paradox is that, ultimately, pressure on water ecosystems will disrupt the productivity and value generation of the same industries that are heavily dependent on water and at the same time responsible for these impacts. Agriculture is the most poignant example. Poor water stewardship on our part is, sadly, in line with the poor stewardship of all other natural resources and services. The water crisis is expanding our awareness of the dependencies, risks and costs associated with the destruction of the ecosystems on which the provision of water depends and calls for a change in the way we value nature . . . as the true source of water.

3.8   Nature, human health and well-being

The understanding of the links between human health and nature is rooted in our deep ancestral relationship with the natural world and our knowledge on how to benefit from its services, knowledge today lost in many parts of the world. This begins with the fact that, as has been the case for millennia, most of the medicines that play an instrumental role in maintaining our quality of life and improving our life expectancy come from chemical compounds found in plants and animals. This includes basic drugs like analgesics and anti-inflammatories such as the ones derived from willow tree and cone snail, as well as life-saving and life-changing antibiotics from fungi, antimalarials from plants like cinchona and artemisia, antifungals from bacteria, antitumorals from the Pacific yew tree and periwinkle and many others. Many Indigenous and local communities around the world still possess a profound knowledge of 'nature's pharmacy' which allows them to use plants and animals with medicinal properties to stay healthier and cure disease.

But nature has a much more profoundly positive influence on us, affecting not just our physical health. E.O. Wilson in his book *Biophilia*, literally 'love for life' [160], suggests the hypothesis that all humans have an innate affinity, curiosity and attraction to the natural world and other life forms. This is something that has evolved to allow us to adapt and survive by having a deep connection and understanding of nature's offerings, as well as its dangers. The more 'connected' and able we are to 'understand' nature, the higher were our chances of survival. That connection to nature still runs deep in all of us. This theory is supported by several psychological and neuroscientific studies, which show how our body and mind respond to nature, and the calming and fulfilling effect that the natural world has on us. Being in nature, even simply walking through an urban park as opposed to traffic, reduces blood pressure, cardiovascular stress, anxiety, rumination and negative thinking [161]. In several countries doctors already prescribe time out in nature or 'nature prescriptions' to their patients to help with both physical and mental health issues.

Visual neuroscientists, meanwhile, have shown that natural landscapes have a powerful neural impact. Images of nature in hospitals have been shown to have a positive impact on patients, including women during labour. 'Pet therapy', the practice of bringing animals into hospitals or private homes, can have a seemingly magical impact on people suffering from depression and even autism.

Even in our most modern and urbanised world, as far away from nature as we could ever have imagined, we bring nature with us, into our homes and into our lives. It is not a coincidence that the more urbanised we become, the more we have pets, gardens, house plants and photographs or paintings of nature in our homes and offices. Most computer screen images are of natural scenes. These are all nature 'surrogates' that make us feel somehow that we are in nature, making us happier in return. In many countries, during the COVID-19 pandemic, one of the few exceptions to lockdown were for people to go out into urban green areas to regenerate their mind, body and spirit. However, as Miles Richardson explains [162], connecting with nature is more than just going for a walk; it is about feeling like we are a part of nature, and that we are connected to her many different components. In other words, establishing a meaningful relationship with the natural world including cognitive, sensorial, emotional and empathetic dimensions.

Nature can be both a saviour and a threat. There are thousands of known bacteria, viruses and fungi that are potentially dangerous to humans. Some 60% of emerging infectious diseases reported globally are of a zoonotic nature, meaning that pathogens have spilled over from animals to humans, and 70% of these cases originated in wildlife [163]. Notorious diseases such as Human Immunodeficiency Virus (HIV), Severe Acute Respiratory Syndrome (SARS), Middle East Respiratory Syndrome (MERS), Nipah virus, Ebola, Lassa fever and rabies, to mention just a few, are all of zoonotic nature.

The recent and catastrophic COVID-19 pandemic also most likely began with the handling or consumption of a traded wild animal [164]. Every year, new zoonotic pathogens are identified. Most of these bacteria and viruses live deep in natural areas, particularly tropical regions, and are 'kept under control' by complex and balanced ecosystems.

In recent decades, humans and our livestock have increasingly encroached on natural habitats, resulting in increased levels of contact with wildlife. The exposure to zoonotic pathogens could occur through direct contact with wildlife (for example through hunting or trading) or indirect contact with infected domestic animals. Wild animals are displaced by the destruction of their habitats and are forced to disperse, increasing the likelihood that they will come into contact with humans. It is believed that an outbreak of the Nipah virus in Malaysia was a direct result of the deforestation of a habitat that was home to a bat colony, which subsequently moved to human settlements, taking the virus with them and spreading it among the human population.

It has also been shown that displaced and distressed animals suffer from a depressed immune system, increasing their viral load and often making them sick and more contagious. While early detection, good hygiene and vaccinations are normally considered methods of prevention, in the aftermath of the COVID-19 pandemic, a panel of experts assembled by WHO, FAO and other agencies highlighted the importance of investing in 'primary prevention' that addresses the root cause of the initial spillover. The problem is that prevention is often not prioritised as, in the absence of a problem, investments are seen as difficult to legitimise and explain to the public, even as an unnecessary waste of resources. On the contrary, large sums are mobilised when a crisis erupts, often too late and with massive additional expenses that could have been prevented at a fraction of the cost by supporting prevention measures [165].

The increasing emergence of zoonotic diseases is linked to large-scale tropical deforestation and conversion for farming, particularly of livestock, alongside modern intensive farming practices where the concentration of animals creates a perfect environment for pathogens to spread. The trade and consumption of high-risk species, meanwhile, is another powerful driver of wildlife-to-human spillover events. With population growth even in remote areas, the consumption of wild meat is growing as a source of protein, while in urban areas it is either a delicacy, part of a person's cultural identity or a status symbol. The illegal trade of live wildlife for the pet and medical testing markets, or of their body parts for food, medicinal or ornamental purposes, is the fourth most profitable illicit activity after drugs, arms and people trafficking and is estimated to be worth $20 billion a year [166, 167].

With the wildlife trade, we are literally 'couriering' pathogens across the world, and virologists warn that unless we change the way we interact with wild spaces and species, the occurrence of another pandemic is not an 'if' but a 'when'. Our knowledge of the links between the degradation of the natural world and human health [168] suggests we need to adopt a 'one health' [169] or 'planetary health' approach [170], which includes humans, animals and the environment and could be the most effective way of preventing the outbreak of new epidemic and pandemic events.

3.9 The business case for Nature (Positive)

"Not a single bee has ever sent you an invoice. And that is part of the problem", said Indian Environmental Economist Pavan Sukhdev in 2008 at the launch of the interim report on the 'The Economics of Ecosystems and Biodiversity' of which he was the study leader and that led the way to a suit of milestone publications [171]. I find this quote perfectly capturing our current failure to economically value and account for Nature's many and irreplaceable contributions to humanity.

While our biological dependence on nature is intuitive, the understanding of the dependence of our economy on the stability of natural ecosystems is still nascent.

In line with the reactive nature of humans, it is the impact on our economy, livelihoods and wellbeing of the decline of nature's services due to climate and ecosystem disruption which is beginning to attract our attention. The unprecedented droughts and floods in recent years, driven by climate change but exacerbated by deforestation, river canalisation, floodplain and wetland reclamation, have affected many regions, depriving millions of people of regular access to drinking water, disrupting agriculture and impacting production in manufacturing sectors. This has cost hundreds of billions of dollars globally and strengthened the business case for combating climate change and, although still less recognised, biodiversity loss. Contrary to climate disruption, the economic impacts of the biodiversity crisis are more subtle and less evident. However, in recent years a sizeable share of business leaders and investors have begun to recognise the extent to which our economy depends on the natural world, including biodiversity. This has triggered the development of a new economic understanding of nature, the risks linked to her loss and the opportunities linked to her conservation and restoration, and the importance of considering natural capital when making investment and development decisions.

The business case for a nature-positive transition begins with the fact that every company has direct or indirect dependencies from and negative impacts on nature and ecosystem services, either in their operations or across their value chain or investment portfolio. In other words, all companies both depend on and impact nature, either directly or indirectly. These dependencies and negative impacts lead to a variety of risks. Unaddressed risks are likely to materialise as costs and losses of various kinds. Moreover, impacts generated by one company or sector can increase risks and costs to others, highlighting the importance of a common global goal like Nature Positive that drives action at the sector and whole economy level.

This is also connected to the emergence of the concept of '*double materiality*': on the one hand, the damage that a business or investment does to nature, and on the other, the risks and costs to the business and investor generated by the loss of nature's services. In essence, nature loss for a company that directly or indirectly depends on goods and services provided by nature equates to a higher risk of decreased productivity and profitability and, therefore, a decrease in its business value. To fulfil their fiduciary responsibility to deliver returns for shareholders, companies and investors' boards are beginning to consider the material implications of nature-negative practices. Our economic growth has been, is and can only be powered by natural capital. As a result of overexploitation and environmental degradation, natural capital is in decline, weakening and destabilising its crucial contribution. Only by sustainably using natural capital can we ensure that the 'cash flow' from nature into our economy continues.

One of the most cited figures used to display the economy's dependence on nature comes from a 2020 report by the WEF in 2020 [172], which estimated that half of the world's GDP was moderately or highly dependent on nature. It makes sense. Think of agriculture, fishing, forestry, tourism, but also mining, energy and manufacturing and how, for example, they depend on regular access to water secured by stable climate but also healthy forests, rivers and wetlands. In reality, the entirety of our economy directly or indirectly depends on nature. Like climate, biodiversity cuts across the whole economy.

Nature's unaccounted contribution to the global economy is larger than global GDP itself and is worth around $150 trillion annually [173], which is in line with earlier calculations [174]. This is not an abstract value but a contribution that is distributed across society and the economy, benefiting everyone, including all businesses. However, it is also a value that has been largely unaccounted for until now, because we could count on nature's resilience and regenerative strength and have not had to worry too much about the impacts we were causing. This is now beginning to change.

The WEF's *Global Risks Report 2020* saw biodiversity appear for the first time in the list of the top 10 global risks facing society and the economy [175], both in terms of likelihood and impact. Since then, it has stubbornly remained near the top of the list. The 2025 edition of the report found extreme weather events and pollution featuring amongst the top ten risks on a two-year horizon, while half of the risks over a 10-year period are environmental, with "biodiversity loss and ecosystem collapse" in second place only after extreme weather events [176]. This new awareness and acknowledgment on the part of society, business and investors has led to many sustainable business and finance initiatives and frameworks, and much stronger corporate engagement in the nature agenda.

Risks and opportunities are the foundation of most decisions made in business and finance, and it is no different when deciding whether to stick to nature-negative practices or to embark on a nature-positive transition. Material risks come from the degradation and depletion of the goods and services from nature on which the productivity of companies ultimately depend. When these goods and services (such as water access and quality, natural commodities and raw materials) are negatively affected, the impact on company operations, productivity and profitability are direct and material. *Physical risks* include supply chain shocks or the disruption of operations resulting from temporary and acute impacts (e.g. crop failure from drought, floods or pest invasions) or more permanent and chronic disruption (e.g. decline of yields because of soil erosion or sterility), putting the very existence of the business at risk. It is important that companies assess and address both their dependencies from and impacts on nature to effectively prevent or mitigate risks.

As biodiversity increasingly features in corporate sector standards of voluntary or regulatory frameworks, and as society increasingly demands that companies address their environmental footprint, stricter environmental legislation is being developed in many countries which will lead to operations and technology restrictions, taxation of environmentally harmful practices or reform of public subsidies, and with that new *compliance, legal and reputational risks*. A good example is the recent EU Corporate Sustainability Reporting Directive and standards [177], which currently requires thousands of companies to report against biodiversity standards included in the EU Taxonomy.

In addition to nature-focused disclosure and target setting frameworks, like the Taskforce on Nature-related Financial Disclosure (TNFD) (https://tnfd.global/) and the Science Based Targets Network (SBTN) (https://sciencebasedtargetsnetwork.org/), several other global voluntary but widely used and highly respected standard frameworks, such as the Global Reporting Initiative (GRI) [178], the International Sustainability Standards Board (ISSB) [179] and the International Organisation for Standardization (ISO) [180], have recently developed or announced that they will develop biodiversity standards. Regulatory, compliance and legal risks are managed by assessing and addressing impacts on nature, allowing businesses to anticipate and comply with upcoming regulations and restrictions. A lack of preparedness could result in sudden increases in operational costs due to the need to adopt new technologies, shift sourcing locations, use alternative raw materials or face permit delays.

One aspect of the business case for nature centres on the double materiality of nature loss, highlighting its risks, dependencies and implications for business resilience. The other aspect focuses on mitigating the negative externalities generated by business activities that impact society, especially given the growing societal expectation for companies and investors to address these issues. Failure – or worse, unwillingness – to tackle negative impacts on nature can lead to significant reputational damage and legal challenges, ultimately affecting a company's social license to operate. It can also influence consumer behaviour, partnerships, investor confidence and, increasingly, employee motivation and productivity. Notably, 69% of employees consider a company's environmental policies important when choosing an employer, while 66% factor in environmental performance when deciding whether to remain with their current employer [181].

While risks are associated with today's nature-negative economic model, embracing a nature-positive transition also represents a huge opportunity. The

WEF estimates that nature-positive transitions in five sectors alone could generate more than $10 trillion in new business revenues and create 395 million jobs by 2030 [182].

Nature-positive transitions need to be seen by businesses and investors as a 'value proposition' rather than as a compliance burden or an additional cost. What we are seeing happening with the transition to decarbonisation and investment in renewable energy technologies will happen with nature-positive transitions in other sectors like regenerative agriculture, sustainable forestry, mining and fishing [183]. Environmental, Social and Governance (ESG)–related assets in the capital market have dramatically grown and are expected to rise from $35 trillion in 2021 to $50 trillion globally by 2025, while sustainability-linked loans and financing exceeded $1.6 trillion in 2021 [184]. Despite political opposition and regulatory rollback in some countries, this trend is expected to continue.

As for emissions, banks will start demanding from their corporate customers that they reduce and even phase out impacts on biodiversity. Governments' procurement policies will increasingly avoid environmental impacts and are estimated to provide a $6 trillion global GDP increase and create 3 million net new jobs by 2050 in the area of net-zero commitments [185].

The '*triple bottom line*' [186] framework brings together financial [187], social and environmental outcomes in the expectation that companies should focus on social and environmental concerns as much as they do on profit, for both business and broader social benefits. Today, an increasing number of regulators are making ESG disclosures mandatory, and ESG ratings are increasingly used to assess investment in companies. As with net zero in climate, nature-positive reporting and disclosure will help financial institutions and governments to identify and invest in companies that are addressing nature risks.

Nature is perhaps the most underdeveloped component of ESG, but not for much longer. Executives, boards and investors' attention to nature-related risks, dependencies and performance has been rapidly growing. Leading companies and investors embracing a nature-positive strategy and reporting will see significant opportunities related to the rise of the ESG and nature agenda. Sustainability-linked finance has already exceeded $1 trillion [188]. Despite the recent rolling back of environmental regulation by some governments, the trend in reporting on corporate environmental sustainability is expected to continue.

Nature-positive commitments and outcomes can create demand-side opportunities in both 'Business-to-Customer' and 'Business-to-Business' contexts. In the former, customers and consumers are increasingly likely to prefer companies, products and services that demonstrate nature-positive outcomes or contributions such as defor-estation-free food, goods sourced from regenerative agriculture or products derived from sustainable fishing practices. In the latter, businesses will face growing demand from their investors or regulatory bodies to align with sustainability standards.

It is known that companies with net-zero emissions strategies and acting to tackle climate change are more successful at motivating their employees and attracting and retaining talent. About half of the employee respondents to a BCG survey said they would not work for organisations that do not reflect their personal beliefs regarding the environment [189]. In essence, reducing impacts and increasing investments in nature will address the material risks of operational disruption resulting from climate change and nature loss, from extreme weather events to soil erosion or scarcity of raw materials, water and other critical natural resources and services. Reducing impacts on biodiversity (ecosystems, species) also contributes directly to the mitigation of climate change.

It will also help mitigate legal, compliance and regulatory risks by anticipating the regulation which will otherwise force sudden and costly operational transitions, restrictions and delays. On the opportunity side, it will secure a social license to operate and unlock new revenue streams as consumers and investors increasingly prioritise companies, products and services which contribute to nature-positive outcomes. Direct opportunities relate to the development of new business lines, technologies and products, from healthier food from regenerative agriculture to low environmental footprint products in a wide range of categories. Moreover, companies that lead on biodiversity will see benefits from better accessing capital as investors are increasingly integrating ESG performance into their decision-making, and market valuations will reflect that.

Finally, embracing a nature-positive transition will drive efficiency, reducing inputs and resource use and waste and reduce costs. As for every transition, there are, of course, associated costs, but the increased business resilience, reduced risks and reputational enhancements make up for those.

Today's local-to-global ecological deficit is economically neglected and invisible, but it is starting to translate into a real financial liability for companies, investors and the global economy. Obscured from governments' budgets, companies' balance sheets and investors' frameworks, climate change and nature loss are generating costs to the economy and society, growing year after year. Every economy, developed or developing, is directly or indirectly dependent on nature's resources and services. Halting and reversing the loss of biodiversity is not just an ethical imperative but a necessity for the resilience of any business and investor. It is about de-risking as much as new value creation. This is the foundation of the new business case for nature and the Nature Positive Global Goal, and the rationale for companies to embark now on a *nature-positive transition*.

References

1 Kopecky, A., 2018. The Globe and Mail, 25 August. Available at: www.theglobeandmail.com/opinion/article-things-have-never-been-so-good-for-humanity-nor-so-dire-for-the/

2 Roser, M., 2019. 'Mortality in the past: Every second child died', Our World in Data. Available at: https://ourworldindata.org/child-mortality-in-the-past (Accessed: November 2024).

3 Volk, A.A. and Atkinson, J.A., 2013. 'Infant and child death in the human environment of evolutionary adaptation', Evolution and Human Behavior, 34(3), pp. 182–192. Available at: https://doi.org/10.1016/j.evolhumbehav.2013.01.003

4 Moatsos, M., 2021. 'Global extreme poverty: Present and past since 1820', In OECD (ed.), How Was Life? Volume II: New Perspectives on Well-Being and Global Inequality since 1820. Paris: OECD Publishing. Available at: https://doi.org/10.1787/3d96efc5-en

5 Dawkins, R., 2006. The Blind Watchmaker. London: Penguin Books.

6 Gari, L., 2006. 'A history of the Hima conservation system', Environment and History, 12(2), pp. 213–228. The White Horse Press.

7 Hampton-Smith, M., Bower, D.S. and Mika, S., 2021. 'A review of the current global status of blast fishing: Causes, implications and solutions', Biological Conservation, 262, p. 109311.

8 Nasi, R., Brown, D., Wilkie, D., Bennett, E., Tutin, C., van Tol, G. and Christophersen, T., 2007. 'Conservation and use of wildlife-based resources: The bushmeat crisis', Secretariat of the Convention on Biological Diversity, Montreal, and Center for International Forestry Research (CIFOR). Available at: www.cbd.int/doc/publications/cbd-ts-33-en.pdf

9 Gombeer, S., Nebesse, C., Musaba, P., et al., 2021. 'Exploring the bushmeat market in Brussels, Belgium: A clandestine luxury business', Biodiversity and Conservation, 30, pp. 55–66. Available at: https://doi.org/10.1007/s10531-020-02074-7

10 Schoon, M. and Cox, M.E., 2018. 'Collaboration, adaptation, and scaling: Perspectives on environmental governance for sustainability', Sustainability, 10(3), p. 679. Available at: https://doi.org/10.3390/su10030679

11 Brundtland, G.H., 1987. Our Common Future: Report of the World Commission on Environment and Development. Oxford: Oxford University Press.

12 Dasgupta, P., 2021. The Economics of Biodiversity: The Dasgupta Review. London: HM Treasury. Available at: https://assets.publishing.service.gov.uk/media/602e92b2e90e07660f807b47/The_Economics_of_Biodiversity_The_Dasgupta_Review_Full_Report.pdf

13 Trucost and TEEB for Business Coalition, 2013. Natural Capital at Risk: The Top 100 Externalities of Business. Available at: https://capitalscoalition.org/wp-content/uploads/2016/08/Trucost-Nat-Cap-at-Risk-Final-Report-web.pdf

14 OECD, 2019. Biodiversity: Finance and the Economic and Business Case for Action. Available at: https://doi.org/10.1787/a3147942-en

15 Meadows, D.H., et al., 1972. The Limits to Growth: A Report for the Club of Rome's Project on the Predicament of Mankind. New York: Universe Books.

16 TEEB, 2010. The Economics of Ecosystems and Biodiversity: Mainstreaming the Economics of Nature: A synthesis of the approach, conclusions and recommendations of TEEB.

17 Diesendorf, M., Davies, G., Wiedmann, T., Spangenberg, J.H. and Hail, S., 2024. 'Sustainability scientists' critique of neoclassical economics', Global Sustainability, 7, e33, pp. 1–13. Available at: https://doi.org/10.1017/sus.2024.36

18 Lenzi, D., Balvanera, P., Arias-Arévalo, P., Eser, U., Guibrunet, L., Martin, A., Muraca, B. and Pascual, U., 2023. 'Justice, sustainability, and the diverse values of nature: Why they matter for biodiversity conservation', Current Opinion in Environmental Sustainability, 64, p. 101353.

19 Gupta, J., Liverman, D.M., Rockström, J., Qin, D., Stewart-Koster, B., et al., 2024. 'A just world on a safe planet: A Lancet Planetary Health–Earth Commission report on Earth-system boundaries, translations, and transformations', The Lancet Planetary Health, 0(0). Available at: www.thelancet.com/journals/lanplh/article/PIIS2542-5196(24)00042-1/fulltext

20 Baber, W.F. and May, J.R. (eds.), 2023. Environmental Human Rights in the Anthropocene. Cambridge: Cambridge University Press. Available at: https://doi.org/10.1017/9781009039642

21 Obura, D.O., DeClerck, F., Verburg, P.H., Gupta, J., Abrams, J.F., Bai, X., Bunn, S., Ebi, K.L., Gifford, L., Gordon, C., Jacobson, L., Lenton, T.M., Liverman, D., Mohamed, A., Prodani, K., Rocha, J.C., Rockström, J., Sakschewski, B., Stewart-Koster, B., van Vuuren, D., Winkelmann, R. and Zimm, C., 2023. 'Achieving a Nature- and People-Positive Future', One Earth, 6(2). Available at: https://doi.org/10.1016/j.oneear.2022.11.013

22 United Nations General Assembly, 2022. The human right to a clean, healthy and sustainable environment: Resolution/adopted by the General Assembly. 76th session, GAOR, Suppl. no. 49. 28 July.

23 Kotzé, L.J., 2014. 'Human rights and the environment in the Anthropocene', The Anthropocene Review, 1(3), pp. 252–275. Available from: https://doi.org/10.1177/2053019614547741

24 H. Slim, 2024. Humanitarianism 2.0. C. Hurst & Co., London, UK

25 Cohen-Shacham, E., Walters, G., Janzen, C. and Maginnis, S. (eds.), 2016. Nature-Based Solutions to Address Global Societal Challenges. Gland, Switzerland: IUCN. Formally adopted by UN Environment Assembly-5, Nairobi, February 2022. Available from: https://doi.org/10.2305/IUCN.CH.2016.13.en

26 CGTN, 2019. 'On the road: Green is gold', 19 October. Available at: https://news.cgtn.com/news/2019-10-19/On-the-Road-Green-is-gold-KUYqvQPzOg/index.html (Accessed: 20 December 2024).

27 Aizen, M.A., Garibaldi, L.A., Cunningham, S.A. and Klein, A.M., 2009. 'How much does agriculture depend on pollinators? Lessons from long-term trends in crop production', Annals of Botany, 103(9), pp. 1575–1583.

28 Stockholm Resilience Centre, Stanford University & EAT, 2021. 'Blue food assessment', Nature. Available at: www.nature.com/immersive/d42859-021-00055-6/index.html#section-wja5pJ2eaN

29 World Economic Forum, 2020. 'Nature risk rising: Why the crisis engulfing nature matters for business and the economy'. Available at: www.weforum.org/publications/nature-risk-rising-why-the-crisis-engulfing-nature-matters-for-business-and-the-economy/

30 Dawkins, R., 2006. The Selfish Gene. Oxford: Oxford University Press.

31 World Commission on Environment and Development, 1987. Our Common Future. Oxford: Oxford University Press.

32 IPBES, 2022. 'Assessment report on diverse values and valuation of nature'. Available at: www.ipbes.net/the-values-assessment

33 Mazzucato, M., 2018. The Value of Everything: Making and Taking in the Global Economy. London: Penguin Books Limited.

34 Harley, N., Shogren, J. and White, B., 2007. Environmental Economics in Theory and Practice. London: Palgrave Macmillan.

35 TEEB, 2010. The Economics of Ecosystems and Biodiversity: Mainstreaming the Economics of Nature: A Synthesis of the Approach, Conclusions and Recommendations of TEEB.

36 World Forum on Natural Capital, 2024. What Is Natural Capital? Available at: https://naturalcapitalforum.com/about/

37 Costanza, R. and Daly, H.E., 1992. 'Natural capital and sustainable development', Conservation Biology, 6(1), pp. 37–46.

38 Day, B., et al., 2024. 'Natural capital approaches for the optimal design of policies for nature recovery', Philosophical Transactions of the Royal Society B, 379(20220327). Available at: http://doi.org/10.1098/rstb.2022.0327

39 Natural Capital Coalition, 2016. Natural Capital Protocol [online]. Available at: www.naturalcapitalcoalition.org/protocol

40 United Nations, 2021. Natural Capital Accounting for Sustainable Macroeconomic Strategies. New York: United Nations, Department of Economic and Social Affairs.

41 United Nations, 1993. Handbook of National Accounting: Integrated Environmental and Economic Accounting – SEEA. New York: United Nations.

42 Ouyang, Z., Song, C., Zheng, H., Polasky, S., Xiao, Y., Bateman, I.J., Liu, J., Ruckelshaus, M., Shi, F., Xiao, Y., Xu, W., Zou, Z. and Daily, G.C., 2020. 'Using gross ecosystem product (GEP) to value nature in decision making', Proceedings of the National Academy of Sciences, 117(25). Available at: https://doi.org/10.1073/pnas.1911439117

43 Zheng, H., Wu, T., Ouyang, Z., Polasky, S., Ruckelshaus, M., Wang, L., Xiao, Y., Gao, X., Li, C. and Daily, G.C., 2023. 'Gross ecosystem product (GEP): Quantifying nature

for environmental and economic policy innovation', Ambio, 52(12), pp. 1952–1967. Available at: https://doi.org/10.1007/s13280-023-01948-8

44 TEEB and Trucost, 2013. Natural Capital at Risk: The Top 100 Externalities of Business. Available at: https://capitalscoalition.org/wp-content/uploads/2016/08/Trucost-Nat-Cap-at-Risk-Final-Report-web.pdf

45 Costanza, R., de Groot, R., Sutton, P., van der Ploeg, S., Anderson, S.J., Kubisze-wski, I., Farber, S. and Turner, R.K., 2014. 'Changes in the global value of ecosystem services', Global Environmental Change, 26, pp. 152–158. Available at: https://doi.org/10.1016/j.gloenvcha.2014.04.002

46 Wilson, F., 2017. 'The gastronomical me by MFK Fisher', The Times, London, 13 May. Available at: www.thetimes.com/life-style/food-drink/article/the-gastronomical-me-by-mfk-fisher-jwsl5z0pv

47 FAO, 2023. World Food and Agriculture – Statistical Yearbook 2023. Rome. Available at: https://doi.org/10.4060/cc8166en

48 FAO, 2018. World Food and Agriculture – Statistical Pocketbook 2018. Rome: FAO. 254 pp. Licence: CC BY-NC-SA 3.0 IGO.

49 Selig, E.R., et al., 2018. 'Mapping global human dependence on marine ecosystems', Conservation Letters, 12, e12617.

50 McIntyre, P.B., Liermann, C.A.R. and Revenga, C., 2016. 'Linking freshwater fishery management to global food security and biodiversity conservation', Proceedings of the National Academy of Sciences of the United States of America, 113, pp. 12880–12885.

51 IPBES, 2019. 'Global assessment report on biodiversity and ecosystem services of the intergovernmental science-policy platform on biodiversity and ecosystem services', In E.S. Brondizio, J. Settele, S. Díaz and H.T. Ngo (eds.), Bonn, Germany: IPBES Secretariat. 1148 pages. Available at: https://doi.org/10.5281/zenodo.3831673

52 Campbell, B.M., Beare, D.J., Bennett, E.M., Hall-Spencer, J.M., Ingram, J.S.I., Jara-millo, F., Ortiz, R., Ramankutty, N., Sayer, J.A. and Shindell, D., 2017. 'Agriculture production as a major driver of the Earth system exceeding planetary boundaries', Ecology and Society, 22(4). Available at: https://doi.org/10.5751/ES-09595-220408

53 WRI, 2021. Just 7 Commodities Replaced an Area of Forest Twice the Size of Germany between 2001 and 2015. World Resources Institute, 11 February. Available at: www.wri.org/insights/just-7-commodities-replaced-area-forest-twice-size-germany-between-2001-and-2015

54 FAO, 2023. Pathways towards Lower Emissions – a Global Assessment of the Green-house Gas Emissions and Mitigation Options from Livestock Agrifood Systems. Rome. Available at: https://doi.org/10.4060/cc9029en

55 Lazarus, O., McDermid, S. and Jacquet, J., 2021. 'The climate responsibilities of in-dustrial meat and dairy producers', Climatic Change, 165, p. 30. Available at: https://doi.org/10.1007/s10584-021-03047-7

56 Ritchie, H., 2021. 'Drivers of deforestation', OurWorldInData.org. Available at: https://ourworldindata.org/drivers-of-deforestation

57 Achieving Peak Pasture, 2019. The Breakthrough Institute. 21 November.

58 Poore, J. and Nemecek, T., 2018. 'Reducing food's environmental impacts through producers and consumers', Science, 360(6392), pp. 987–992. Available at: https://doi.org/10.1126/science.aaq0216

59 Klaura, J., Breeman, G. and Scherer, L., 2023. 'Animal lives embodied in food loss and waste', Sustainable Production and Consumption. Available at: www.universiteitleiden.nl/en/news/2023/11/18-billion-animals-a-year-they-do-die-but-never-end-up-on-our-plate

60 Branthomme, A., Merle, C., Kindgard, A., Lourenço, A., Ng, W.-T., D'Annunzio, R. and Shapiro, A., 2023. How Much Do Large-Scale and Small-Scale Farming Contrib-ute to Global Deforestation? Results from a Remote Sensing Pilot Approach. Rome: FAO. Available at: https://doi.org/10.4060/cc5723en

61 Yang, L., He, X., Ru, S., et al., 2024. 'Herbicide leakage into seawater impacts primary productivity and zooplankton globally', Nature Communications, 15, p. 1783. Available at: https://doi.org/10.1038/s41467-024-46059-4

62 Dryden, H. and Duncan, D., 2022. 'Climate disruption caused by a decline in marine biodiversity and pollution', International Journal of Environment and Climate Change, 12(11), pp. 3414–3436.

63 FAO, IFAD, UNICEF, WFP and WHO, 2024. The State of Food Security and Nutrition in the World 2024 – Financing to End Hunger, Food Insecurity and Malnutrition in All Its Forms. Rome: FAO.

64 Gustavsson, J., Cederberg, C., Sonesson, U., Otterdijk, R. and Mcybeck, A., 2011. Global Food Losses and Food Waste: Extent, Causes and Prevention. Rome: FAO. Available at: www.fao.org/3/mb060e/mb060e00.htm

65 United Nations Environment Programme, 2024. Food Waste Index Report 2024. Think Eat Save: Tracking Progress to Halve Global Food Waste. Available at: https://wedocs.unep.org/20.500.11822/45230

66 Boeckel, T.P.V., Glennon, E.E., Chen, D., Gilbert, M., Robinson, T.P., Grenfell, B.T., Levin, S.A., Bonhoeffer, S. and Laxminarayan, R., 2017. 'Reducing antimicrobial use in food animals', Science, 357, pp. 1350–1352. Available at: 10.1126/science.aao1495

67 Murray, C.J.L., et al., 2022. 'Global burden of bacterial antimicrobial resistance in 2019: A systematic analysis', The Lancet, 399(10325), pp. 629–655.

68 McIntosh, W.L., Spies, E., Stone, D.M., Lokey, C.N., Trudeau, A.T. and Bartholow, B., 2016. 'Suicide rates by occupational group', MMWR Morbidity and Mortality Weekly Report. Available at: https://doi.org/10.15585/mmwr.mm6525a1

69 Beard, J. D., Umbach, D. M., Hoppin, J. A., Richards, M., Alavanja, M.C.R., Blair, A., Sandler, D. P. and Kamel, F., 2014. 'Pesticide exposure and depression among male private pesticide applicators in the agricultural health study', Environmental Health Perspectives, 122(9). https://doi.org/10.1289/ehp.1307450

70 Ruggeri Laderchi, C., Lotze-Campen, H., DeClerck, F., Bodirsky, B.L., Collignon, Q., Crawford, M.S., Dietz, S., Fesenfeld, L., Hunecke, C., Leip, D., Lord, S., Lowder, S., Nagenborg, S., Pilditch, T., Popp, A., Wedl, I., Branca, F., Fan, S., Fanzo, J., Ghosh, J., Harriss-White, B., Ishii, N., Kyte, R., Mathai, W., Chomba, S., Nordhagen, S., Nugent, R., Swinnen, J., Torero, M., Laborde Debouquet, D., Karfakis, P., Voegele, J., Sethi, G., Winters, P., Edenhofer, O., Kanbur, R. and Songwe, V., 2024. The Economics of the Food System Transformation. Food System Economics Commission (FSEC), Global Policy Report.

71 FAO, UNDP and UNEP, 2021. A Multi-Billion-Dollar Opportunity – Repurposing Agricultural Support to Transform Food Systems. Rome: FAO. Available at: https://doi.org/10.4060/cb6562en

72 Ding, H., Markandya, A., Barbieri, R., Calmon, M., Cervera, M., Duraisami, M., Singh, R., Warman, J. and Anderson, W., 2021. Repurposing agricultural subsidies to restore degraded farmland and grow rural prosperity. Washington. Available at: https://doi.org/10.46830/wrirpt.20.00013

73 Potapov, P., Turubanova, S., Hansen, M.C., et al., 2022. 'Global maps of cropland extent and change show accelerated cropland expansion in the twenty-first century', Nature Food, 3, pp. 19–28. Available at: https://doi.org/10.1038/s43016-021-00429-z

74 Benton, T.G. and Bailey, R., 2019. 'The paradox of productivity: Agricultural productivity promotes food system inefficiency', Global Sustainability, 2, e6. Available at: https://doi.org/10.1017/sus.2019.3

75 Schipanski, M.E., MacDonald, G.K., Rosenzweig, S., Chappell, M.J., Bennett, E.M., Bezner Kerr, R., Blesh, J., Crews, T., Drinkwater, L., Lundgren, J.G. and Schnarr, C., 2016. 'Realizing resilient food systems', BioScience, 66(7), pp. 601–610. Available at: https://doi.org/10.1093/biosci/biw052

76 Rodale, R., 1983. 'Breaking new ground: The search for a sustainable agriculture', Futurist, 17.

77 Phalan, B., Onial, M., Balmford, A. and Green, R.E., 2011. 'Reconciling food production and biodiversity conservation: Land sharing and land sparing compared', Science, 333(6047), pp. 1289–1291. Available at: 10.1126/science.1208742

78 LaCanne, C.E. and Lundgren, J.G., 2018. 'Regenerative agriculture: Merging farming and natural resource conservation profitably', PeerJ, 6, e4428. Available at: https://doi.org/10.7717/peerj.4428

79 Seufert, V., Ramankutty, N. and Foley, J., 2012. 'Comparing the yields of organic and conventional agriculture', Nature, 485, pp. 229–232. Available at: 10.1038/nature11069

80 FAO, 2020. The State of World Fisheries and Aquaculture 2020: Sustainability in Action. Rome: FAO. Available at: https://doi.org/10.4060/ca9229en

81 Selig, E.R., et al., 2018. 'Mapping global human dependence on marine ecosystems', Conservation Letters, 12.

82 McIntyre, P.B., Liermann, C.A.R. and Revenga, C., 2016. 'Linking freshwater fishery management to global food security and biodiversity conservation', Proceedings of the National Academy of Sciences of the United States of America (PNAS), 113.

83 Žydelis, R., Wallace, B.P., Gilman, E.L. and Werner, T.B., 2009. 'Conservation of marine megafauna through minimization of fisheries bycatch', Conservation Biology, 23, pp. 608–616. Available at: https://doi.org/10.1111/j.1523-1739.2009.01172.x

84 Wallace, B.P., Lewison, R.L., McDonald, S.L., McDonald, R.K., Kot, C.Y., Kelez, S., Bjorkland, R.K., Finkbeiner, E.M., Helmbrecht, S. and Crowder, L.B., 2010. 'Global patterns of marine turtle bycatch', Conservation Letters, 3, pp. 131–142. Available at: https://doi.org/10.1111/j.1755-263X.2010.00105.x

85 Pérez Roda, M.A. (ed.), Gilman, E., Huntington, T., Kennelly, S.J., Suuronen, P., Chaloupka, M. and Medley, P., 2019. 'A third assessment of global marine fisheries discards', FAO Fisheries and Aquaculture Technical Paper No. 633. Available at: https://openknowledge.fao.org/handle/20.500.14283/ca2905en

86 Davies, R.W.D., Cripps, S.J., Nickson, A. and Porter, G., 2009. 'Defining and estimating global marine fisheries bycatch', Marine Policy, 33(4), pp. 661–672. Available at: https://doi.org/10.1016/j.marpol.2009.01.003

87 Cashion, T., 2016. 'The end use of marine fisheries landings', Fisheries Centre Research Reports, 24, p. 108.

88 Roberts, S., Jacquet, J., Majluf, P. and Hayek, M.N., 2024. 'Feeding global aquaculture', Science Advances, 10, p. 9698. Available at: https://doi.org/10.1126/sciadv.adn9698

89 FAO, 2020. The State of World Fisheries and Aquaculture 2020: Sustainability in Action. Rome: FAO.

90 Edgar, G.J., et al., 2024. 'Stock assessment models overstate sustainability of the world's fisheries', Science, 385. Available at: https://doi.org/10.1126/science.adl6282

91 Zupan, M., Fragkopoulou, E., Claudet, J., Erzini, K., Horta e Costa, B. and Gonçalves, E.J., 2018. 'Marine partially protected areas: Drivers of ecological effectiveness', Frontiers in Ecology and the Environment, 16, pp. 381–387.

92 Marine Conservation Institute, 2022. The Marine Protection Atlas. Available at: https://mpatlas.org

93 Sala, E., Mayorga, J., Bradley, D., Cabral, R.B., Atwood, T.B., Auber, A., Cheung, W., Costello, C., Ferretti, F., Friedlander, A.M., Gaines, S.D., Garilao, C., Goodell, W., Halpern, B.S., Hinson, A., Kaschner, K., Kesner-Reyes, K., Leprieur, F., McGowan, J., Morgan, L.E., Mouillot, D., Palacios-Abrantes, J., Possingham, H.P., Rechberger, K.D., Worm, B. and Lubchenco, J., 2021. 'Protecting the global ocean for biodiversity, food and climate', Nature, 592, pp. 397–402.

94 Erguden, D. and Turan, C., 2005. 'Growth properties of sea bass (Dicentrarchus labrax (L., 1758), Perciformes: Moronidae) live in Iskenderun Bay', Pakistan Journal of Biological Sciences, 8, pp. 1584–1587.

95 Lubchenco, J. and Gaines, S.D., 2019. 'A new narrative for the ocean', Science, 364(6444), p. 911. Available at: https://doi.org/10.1126/science.aay224

96 Brashares, J.S., Golden, C.D., Weinbaum, K.Z., Barrett, C.B. and Okello, G.V., 2011. 'Economic and geographic drivers of wildlife consumption in rural Africa', Proceedings of the National Academy of Sciences, 108, pp. 13931–13936. Available at: https://doi.org/10.1073/pnas.1011526108

97 Swamy, V. and Pinedo-Vasquez, M., 2014. 'Bushmeat harvest in tropical forests: Knowledge base, gaps and research priorities', Occasional Paper 114. Bogor, Indonesia: CIFOR.

98 Fa, J.E., Peres, C.A. and Meeuwig, J., 2002. 'Bushmeat exploitation in tropical forests: An intercontinental comparison', Conservation Biology, 16(1), pp. 232–237. Available at: https://doi.org/10.1046/j.1523-1739.2002.00275.x

99 Eurostat, 2024. Agricultural Production – Livestock and Meat. Data Extracted in October 2024. Planned article update: 18 October 2025. Available at: https://ec.europa.eu/eurostat

100 Cawthorn, D.M. and Hoffman, L.C., 2015. 'The bushmeat and food security nexus: A global account of the contributions, conundrums and ethical collisions', Food Research International, 76, pp. 906–925. Available at: https://doi.org/10.1016/j.foodres.2015.03.025

101 Chaber, A.-L., Allebone-Webb, S., Lignereux, Y., et al., 2010. 'The scale of illegal meat importation from Africa to Europe via Paris', Conservation Letters, 3, pp. 317–321.

102 Kurpiers, L.A., Schulte-Herbrüggen, B., Ejotre, I. and Reeder, D.M., 2016. 'Bushmeat and emerging infectious diseases: Lessons from Africa', In F. Angelici (ed.), Problematic Wildlife. Springer. Available at: https://doi.org/10.1007/978-3-319-22246-2_24

103 Peros, C.S., Dasgupta, R., Kumar, P. and Johnson, B.A., 2021. 'Bushmeat, wet markets, and the risks of pandemics: Exploring the nexus through systematic review of scientific disclosures', Environmental Science & Policy, 124. https://doi.org/10.1016/j.envsci.2021.05.025

104 Cawthorn, D.M. and Hoffman, L.C., 2015. 'The bushmeat and food security nexus: A global account of the contributions, conundrums and ethical collisions', Food Research International, 76, pp. 906–925. Available at: https://doi.org/10.1016/j.foodres.2015.03.025

105 Godfray, H.C.G., Aveyard, P., Garnett, T., Hall, J.W., Key, T.J., Lorimer, J., Pierrehumbert, R.T., Scarborough, P., Springmann, M. and Jebb, S.A., 2018. 'Meat consumption, health, and the environment', Science, 361(6399), eaam5324. Available at: https://doi.org/10.1126/science.aam5324

106 Cassidy, E.S., West, P.C., Gerber, J.S. and Foley, J.A., 2013. 'Redefining agricultural yields: From tonnes to people nourished per hectare', Environmental Research Letters, 8(3), 034015. Available at: https://doi.org/10.1088/1748-9326/8/3/034015

107 Ipsos MORI, 2018. An Exploration into Diets around the World. Available at: www.ipsos.com/sites/default/files/ct/news/documents/2018-09/an_exploration_into_diets_around_the_world.pdf

108 EAT-Lancet Commission, 2019. The Planetary Health Diet. Available at: https://eatforum.org/eat-lancet-commission/the-planetary-health-diet-and-you/

109 IPCC, 2022. 'Summary for policymakers', In P.R. Shukla, J. Skea, A. Reisinger, R. Slade, R. Fradera, M. Pathak, A. Al Khourdajie, M. Belkacemi, R. van Diemen, A. Hasija, G. Lisboa, S. Luz, J. Malley, D. McCollum, S. Some and P. Vyas (eds.), Climate Change 2022: Mitigation of Climate Change. Contribution of Working Group III to the Sixth Assessment Report of the Intergovernmental Panel on Climate Change. Cambridge, UK and New York, NY: Cambridge University Press. Available at: https://doi.org/10.1017/9781009157926.001

110 Humpenöder, F., Popp, A., Merfort, L., Luderer, G., Weindl, I., Bodirsky, B.L., Stevanović, M., Klein, D., Rodrigues, R., Bauer, N., Dietrich, J.P., Lotze-Campen, H.

and Rockström, J., 2024. 'Food matters: Dietary shifts increase the feasibility of 1.5°C pathways in line with the Paris Agreement', Science Advances, 10(13), eadj3832. Available at: www.science.org/doi/10.1126/sciadv.adj3832

111 Sun, Z., Scherer, L., Tukker, A., et al., 2022. 'Dietary change in high-income nations alone can lead to substantial double climate dividend', Nature Food, 3, pp. 29–37.

112 Tilt Collective and Systemiq, 2024. Transforming the Global Food System: A Philanthropic Return on Investment Analysis. Available at: https://tiltcollective.org/wp-content/uploads/2024/09/Transforming-the-global-food-system_190924.pdf

113 Food and Agriculture Organization of the United Nations, 2023. Statistical Yearbook 2023: World Food and Agriculture. Available at: https://openknowledge.fao.org/server/api/core/bitstreams/28cfd24e-81a9-4ebc-b2b5-4095fe5b1dab/content/cc8166en.html

114 Naylor, R., et al., 2005. 'Losing the links between livestock and land', Science, 310, pp. 1621–1622.

115 Alexandratos, N. and Bruinsma, J., 2012. World Agriculture towards 2030/2050: The 2012 Revision. FAO, ESA Working Paper No. 12–03.

116 Cassidy, E.S., West, P.C., Gerber, J.S. and Foley, J.A., 2013. 'Redefining agricultural yields: From tonnes to people nourished per hectare', Environmental Research Letters, 8(3), 034015. Available at: https://doi.org/10.1088/1748-9326/8/3/034015

117 Food and Agriculture Organization of the United Nations (FAO), 2014. Family Farming. Available at: www.fao.org/family-farming/detail/en/c/273649/

118 United Nations, Department of Economic and Social Affairs, Population Division, 2024. World Population Prospects 2024 Revision. Available at: https://population.un.org/wpp/

119 Garnett, T., Appleby, M.C., Balmford, A., Bateman, I.J., Benton, T.G., Bloomer, P., Burlingame, B., Dawkins, M., Dolan, L., Fraser, D., Herrero, M., Hoffmann, I., Smith, P., Thornton, P.K., Toulmin, C., Vermeulen, S.J. and Godfray, H.C.J., 2013. 'Sustainable intensification in agriculture: Premises and policies', Science, 341, pp. 33–34.

120 Benton, T.G., 2015. 'Sustainable intensification', In B. Pritchard, R. Ortiz and M. Shekar (eds.), Routledge Handbook of Food and Nutrition Security. Chapter 6. Abingdon: Routledge.

121 Land Carbon Lab, n.d. Land Carbon Lab. Available at: https://landcarbonlab.org

122 Ruggeri Laderchi, C., Lotze-Campen, H., DeClerck, F., Bodirsky, B.L., Collignon, Q., Crawford, M.S., Dietz, S., Fesenfeld, L., Hunecke, C., Leip, D., Lord, S., Lowder, S., Nagenborg, S., Pilditch, T., Popp, A., Wedl, I., Branca, F., Fan, S., Fanzo, J., Ghosh, J., Harriss-White, B., Ishii, N., Kyte, R., Mathai, W., Chomba, S., Nordhagen, S., Nugent, R., Swinnen, J., Torero, M., Laborde Debouquet, D., Karfakis, P., Voegele, J., Sethi, G., Winters, P., Edenhofer, O., Kanbur, R. and Songwe, V., 2024. The Economics of the Food System Transformation. Food System Economics Commission (FSEC), Global Policy Report.

123 Chafin, C., 2022. The Dawn of the Climavores. Kearney. Available at: www.kearney.com/industry/consumer-retail/article/-/insights/dawn-of-the-climavores

124 Godfray, H.C.J., Aveyard, P., Garnett, T., Hall, J.W., Key, T.J., Lorimer, J., Pierrehumbert, R.T., Scarborough, P., Springmann, M. and Jebb, S.A., 2018. 'Meat consumption, health, and the environment', Science, 361(6399), eaam5324. Available at: https://doi.org/10.1126/science.aam5324

125 Lal, R., Smith, P., Jungkunst, H.F., Mitsch, W.J., Lehmann, J., Ramachandran Nair, P.K., McBratney, A.B., de Moraes Sá, J.C., Schneider, J., Zinn, Y.L., Skorupa, A.L.A., Zhang, H.-L., Minasny, B., Srinivasrao, C. and Ravindranath, N.H., 2018. 'The carbon sequestration potential of terrestrial ecosystems', Journal of Soil and Water Conservation, 73(6), pp. 145A–152A. Available at: https://doi.org/10.2489/jswc.73.6.145A

126 Pan, Y., Birdsey, R.A., Fang, J., Houghton, R., Kauppi, P.E., Kurz, W.A., Phillips, O.L., Shvidenko, A., Lewis, S.L., Hayes, D., et al., 2011. 'A large and persistent carbon

sink in the world's forests', Science, 333(6045), pp. 988–993. Available at: https://doi.org/10.1126/science.120160

127 Noon, M.L., Goldstein, A., Ledezma, J.C., et al., 2022. 'Mapping the irrecoverable carbon in Earth's ecosystems', Nature Sustainability, 5, pp. 37–46. Available at: https://doi.org/10.1038/s41893-021-00803-6

128 Gauci, V., Pangala, S.R., Shenkin, A., et al., 2024. 'Global atmospheric methane uptake by upland tree woody surfaces', Nature, 631, pp. 796–800. Available at: https://doi.org/10.1038/s41586-024-07592-w

129 Crowther, T., Glick, H., Covey, K., et al., 2015. 'Mapping tree density at a global scale', Nature, 525, pp. 201–205. Available at: https://doi.org/10.1038/nature14967

130 Tyukavina, A., Potapov, P., Hansen, M.C., Pickens, A.H., Stehman, S.V., Turubanova, S., Parker, D., Zalles, V., Lima, A., Kommareddy, I., Song, X.-P., Wang, L. and Harris, N., 2022. 'Global trends of forest loss due to fire from 2001 to 2019', Frontiers in Remote Sensing, 3 [online]. Available at: https://doi.org/10.3389/frsen.2022.825190

131 Ke, P., Ciais, P., Sitch, S., Li, W., Bastos, A., Liu, Z., Xu, Y., Gui, X., Bian, J., Goll, D.S., Xi, Y., Li, W., O'Sullivan, M., Goncalves de Souza, J., Friedlingstein, P. and Chevallier, F., 2024. 'Low latency carbon budget analysis reveals a large decline of the land carbon sink in 2023', arXiv Cornell University [online]. Available at: https://arxiv.org/abs/2407.12447

132 Luonnonvarakeskus(Luke), 2023. 'Greenhouse gas inventory 2022', Monitoring Release. Available at: www.luke.fi/en/monitorings/maatalous-ja-lulucfsektorin-kasvihuonekaasuinventaario/greenhouse-gas-inventory-2022-no-significant-changes-in-the-final-results-for-the-agriculture-and-lulucf-sectors-compared-to-the-preliminary-data-published-in-december-2023

133 Ramsar Convention on Wetlands, 2018a. Global Wetland Outlook: State of the World's Wetlands and their Services to People. Gland, Switzerland: Ramsar Convention Secretariat.

134 Were, D., Kansiime, F., Fetahi, T., et al., 2019. 'Carbon sequestration by wetlands: A critical review of enhancement measures for climate change mitigation', Earth Systems and Environment, 3, pp. 327–340. Available at: https://doi.org/10.1007/s41748-019-00094-0

135 Anisha, N.F., Mauroner, A., Lovett, G., Neher, A., Servos, M., Minayeva, T., Schutten, H. and Minelli, L., 2020. Locking Carbon in Wetlands: Enhancing Climate Action By Including Wetlands in NDCs. Alliance for Global Water Adaptation and Wetlands International.

136 Parish, F., Sirin, A., Charman, D., Joosten, H., Minayeva, T., Silvius, M. and Stringer, L. (eds.), 2008. Assessment on Peatlands, Biodiversity and Climate Change. Global Environment Centre and Wetlands International.

137 Joosten, H., Tapio-Biström, M.-L. and Tol, S. (eds.), 2012. Peatlands – Guidance for Climate Change Mitigation By Conservation, Rehabilitation and Sustainable Use. Mitigation of Climate Change in Agriculture Series 5. FAO and Wetlands International.

138 Tarnocai, C., Canadell, J.G., Schuur, E.A.G., Kuhry, P., Mazhitova, G. and Zimov, S., 2009. 'Soil organic carbon pools in the northern circumpolar permafrost region', Global Biogeochemical Cycles, 23(2). AGU. Available at: https://doi.org/10.1029/2008GB003327

139 Schuur, E.A.G. and Abbott, B., 2011. 'High risk of permafrost thaw', Nature, 480, p. 32. Available at: https://doi.org/10.1038/480032a

140 White, R.P., Murray, S. and Rohweder, M., 2000. Pilot Analysis of Global Ecosystems: Grassland Ecosystems. World Resources Institute.

141 Bai, Y. and Cotrufo, M.F., 2022. 'Grassland soil carbon sequestration: Current understanding, challenges, and solutions', Science, 377. Available at: https://doi.org/10.1126/science.abo2380

142 Spalding, M.D. and Leal, M. (eds.), 2021. The State of the World's Mangroves 2021. Global Mangrove Alliance. Available at: www.mangrovealliance.org/wp-content/uploads/2021/07/The-State-of-the-Worlds-Mangroves-2021-FINAL-1.pdf

143 Naselli-Flores, L. and Padisák, J., 2023. 'Ecosystem services provided by marine and freshwater phytoplankton', Hydrobiologia, 850(12–13), pp. 2691–2706. Available at: https://doi.org/10.1007/s10750-022-04795-y

144 Khatiwala, S., Primeau, F. and Hall, T., 2009. 'Reconstruction of the history of anthropogenic CO_2 concentrations in the ocean', Nature, 462, pp. 346–349. Available at: https://doi.org/10.1038/nature08526

145 Chami, R., Cosimano, T., Fullenkamp, C. and Oztosun, S., 2019. 'Nature's solution to climate change', Finance and Development, 56, pp. 34–38.

146 IPCC, 2019. IPCC Special Report on the Ocean and Cryosphere in a Changing Climate. Cambridge University Press. Available at: https://doi.org/10.1017/9781009157964

147 Gruber, N., et al., 2019. 'The oceanic sink for anthropogenic CO_2 from 1994 to 2007', Science, 363, pp. 1193–1199.

148 IPCC, 2018. 'Summary for policymakers', In V. Masson-Delmotte, et al. (eds.), Global Warming of 1.5°C: An IPCC Special Report on the Impacts of Global Warming of 1.5°C Above Pre-Industrial Levels and Related Global Greenhouse Gas Emission Pathways, in the Context of Strengthening the Global Response to the Threat of Climate Change, Sustainable Development, and Efforts to Eradicate Poverty. Cambridge, UK and New York, NY: Cambridge University Press, pp. 3–24. Available at: https://doi.org/10.1017/9781009157940.001

149 Grill, G., Lehner, B., Thieme, M., et al., 2019. 'Mapping the world's free-flowing rivers', Nature, 569, pp. 215–221. Available at: https://doi.org/10.1038/s41586-019-1111-9

150 Convention on Wetlands, 2021. Global Wetland Outlook: Special Edition 2021. Secretariat of the Convention on Wetlands, Gland, Switzerland.

151 WWF and ZSL, 2024. Living Planet Report. Available at: https://wwf.panda.org/wwf_news/?12179466/LPR-2024

152 World Bank, 2022. Water in Agriculture. Available at: www.worldbank.org/en/topic/water-in-agriculture

153 WWF, 2021. Rivers of Food: How Healthy Rivers Are Central to Feeding the World. Available at: https://rivers-of-food.panda.org/?sf152152046=1#intro

154 ETH, 2023. Water-Energy Nexus. Available at: https://rre.ethz.ch/research/research-pillars/interdependent-energy-chemical-networks/water-energy-nexus.html

155 Wetlands International, 2020. Urban Wetlands for Cooler Cities. Available at: www.wetlands.org/case-study/urban-wetlands-for-cooler-cities

156 WWF, 2021. The World's Forgotten Fishes. Available at: https://wwfint.awsassets.panda.org/downloads/world_s_forgotten_fishes__final_april9_.pdf

157 WWF, 2023. High Cost of Cheap Water: The True Value of Water and Freshwater Ecosystems to People and Planet. Available at: https://wwfint.awsassets.panda.org/downloads/wwf-high-cost-of-cheap-water--final-lr-for-web-.pdf

158 Bowman, K.W., Dale, S.A., Dhanani, S., Nehru, J. and Rabishaw, B.T., 2022. 'The degradation of the Amazon rainforest: Regional and global climate implications', In V. Ongoma and H. Tabari (eds.), Climate Impacts on Extreme Weather. Chapter 13. Elsevier. Available at: https://doi.org/10.1016/B978-0-323-88456-3.00011-3

159 Sheil, D., 2018. 'Forests, atmospheric water and an uncertain future: The new biology of the global water cycle', Forest Ecosystems, 5, 19. Available at: https://doi.org/10.1186/s40663-018-0138-y

160 Schneider, U., Finger, P., Meyer-Christoffer, A., Rustemeier, E., Ziese, M. and Becker, A., 2017. 'Evaluating the hydrological cycle over land using the newly-corrected precipitation climatology from the global precipitation climatology centre (GPCC)', Atmosphere, 8, pp. 1–17.

161 Global Commission on the Economics of Water, 2023. Turning the Tide: A Call to Collective Action. Available at: https://watercommission.org/wp-content/uploads/2023/03/Turning-the-Tide-Report-Web.pdf

162 Wilson, E.O., 1984. Biophilia. Harvard University Press.

163 Bratman, G., Hamilton, J.P., Hahn, K., Daily, G. and Gross, J., 2015. 'Nature experience reduces rumination and subgenual prefrontal cortex activation', Proceedings of the National Academy of Sciences (PNAS), 112, pp. 8567–8572. Available at: www.ncbi.nlm.nih.gov/pmc/articles/PMC4507237/

164 Richardson, M., 2023. Reconnection: Fixing Our Broken Relationship with Nature. Pelagic.

165 Jones, K., Patel, N., Levy, M., et al., 2008. 'Global trends in emerging infectious diseases', Nature, 451, pp. 990–993. Available at: https://doi.org/10.1038/nature06536

166 Mallapaty, S., 2024. 'COVID pandemic started in Wuhan market animals after all, suggests latest study', Nature, 634. Available at: https://doi.org/10.1038/d41586-024-03026-9

167 Bernstein, A.S., Ando, A.W., Loch-Temzelides, T., et al., 2022. 'The costs and benefits of primary prevention of zoonotic pandemics', Science Advances, 8. Available at: https://doi.org/10.1126/sciadv.abm7571

168 UNODC, 2024. World Wildlife Crime Report: Summary, Conclusions & Policy Implications. Available at: www.unodc.org/documents/data-and-analysis/wildlife/2024/Wildlife_SpecialPoint_Final.pdf

169 Nellemann, C. (Editor in Chief), Henriksen, R., Kreilhuber, A., Stewart, D., Kotsovou, M., Raxter, P., Mrema, E. and Barrat, S. (eds)., 2016. The Rise of Environmental Crime–A Growing Threat to Natural Resources Peace, Development and Security. A UNEP and INTERPOL Rapid Response Assessment. Available at: https://www.unep.org/resources/report/rise-environmental-crime-growing-threat-natural-resources-peace-development-and

170 Pfenning-Butterworth, A., Buckley, L.B., Drake, J.M., Farner, J.E., Farrell, M.J., Gehman, A.-L.M., Mordecai, E.A., Stephens, P.R., Gittleman, J.L. and Davies, T.J., 2024. 'Interconnecting global threats: Climate change, biodiversity loss, and infectious diseases', The Lancet Planetary Health, 8(4). Available at: https://doi.org/10.1016/S2542-5196(24)00021-4

171 Behravesh, C., 2016. 'One health: People, animals, and the environment', Emerging Infectious Diseases, 22(4), pp. 766–767. Available at: https://doi.org/10.3201/eid2204.151887

172 Myers, S. and Frumkin, H., 2020. Planetary Health: Protecting Nature to Protect Ourselves. Island Press.

173 The Economics of Ecosystems and Biodiversity, 2008. TEEB Interim Report. Available at: https://teebweb.org/publications/other/teeb-interim-report/

174 World Economic Forum, 2020. Nature Risk Rising: Why the Crisis Engulfing Nature Matters for Business and the Economy. Available at: www.weforum.org/publications/nature-risk-rising-why-the-crisis-engulfing-nature-matters-for-business-and-the-economy/

175 Kurth, T., Wübbels, G., Portafaix, A., Meyer zum Felde, A. and Zielcke, S., 2021. The Biodiversity Crisis Is a Business Crisis. Boston Consulting Group. Available at: www.bcg.com/publications/2021/biodiversity-loss-business-implications-responses

176 Costanza, R., de Groot, R., Sutton, P., van der Ploeg, S., Anderson, S.J., Kubiszewski, I., Farber, S. and Turner, R.K., 2014. 'Changes in the global value of ecosystem services', Global Environmental Change, 26, pp. 152–158. Available at: https://doi.org/10.1016/j.gloenvcha.2014.04.002

177 World Economic Forum (WEF), 2020. Global Risks Report 2020. Available at: www3.weforum.org/docs/WEF_Global_Risk_Report_2020.pdf

178 World Economic Forum (WEF), 2025. The Global Risks Report 2025. Available at: https://reports.weforum.org/docs/WEF_Global_Risks_Report_2025.pdf

179 Ernst & Young, 2023. European Sustainability Reporting Standards (ESRS) in a Nutshell. Available at: https://denkstatt.eu/esrs-standards-explained/#:~:text=The%20five%20environmental%20standards%20cover,circular%20economy%20(ESRS%20E5)

180 GRI, 2024. GRI 101: Biodiversity 2024. Available at: www.globalreporting.org/standards/standards-development/topic-standard-project-for-biodiversity/

181 IFRS, 2024. ISSB to Commence Research Projects About Risks and Opportunities Related to Nature and Human Capital. Available at: www.ifrs.org/news-and-events/news/2024/04/issb-commence-research-projects-risks-opportunities-nature-human-capital/

182 ISO, 2024. ISO Standards in Support of the Global Biodiversity Framework. Available at: www.iso.org/biodiversity

183 PwC, 2024. PwC's Global Workforce ESG Preferences Study. Available at: www.pwc.com/gx/en/issues/workforce/pwcs-global-workforce-sustainability-study.html

184 World Economic Forum, 2020. The Future of Nature and Business. Available at: www3.weforum.org/docs/WEF_The_Future_Of_Nature_And_Business_2020.pdf

185 Young, D. and Beck, S., 2022. The Strategic Race to Sustainability. Boston Consulting Group. Available at: www.bcg.com/publications/2022/winning-strategic-race-to-sustainability

186 Kishan, S., 2022. ESG by the Numbers: Sustainable Investing Set Records in 2021. Bloomberg. Available at: www.bloomberg.com/news/articles/2022-02-03/esg-by-the-numbers-sustainable-investing-set-records-in-2021

187 World Economic Forum, 2022. Green Public Procurement: Catalysing the Net-Zero Economy. WEF White Papers. Available at: www.weforum.org/publications/green-public-procurement-catalysing-the-net-zero-economy

188 Spreckley, F. and Freer, C., 1981. Social Audit: A Management Tool for Co-Operative Working. Beechwood College.

189 Elkington, J., 1999. Cannibals with Forks: The Triple Bottom Line of 21st Century Business. Oxford: Capstone.

190 Cochelin, P. and Popoola, B., 2024. Sustainable Bond Issuance to Approach $1 Trillion in 2024. Standard & Poor's Global, 13 February. Available at: www.spglobal.com/_assets/documents/ratings/research/101593071.pdf

191 Strack, R., Kovács-Ondrejkovic, O., Baier, J., Antebi, P., Kavanagh, K. and López Gobernado, A., 2021. Decoding Global Ways of Working. Boston Consulting Group. Available at: www.bcg.com/publications/2021/advantages-of-remote-work-flexibility

4 Becoming Nature Positive

The 'Great Transition' from tipping to turning point

Marco Lambertini

'The measure of intelligence is the ability to change', is a reflection often attributed to German theoretical physicist Albert Einstein. Change is one of the universal laws of nature and a powerful dimension of natural evolution, linked to the need to adapt and respond to evolving conditions. Importantly, it can be either the result of choice or forced upon us by external variables.

We know the problem and the threat. We have the knowledge to understand the disastrous consequences of inaction and the many benefits of a transition. We also know the solutions, and we have agreed the destination through major international agreements: a carbon-neutral, nature-positive and equitable society and economy. Now, the focus should be on the journey. We know we are on a train going in the wrong direction. The longer we wait the more costly the return ticket will be. The more we wait the greater the chance that we will miss the connection that will take us to where we need to go.

As biodiversity loss is increasingly considered one of the most pressing macro-level risks for the economy and society [1, 2], the moral, social and economic case for a nature-positive transition has never been stronger. It is about avoiding planetary tipping points and reaching a turning point in our relationship with nature. It is about agreeing on the need for change, what that change looks like, where we want to go and how we will get there. It is inevitably a journey, and it all starts from defining the destination, the goal that will unite us along the journey: the Nature Positive goal.

4.1 From aspiration to a measurable global goal for nature: the definition of Nature Positive

Nature Positive is a global societal goal defined as 'Halt and Reverse Nature Loss by 2030 on a 2020 baseline, and achieve full recovery by 2050'. To put this more simply, it means ensuring more nature in the world in 2030 than in 2020 and continued recovery after that. Delivering the Nature Positive goal requires measurable net-positive biodiversity outcomes through the improvement in the abundance, diversity, integrity and resilience of species, ecosystems and natural processes. The Nature Positive goal is designed to drive society to deliver a measurable absolute improvement in the state of nature against a defined baseline, which will in turn improve nature's ability to contribute to human wellbeing [3].

DOI: 10.4324/9781003474043-5

Humanity is facing a '*polycrisis*' made of many challenges, but three are existential and interconnected: *nature loss, climate change* and *inequality*. All these crises are of our own making, and, if we act swiftly, we have the ability to address them.

The stimulus to develop a global goal for nature was inspired by the high-level goal to combat climate change in the Paris Agreement, which was signed by 196 parties in 2015. These countries pledged to "aim to reach global peaking of greenhouse gas emissions as soon as possible . . . as to achieve a balance between anthropogenic emissions by sources and removals by sinks of greenhouse gases in the second half of this century", effectively embracing a *carbon-neutral goal* that would give us a reasonable chance to keep global average temperatures well below 2°C.

Three years later the Intergovernmental Panel on Climate Change (IPCC) Special Report [4], *Global Warming of 1.5°C*, stated that, "limiting temperature rise to around 1.5°C . . . implies reaching net zero emissions of CO_2 by mid-century along with deep reductions of non-CO_2 emissions", effectively setting a measurable *net-zero emission goal* for action. Political and civil society initiatives like the Glasgow Climate Pact, the Race to Net-Zero, the Green House Gas protocol, the Science Based Targets Initiative on climate, amongst others, boosted the impetus towards delivering the net-zero emissions goal by 2050 and the 2030 milestone of halving global carbon dioxide emissions relative to 2010 levels. Net-zero emissions is the 'response target' to the main anthropogenic pressure on climate – GHG emissions, while 1.5°C is the overall 'outcome goal'.

The effect of setting a climate Global Goal has been transformational in terms of inspiring governments and companies to align with the global ambition of achieving net-zero emissions. One year after the IPCC Special Report, net-zero pledges already covered one-sixth of global GDP, continuing to grow to nine-tenths in 2022 [5].

Of course, pledges are only as good as they are implemented, and emissions, even though they have slowed down in several countries, have continued to rise globally since the Paris Agreement was signed, indicating the need to accelerate decarbonisation as well as protecting natural carbon sinks. By 2023 nearly 40% of listed companies had set decarbonisation targets that aimed to achieve net-zero emissions, and over half of listed companies had disclosed an emissions reduction commitment. However, only 11% of listed companies are on a pathway aligned with a 1.5°C temperature rise and 38% to 2°C. The world's listed companies are still on a path to warm the planet by 3°C above pre-industrial levels this century [6]. While global GHG emissions are at a record high, the share of fossil fuels in the global energy mix is slowly falling [7].

However, without the transparency and accountability that comes with having clarity on a common ambition, we could confidently assume that progress would have lagged much further. We now know how far off we are in mitigating climate change, which increases accountability and can in turn spur more action.

Approved on same year as the Paris Agreement, the Sustainable Development Goals (SDGs) provided a set of 17 socio-economic and environmental goals to achieve high standards of human welfare and long-term viability of the natural systems to secure a more equitable and sustainable world by 2030 [8]. They address all the most pressing challenges our society faces today, from poverty to malnutrition, health and environmental degradation.

After Paris it not only became clear that we needed a similar, complementary and measurable global goal for nature, but also that it was possible. In December 2019, a group of CEOs from environmental organisations, sustainable business platforms and research institutes gathered under the Global Goal for Nature Group to begin work to identify an ambitious, science-based and measurable global goal for nature, and to advocate for this goal to be adopted internationally [9]. The group recognised the need for an overarching goal equivalent to the one agreed in Paris for climate that would set a common ambition on nature and biodiversity and would be mutually supportive to both the Paris Agreement goal of 1.5 °C (and the net-zero emissions pathway to deliver it) and the SDGs.

This was agreed to be a *net-positive*, Nature Positive Global Goal, emphasising the need to halt and reverse nature loss. The lack of a clear and measurable goal for addressing the nature loss crisis was considered a major obstacle to aligning on ambition, driving progress and increasing accountability in the effort to address nature loss. This was reinforced by the *Global Biodiversity Outlook 5* [10], released ahead of the approaching UN Convention on Biological Diversity (UN CBD), which concluded that none of the 20 biodiversity targets agreed at the UN Convention on Biological Diversity in Aichi in 2010 had been fully met at the global level. The approaching Conference of the Parties of the UN CBD, set to agree the next 10-year plan, was identified as the opportunity for the world to adopt and codify such a new global goal for nature. It followed an intense, coordinated advocacy effort mobilising over 300 civil society organisations from the environmental, humanitarian and development sectors, faith groups and Indigenous organisations [11], hundreds of companies through sustainable business platforms like the World Economic Forum (WEF), the World Business Council for Sustainable Development (WBCSD) and Business for Nature, amongst others, and 96 countries and the European Union who were all signatories, at Head of State level, to a *Leaders' Pledge for Nature* [12].

After the delays caused by the COVID-19 pandemic, in December 2022, the 188 nations attending the UN CBD COP15 adopted the Kunming-Montreal Global Biodiversity Framework (GBF), the mission of which is to "halt and reverse biodiversity loss by 2030". Although the phrase 'Nature Positive' is not used in the GBF text, the language "halt and reverse by 2030" is fully consistent with the original definition of the Nature Positive Global Goal and, for the first time, sets a north star or southern cross for the world's nature action agenda, a global goal for nature to be collectively delivered, equivalent to and supportive of the 1.5°C Global Goal for climate.

The Nature Positive Global Goal the Nature Positive Global Goal includes both a *deadline* and a *baseline* to ensure its measurability and to increase accountability in all those responsible for delivering it. By the 2030 deadline, we should have halted and reversed nature loss by achieving a global net-positive biodiversity outcome. This signifies the science-based ambition and urgent necessity to halt and reverse nature loss and bring our society and economy back within the safe planetary boundaries of biosphere integrity. The 2020 baseline, meanwhile, adds realism and feasibility to this ambition. The further back we set the baseline, the more monumental the effort to reverse nature loss and achieve a net-positive outcome by the 2030 deadline. A clear 2020 baseline was preferred over less quantifiable and more discretionary options such as 'pristine or high integrity conditions'. Both the 2020 baseline and 2030 deadline were also chosen to align with the timeline of the UN CBD Kunming-Montreal agreement. But of course, the nature-positive journey doesn't stop at 2030, and the Nature Positive Global Goal definition sets a further ambition to achieve 'full' recovery of nature by 2050 in line with the GBF 2050 vision. At the time of coining the Nature Positive goal, the focus was on the 2030 deadline, and the concept of 'full recovery by 2050' was left as an aspiration and was not more precisely defined. As we approach the first deadline of 2030, it will require a clear articulation.

The term '*Nature Positive*' was chosen over more technical wording to facilitate an intuitive understanding in people, and to foster greater communications and uptake. Following the approval of the GBF in 2022, a larger and more diverse group of global organisations came together to establish the *Nature Positive Initiative* (NPI) to preserve the integrity of the Nature Positive definition, align on guidance to deliver genuine nature-positive outcomes and contributions, and to coordinate further joint advocacy efforts. The NPI is a convening platform to align on providing guidance and to coordinate support for the delivery of the Nature Positive goal across the Initiative's members, and the broader sustainability ecosystem of organisations and stakeholders.

In 2023, the initiative published a more concise and nuanced definition paper [13], although it remained totally aligned to the 2020 original. This consensus definition, endorsed by the 27 leading organisations that are core members of the NPI, removed any lingering confusion over the meaning of the Nature Positive goal. This should now be considered *the reference definition of Nature Positive*.

The definition paper underscores that the Nature Positive goal is measured by net-positive biodiversity outcomes, thus clarifying the alignment between the goal itself and the mission of the UN Global Biodiversity Framework. Along with the growing awareness of the risks associated with nature loss, the concept of Nature Positive has proven successful, and its uptake has been widespread. The intuitive nature of the term has, however, sometimes led to a superficial use and interpretation, and at times it is used as a catchphrase and generic aspiration rather than a measurable global goal. This defies the scope and ambition of the Nature Positive goal to be measurable and drive action and accountability, recognising that it is no longer enough to do 'less harm' or something good for nature, and we must deliver or contribute to the delivery of a quantifiable, net-positive biodiversity outcome.

Not every action that benefits biodiversity equates to a nature-positive outcome, nor can a company, a government or any other actor that does 'something positive' for nature claim to deliver a nature-positive outcome. Nature-positive outcomes need to demonstrate a quantifiable net-positive improvement in the state of biodiversity by 2030 against the 2020 baseline and at a defined scale, also ensuring that it has not resulted in a displacement of impacts somewhere else.

The Nature Positive goal was designed to move us away from insufficient actions and generic claims to demonstrable outcomes. It moves us towards the development and implementation of appropriate actions and structured plans that will deliver what is necessary to halt and reverse the loss of nature and demonstrate measurable net-positive biodiversity outcomes. Within this resides the measurable ambition and the disruptive nature of the Nature Positive concept that must be preserved [14, 15]. The preservation of the integrity of the meaning of the Nature Positive goal and its applications are the core objectives of the NPI coalition.

The rigour applied to the Nature Positive definition should not discourage companies or governments. Most companies should start with designing nature-positive strategies and measure the progress made in their sphere of control (operations) or influence (value chain, procurements, etc.) aiming to improve coverage and granularity over time. We will discuss corporate strategies in more detail later in this chapter.

Anything a government, business or investor does for nature, albeit welcome, cannot simplistically be labelled 'nature positive', but rather a contribution to the overall global goal. Then the key question is whether the contribution is sufficient or adequate vis-à-vis the negative impacts on nature of the entity. For this reason, the Nature Positive Global Goal was designed as a measurable goal through the adoption of a defined baseline and deadline, and every stakeholder should develop plans and targets to deliver or contribute to the delivery of net-positive biodiversity outcomes through a holistic nature-positive strategy, complemented by transparent disclosure and reporting.

4.2   We need a world with more nature, not less

Nature Positive is an *intuitive concept*. In simple terms, it means more nature, not less. More nature by 2030 than in 2020. More natural forests, more wetlands and natural floodplains, wild fish in the ocean, rivers and lakes, more soil biodiversity and more pollinators. More biodiversity worldwide, from wild habitats to productive landscapes and the urban environment. From these simple examples it is immediately apparent that a Nature Positive Global Goal will bring huge benefits to society, the economy and our well-being in addressing issues like climate change, water and food security and other key challenges. The goal embodies an ambition to move from today's nature-negative society and economy to a nature-positive one, which will halt and reverse nature and biodiversity decline by the end of this decade and continue the recovery beyond this timeframe.

Nature Positive is not a slogan, but a *measurable goal*, comparable to carbon neutral. The carbon-neutral goal is achieved by net-zero emissions by 2050 with a milestone of 45% reduction by 2030 from 2010 levels. The Nature Positive goal is achieved by a net-positive biodiversity outcome by 2030 on a 2020 baseline. The overall climate goal is to contain average global atmospheric temperatures below 1.5 °C, while the overall biodiversity goal is to halt and reverse biodiversity loss.

Nature Positive is a *necessary goal* because we have lost so much biodiversity, thus weakening the resilience, functionality and productivity of many essential ecosystems and driving them towards dangerous and irreversible ecological tipping points and heavily disrupting the liveability of the planet as we know it. This is why a net-zero outcome for nature would not be enough. We need to halt as well as reverse nature loss, by conserving, restoring and sustainably managing natural resources and biodiversity. Achieving the Nature Positive goal is imperative because humanity cannot thrive in the future if biodiversity continues to decline and ecosystem services continue to weaken.

The evidence has never been clearer. The impacts of climate change and the unfolding biodiversity crisis on humanity are already profound. Furthermore, the consequences are not felt equally across society, and socio-economic inequality is being exacerbated.

Nature Positive is an ambitious but *achievable* goal because we have the knowledge, technologies and means to deliver it and because nature's intrinsic regenerative capacity is formidable. Contrary to climate, biodiversity can often bounce back, and much more quickly, when it is given a chance. There are so many examples from around the world, such as the way most whale populations were able to rebound following the end of large-scale commercial whaling, the comeback of salmon after the de-damming of rivers, the exponential increase of pollinators and soil biodiversity in regenerative agriculture, the return of predators like wolves, lynx and beavers in Europe once they stopped being persecuted, the recovery of tiger populations in India, forest elephants in Gabon, black rhinos in Kenya, mountain gorillas in Rwanda, waterfowls in North America and tuna in the Mediterranean, to name just a few. Not to mention many local examples of ecosystems like forests, wetlands and mangroves which have bounced back across the world. These are the result of better protection, habitat conservation and changes in human attitudes. Various studies show how conservation efforts have made a difference in improving the status or in slowing down the decline of biodiversity, and how many areas have witnessed an extraordinary 'rewilding' – whether spontaneous or assisted – even in industrialised and highly populated regions [16, 17].

Finally, Nature Positive is a positively *disruptive goal* driving systemic transitions that de-risk our society and economy. Becoming Nature Positive implies both cultural and structural changes to the way we live our lives and run our society and economy. While recognising the necessity of building a nature-positive future and the huge benefits that will derive from it, we also need to accept that its ambition to halt and reverse biodiversity loss by 2030 requires deep transformations in both our society and our economy, with implications for the way we value nature, as well as the way we produce and consume, allocate financial flows, decide on land use, manage seascapes and marine resources and how we relate to nature more generally. As the carbon/neutral and net-zero emissions goal has disrupted the energy sector, the nature-positive and net-positive biodiversity goal is designed to disrupt the key economic nature-negative sectors that are today main drivers of nature loss and instead drive the adoption of sustainable practices. Top of the list: agriculture, fishing, extractives, forestry and infrastructure but also energy, manufacturing, tourism and others. Every sector has a contribution to make to the Nature Positive goal.

Nature Positive is intuitive, measurable, necessary, achievable and disruptive. By embarking on this journey, we should anchor our resolve in the rationale behind the need to become Nature Positive and the benefits it will deliver. But if the intuitive elements of Nature Positive are at the foundation of its success, then its definition as a measurable goal raises fundamental questions: what does 'more nature' or 'net-positive biodiversity' mean, and how can it be measured?

4.3   The inevitability of the 'net' concept

Before addressing the measurability of Nature Positive, it's necessary to focus on a fundamental dimension of its definition, the notion of '*net*'. A 'net-outcome' is the result of a series of negative and positive events. For example, in simple financial terms, a net result is the difference between income and expenditure. It would be nice, but highly unrealistic and in fact socially unacceptable in many regions of the world, to think that we can reduce our impact on nature to zero, just as it is unrealistic to think we can achieve zero emissions in the near future. That's why the concept of net-zero is embedded in the Global Climate Action Agenda, recognising that in the short and medium term, some emissions are inevitable but that overall they should be neutralised. Humanity's current stage of development and the fact that many basic needs are yet to be fulfilled in large parts of the world simply does not allow for a 'zero-impact' goal.

The concept of 'net' in the context of biodiversity acknowledges that some loss or degradation of nature is an inevitable result of human use and pressures, the result of demographic growth and development. In developing economies in particular, infrastructure needs to be built, agriculture needs to be expanded and manufacturing capacity to be developed, and this will generate an inevitable degree of impact on nature. Today, Africa has a mere 1% of the world's railways, and less than half of the rural population has access to electricity [18].

However, the overall 'net-outcome' must be positive in biodiversity terms and at the appropriate scale to reflect the intrinsic local characteristics and distribution of biodiversity. Unlike the financial industry, where transactions are standardised using monetary units normalised through exchange rates, or greenhouse gasses through a CO_2 equivalent metric, biodiversity is inherently local and lacks *equivalence* across ecosystems and species, adding to its complexity. Achieving a genuine net outcome requires a stringent application of a process of assessment and mitigation. The '*mitigation hierarchy*' is a tool designed to help investors, companies and developers reduce the negative impacts on nature and biodiversity and achieve a genuine net-zero loss or net-positive outcome [19, 20].

The mitigation hierarchy includes a series of logical, sequential, iterative steps that begin with *assessing* the impacts and then *avoiding*, *minimising*, *restoring* and a final action of *compensation*, applicable exclusively to truly unavoidable impacts. The aim is to first achieve a *net-zero loss* (no-net loss) and subsequently a *net-positive outcome* (net-gain or nature-positive). The net principle recognises that some loss of nature may be unavoidable vis-à-vis human development, but through the application of the mitigation hierarchy, it aims to reduce to an absolute minimum the impacts on nature and, through a set of corrective measures, halt and reverse biodiversity loss while allowing for socio-economic development.

A net-positive outcome for biodiversity means that the negative impacts of an activity are outweighed by the results of actions taken to avoid, minimise, restore and compensate those impacts, ultimately resulting in an improvement of the state of biodiversity in a defined geographic area and timescale, while also avoiding '*leakage*' of negative impacts elsewhere. The approach can be applied at different scales, from project, site and operation to organisation, sector, landscape or jurisdiction level for cumulative impacts. The reality of our current economic model, until now largely oblivious to environmental externalities, means that there is huge room for improvement in terms of reducing our impacts and improving the state of nature. The new Nature Positive goal and ambition is not to simply 'do less harm' but to deliver a positive outcome for nature.

The complexity of nature, the lack of equivalence between species and ecosystems, and the many potential impacts associated with development projects mean that the application of the net concept comes with significant integrity risks and raises legitimate concerns. The main concern, witnessed endless times in the application of the 'net-zero' approach to climate action, is that companies resort to compensation 'too much and too soon', rather than first reducing their impacts. In fact, in the absence of regulation and an adequate price on carbon, it's often cheaper to negotiate the purchase of carbon offsets rather than invest in transforming systems of production. The mitigation hierarchy was designed to avoid companies resorting to compensation before taking necessary avoidance and mitigation actions.

There are a number of fundamental elements that can guarantee the high integrity application of the mitigation hierarchy. It all begins with the company's public *commitment* to contribute to the Nature Positive goal and an honest and comprehensive *assessment* of its impacts on nature as a crucial first step that will determine the rigour of the process and credibility of the outcomes.

The assessment should aim to be as comprehensive as possible and leave no stone unturned when looking at impacts both in direct operations and along the value chain. It is important to recognise that companies with complex supply chains may need time to assess the full range of their impact on nature. Therefore, it would be counterproductive to the overall nature-positive outcome for companies to wait until the full impact assessment is complete. Following a public commitment to a nature-positive outcome, the company should immediately start addressing known impacts and those of greatest materiality, affecting priority sites, species and natural processes. The assessment should be periodically reviewed and the response plan adapted as conditions change and new elements emerge. It is also critical that assessments and corrections are publicly disclosed for transparency and to allow scrutiny from all stakeholders and ensure credibility and recognition.

Secondly, the same rigour and transparency need to apply to plans to *avoid and minimise* the identified impacts. This may require bold decisions, such as avoiding exploitation or development in areas with significant biodiversity. Overall, it will need to avoid and minimise impacts to reduce the need for compensation to an absolute minimum.

Thirdly, the most delicate and controversial implication of a net approach relates to the final and potential act of *compensation*. We should stress that the primary focus should firstly be on reducing impacts within the company's operations and value chain as well as restoring and improving the state of nature (also referred to as '*insetting*') before resorting to funding or driving positive outcomes outside the sphere of control or influence of the company (*offsetting*). The most important step is to ensure that the impacts assessed as unavoidable are truly that. When the unavoidability of the impact is ascertained (and disclosed), compensation becomes a necessary last resort. In situations with some levels of truly unavoidable impacts, compensation is needed to achieve overall net-positive outcomes or contributions. All this will need to be periodically reviewed as impacts deemed unavoidable today may become avoidable tomorrow thanks to new technologies or practices.

The complexity of biodiversity necessitates adherence to several essential principles when applying compensation measures. Contrary to the mainstream climate metric of metric tonne of carbon dioxide equivalent, biodiversity has unique localised features that are not fungible or equivalent across typologies and geographies. While the tonnes of carbon emitted by a power plant can be compared in pure quantitative terms to those absorbed by restoring a forest, in biodiversity terms two forests just a few kilometres apart (for example along a mountain slope or on different soil) can be very different in structure, plant and animal communities.

Therefore, conservation and restoration measures intended to compensate for unavoidable impacts on nature must take place within the same ecosystem type that has been impacted, to fulfil as much as possible the principle of '*equivalence*' or 'like-for-like'. Compensatory actions on species or ecosystems other than those equivalent to the one impacted are flawed and unacceptable. This makes compensatory actions on biodiversity particularly delicate and radically different in practical terms from the ones on carbon. Compensation of biodiversity impacts need to be hyper-local so that like-for-like is attained, and international offsetting or offsetting across large geographies is not acceptable. Only by ensuring such a high-integrity approach can we ensure that any compensation does not legitimise but, on the contrary, comes at the end of a thorough process of avoidance and mitigation and drives a reduction of impacts on nature, ultimately delivering genuine net-positive outcomes. This is the true spirit and the objective of applying the mitigation hierarchy.

Additional voluntary actions which do not directly compensate for project or company impacts can help achieve a greater contribution to the Nature Positive goal. However, the main focus should primarily be on avoiding, reducing, restoring and compensating the negative impacts. Being disassociated from the company's impacts, these voluntary additional contributions do not need to follow the equivalence principle but also cannot be counted as part of the compensatory actions. They can, however, contribute to an overall and higher net-positive outcome once, as a minimum, a 'net-zero loss' outcome is achieved within the company's operations and value chain. Safeguarding principles for net-positive outcomes have already been developed [21, 22].

Sites with high conservation value, characterised by significant biodiversity, unique and intact ecosystems or ecosystems which provide crucial services, such as sites crucial to the local hydrological cycle or the survival of endemic or threatened biodiversity (such as Key Biodiversity Areas or Alliance for Zero Extinction sites), must be strictly preserved, and impacts must be avoided or highly minimised. This principle is paramount. Any development project or investment should first seek to avoid and minimise impacts to critical natural habitats and species to the maximum extent possible, and alternatives should be seriously considered and pursued. It is not a coincidence that the Nature Positive goal and GBF mission mention 'halt' first and then 'reverse'!

The principle of equivalence should also be applied to '*ecosystem integrity*': the loss of intact ecosystems cannot normally be compensated through the restoration of degraded ones, even of the same typology, unless the impacts are truly unavoidable. Intact ecosystems are the product of innumerable interactions between their biotic and abiotic components which have evolved over thousands, even millions, of years. Most of these ecosystems are difficult and often impossible to restore to their original condition during a human timescale. Avoiding impacts on these intact ecosystems should always be a top priority. Restoration should only be acceptable for already-degraded ecosystems or limited portions of a higher integrity site or landscape. In most cases, impacts to highly intact ecosystems cannot be fully compensated through restoration.

To put it simply, we must 'protect the best and improve the rest'. However, if safeguarding the highest conservation value areas is a priority, this does not mean that other areas that are not intact or that do not qualify as high biodiversity value but are still in a natural or seminatural state can be converted or unsustainably exploited. Biodiversity is ubiquitous across geographies and habitats, and even somewhat degraded ecosystems host significant levels of biodiversity and may still provide important services, not to mention their potential to recover over time. We have already lost too much nature and urgently need to restore many degraded ecosystems to conserve biodiversity, stabilise the climate and secure nature's many services to human development and well-being. Science suggests that to help prevent biodiversity loss and preserve Earth's ecological resilience, it is necessary to conserve, in their 'natural state', between 50% and 70% of the planet [23–26].

Because biodiversity is very specific in location and distribution and not comparable to a mixture of heat trapping gases in a contiguous and shared atmosphere, it is commonly perceived as an obstacle to high integrity compensation action for biodiversity loss. But what if this lack of equivalence could in fact be a blessing in disguise? Once we establish that compensation must be 'like-for-like' and based on ecological equivalence, the local nature and distribution of biodiversity means that any compensation must firstly focus on the same ecosystem or landscape where the impacts and losses happened. This ecosystem and landscape focus will make compensation for biodiversity loss more traceable and verifiable, addressing a major weakness and risk in other compensation schemes. It would also allow for greater transparency, allowing stakeholders to scrutinise and better assess benefits to biodiversity as well as local communities, avoiding the risk of impact displacement, uneven trade-offs and inequity. The 'net approach', despite its inevitability, is not without controversy and, without the clear safeguards and a strong mechanism for enforcement, could well be gamed by bad actors. However, if guided by strong principles and guidance and applied within regulatory frameworks, compensation schemes have the potential to rectify the current unacceptable situation where nature is destroyed without any liability, responsibility or consequence for the perpetrators.

4.4   Measuring Nature Positive: let's not make perfect the enemy of good

It's worth reiterating. Nature Positive is not a slogan or a vague aspiration, but a measurable goal. As such, measuring the actions taken to deliver it as well as the final outcome is crucial for the effectiveness, credibility, accountability, recognition and, most importantly, to deliver genuine improvements in the state of nature. The biggest challenge for both the public and private sector in developing and implementing a nature-positive strategy is figuring out what are the impacts and what indicators, metrics and data to use to credibly measure progress against the net-positive outcome. This is an area where more clarity and alignment are still needed.

Firstly, let's clarify a few key terms. An *indicator* is a quantitative or qualitative variable that provides a simpler and relevant way to measure a condition, a level or a performance. More specifically, regarding the state of nature, an indicator can also be defined as a quantitative or qualitative factor contributing to understanding the state of nature, including positive upwards or downwards trends (e.g. ecosystem extent, ecosystem condition, species richness and abundance). A '*metric*' is the standard of measurement used by an indicator and to measure the current state or condition and compare changes over time. Metrics are based on reliable, standardised and comparable '*data*'. Data is collected through various *methodologies* and *technologies*. To use a straightforward example, 'ecosystem extent' is a biodiversity spatial indicator, and 'area' is the metric to measure loss or gain of this extent using standardised units such as hectares and with data derived, for example, from satellite imagery technology.

The term '*biodiversity*' refers to the diversity within and among species and ecosystems (the biosphere). The term 'nature' is broader, as it commonly refers to both living components (biosphere) and non-living components (geosphere/lithosphere, atmosphere and hydrosphere/cryosphere). The interactions between these realms co-evolved and are deep, intricate and not fully understood. Biodiversity (species, ecosystems) has adapted to geophysical systems and can influence the resilience and behaviour of natural processes as well as being affected by their alteration. Ultimately biodiversity, the 'living component of nature', and the overall liveability of the planet depends on the health and stability of all other abiotic realms. For this reason, biodiversity can be considered a reliable proxy for the overall state of nature. The decline of biodiversity is also a warning system for humanity, as our well-being is dependent on the same environmental conditions. If biodiversity declines it is bad news for us too. The achievement of the Nature Positive goal is therefore ultimately best measured against net-positive biodiversity outcomes. The Global Biodiversity Framework, adopted internationally under UN CBD, also refers to the need to 'halt and reverse biodiversity loss by 2030', a language fully consistent with the definition and measurability of the Nature Positive goal [27].

There are three broad categories of metrics necessary to drive nature-positive outcomes, and they are equally important. The first two categories include '*pressure/impact metrics*' and '*response metrics*' and respectively measure the ecological footprint of a company and the progress they make in reducing it. The 'pressure/impact metrics' are inevitably specific to each company's impacts and nature-positive plan. Logically, the impact on nature of fishing, mining, manufacturing or agricultural operations are very different, therefore so is the way to measure it. Getting these metrics right is critical to ensure high integrity and accuracy in both assessing and addressing the company's negative impacts on nature, as well as measuring the positive ones. The 'response metrics', which measure the actions taken to avoid, reduce and compensate the impacts, are also specific to each company's impacts and context. The other, equally crucial category relates to '*outcome/state metrics*', which demonstrate the achievement of,

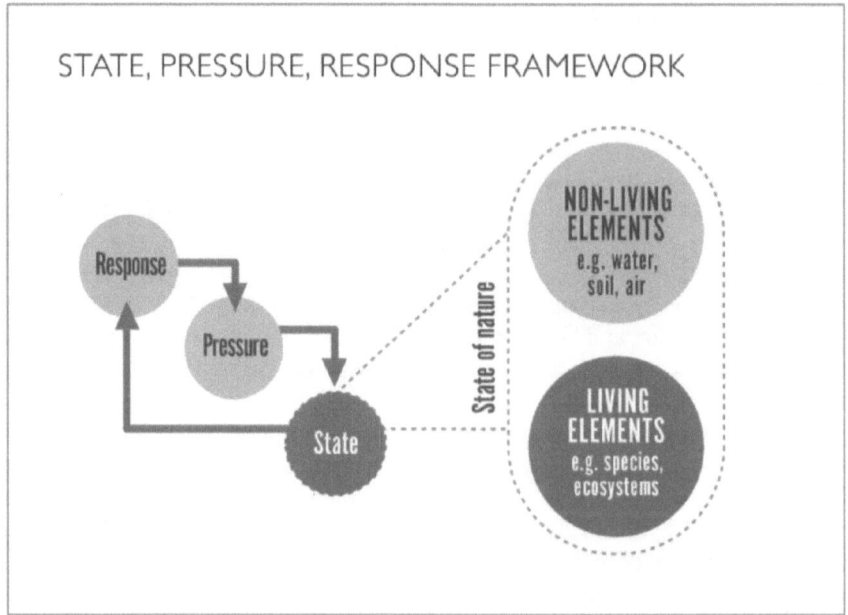

Figure 4.1 State, pressure and response metrics are all essential to guide strategies to deliver net-positive biodiversity outcomes.

Source: Nature Positive Initiative

or contribution to, an absolute improvement in the state of nature (net-positive biodiversity outcomes) as a result of implementing a corporate nature-positive strategy or national biodiversity plan for governments. Nature-positive outcome metrics refer to the state of biodiversity but indicate the overall result of impacts and responses. These metrics are essential to validate the adequacy of nature-positive strategies and if they have delivered nature-positive (net-positive biodiversity) outcomes, failing which means that either the responses are not adequate and/or external factors are at play.

There are several methodologies and frameworks focussing on pressure/impacts and response that are available to companies and investors, with the Taskforce on Nature-related Financial Disclosures (TNFD), the Science Based Target Network (SBTN), the Global Reporting Initiative (GRI), Nature Action 100 and WBCSD among the most recognised and increasingly aligned for interoperability. Several consultancies have developed guidance on how to deliver nature-positive corporate strategies as companies and investors find themselves under an increasing amount of pressure to adopt credible action plans and standards. We are beginning to see regulators require the private sector to disclose their biodiversity-related risks and strategies, such as through the EU Sustainable Finance Disclosure Regulation and Corporate Sustainability Reporting Standards focussing on impacts and dependencies [28].

This is how the pressure, response and state metrics connect. The pressure/impact and response metrics measure the impacts and the corrective actions, while the 'outcome metrics' or 'state of nature metrics' measure the ultimate result in terms of change in state. The need for standardised *State of Nature Metrics* lies in the fact that without them, no actor would be able to know if their nature-positive strategy is fit for purpose and able to credibly demonstrate genuine nature-positive outcomes in a comparable way. The lack of meaningful metrics will lead to various ways of assessing and reporting the results of a company's nature-positive strategy, risking accusations of lack of transparency, poor ambition and effectiveness and even greenwashing.

Biodiversity is distributed in every habitat; natural, semi-natural or anthropic. Although intact natural habitats rank at the top of the conservation priority list because they host an immense diversity of life, contribute to the stability of important natural processes and provide powerful direct services to people, we must strive to preserve biodiversity everywhere on the planet. Biodiversity's contributions to ecological stability and human well-being are important everywhere, so this should include modified landscapes like farmland and urban environments.

The *Three Global Conditions* provides a framework for biodiversity conservation by identifying three main landscape patterns [29]: large wild areas, shared lands (mosaic landscapes due to human intervention) and farms and cities with a high degree of human alteration. Each condition requires different strategies for biodiversity conservation and sustainable use. An important principle in this framework is that, although with different level of richness, abundance and significance, biodiversity is present and requires conserving in all three conditions. The need to measure net-positive biodiversity outcomes applies across all three conditions.

Biodiversity is undoubtedly a complex concept, with many dimensions and drivers of loss that interact with an equal number of abiotic variables. For this reason, biodiversity loss is often thought of as much more complex than climate change. We often come to this conclusion by comparing the many, localised manifestations of biodiversity, as opposed to the ubiquitous nature of a few key greenhouse atmospheric gases. In reality, the climate is also a very complex system with many drivers interacting and generating mutual feedbacks, not to mention the bioclimatic dimensions that involve biodiversity.

In measurement terms, biodiversity is undoubtedly complex, but rather than making perfect the enemy of good, the biodiversity community should agree on what are 'good enough' metrics. These must strike the right balance in being science based and credible, but also accessible, applicable, comparable and affordable. As technology evolves, more methodologies, metrics and data become available, allowing us to measure and monitor in more sophisticated and holistic ways the state of nature.

Some of these technologies and methodologies are already available and evolving such as remote sensing, e-DNA and bio-acoustics. Companies are encouraged to use these metrics to deepen the understanding of their impacts. However, as a societal goal, outcomes and contributions to Nature Positive must be measured in a comparable way across scales and users, and for a variety of use cases.

– *Nature Positive metrics: the 'proof is in the state'*

Measuring changes in the state of nature in a standardised and comparable way is essential. It allows for credible demonstrations of progress in addressing negative impacts and achieving net-positive biodiversity (nature-positive) outcomes and aligns with the Kunming-Montreal Global Biodiversity Framework's mission to "halt and reverse biodiversity loss by 2030".

Nature is complex and so is biodiversity, the living component of nature. Biodiversity encompasses a wide variety of species, ecosystems and the intricate relationships and natural processes that connect biotic and abiotic elements of the environment. We have made amazing progress in measuring many dimensions of nature, and new technologies and methodologies are continuously improving. But, let's face it, we will never be able to accurately measure all elements of nature at every scale and everywhere in the world, at least for the foreseeable future. And we don't have time. To begin with, we know only a fraction of the planet's animals, plants, fungi and microorganisms, and even less about the interactions governing the planetary ecological system. Measuring every aspect of nature is not feasible or practical. Therefore, we should identify a small but meaningful set of metrics that can act as an indication or proxy of nature's overall health to gauge whether it is in recovery or not and inform our actions.

We have many possible indicators and metrics to choose from, some of which are already used effectively for important research and specific purposes, locations, species and ecosystems. What is missing is consensus on a set of credible yet practical and accessible metrics that are not perfect but good enough to measure change in the state of nature vis-à-vis human pressures and to assess nature-positive outcomes. Metrics which are both science-based and credibly cover key dimensions of biodiversity but are also practical, affordable, applicable and comparable at different levels.

This lack of consensus hinders engagement, mobilisation, accountability, recognition, disclosure and, most importantly, progress in halting and reversing nature loss. Nature is complex, and no single indicator and metric can fully capture the state of nature. But with more than 600 available nature metrics (and counting), it is truly challenging for organisations, companies, financial institutions as well as national and sub-national governments to determine what to measure and how to do it in a relevant, credible and consistent way. This leads to confusion, lack of confidence and, ultimately, inaction. Agreeing on a set of SON metrics is important to effectively know whether our actions are truly contributing to nature's recovery. These metrics are a central component of any nature strategy and transition plan, complementing and informing the other critical metrics that measure pressure and response.

The starting point for every company or financial institution is to identify and measure their pressures and impacts on nature. The second step is, of course, developing a plan to respond to those pressures and impacts. Measuring the state of nature means to assess the outcomes of the responses to the pressures and impacts. For example, a company must measure their pressures (e.g. wastewater volume) and their responses (e.g. volume of wastewater treated) using specific, relevant metrics. Yet, the proof of an effective response can only ultimately come from measuring the effects on the abundance, diversity, integrity and resilience of ecosystems and species (i.e. aquatic biodiversity) – the best representation of the overall 'state of nature'.

Having a set of measurable indicators and metrics that capture the effectiveness of our efforts to halt nature loss and set it on a path to recovery, thereby delivering nature-positive outcomes, is the fundamental basis to ensure transparency, accountability, recognition and, most importantly, to drive the right action at the right scale and speed. This is particularly relevant to equip the private sector (where most of the footprint on nature comes from) to act and be both accountable and recognised for taking action. But the use cases for SON metrics do not stop with the private sector. These metrics are also critical to inform actions and reporting by national and subnational governments, cities, private landowners, land stewards and also as advocacy tools for civil society.

The NPI in 2024 has convened a process to build broad consensus on a small practical and meaningful universal set of State of Nature Metrics, starting from terrestrial biodiversity. In selecting a set of State of Nature Metrics it is important to define the priority *use-cases*, to ensure the metrics are useful and practical in supporting decision-making across key users and scales. The objective is not to measure nature's elements for the sake of measuring but to inform decisions that will lead to nature-positive outcomes. This will allow for aggregation and comparison, something that is missing today. Three overarching use-case dimensions were identified: *user* (e.g. companies, financial institutions, governments, cities, landowners, land stewards), s*cale* (e.g. site, landscape, jurisdiction, country, global, value chain, investment portfolio) and *purpose* (e.g. strategy and target setting, disclosure, reporting, assurance). Practical but meaningful State of Nature Metrics should be selected using *criteria* to test their universal applicability and viability for mass adoption, including whether they are science-based, responsive, comparable, accessible, affordable and auditable. The process also recognised the need to develop a complementary framework of 'measures' based on the depth of traditional and local knowledge, aspirations, priorities and values.

Some State of Nature Metrics could be chosen to be *universal metrics* – a minimum set of indicators and metrics that should be measured by all users and in multiple use cases, and others to be *case-specific metrics* – a set of additional indicators and metrics triggered under certain conditions. This approach would not only simplify the approach, reduce the number of indicators to be measured by everyone, but also ensure that important dimensions are not overlooked but measured only when necessary. Both types of metrics could be organised along a *Metric Maturity Scale*, which includes an 'entry' level, and subsequent more granular ones. This structure aims to provide a clear and accessible pathway for users with varying levels of experience and capacity to begin the process and improve over time. This way, universal and case-specific metrics can be applied at different levels of granularity. This 'tiered approach' could help users of varying sizes and capabilities to get involved, begin the journey towards developing and implementing a nature-positive strategy and moving towards more sophisticated biodiversity monitoring over time. Different levels of granularity may also be appropriate with reference to different use cases. This approach is designed to encourage entities to begin their nature-positive journey while also promoting a 'race to the top', in terms of biodiversity monitoring.

The maturity scale could help to guide users on how to continuously improve their data quality, granularity and methodologies over time, strengthen their commitment to action and improve the credibility of their disclosures and reporting. When selecting a minimum set of indicators and metrics to provide a comprehensive view of the state of nature, the focus is on essential elements such as ecosystem extent and condition at landscape and site level and species extinction risk. At this stage in the nature-positive journey – where many stakeholders are still learning and are often deterred by the perceived complexity of measuring biodiversity – the goal is not to identify a long list of metrics nor to search for a 'silver bullet metric' or to develop entirely new ones. Instead, the priority is to leverage what is already available and practical for universal use, while acknowledging the potential of emerging technologies and methodologies. The indicators and metrics that will be chosen should be good enough to represent the overall state of biodiversity effectively.

The rationale behind a 'minimum but meaningful' set of universally usable metrics is the following:

1. *Extent of natural ecosystems.* Biodiversity requires an adequate area of appropriate habitat of a certain ecosystem for it to be maintained, and so measurement

of extent of natural ecosystems is the most essential metric and is of paramount importance for tracking nature-positive outcomes. After all, if we lose habitat, we lose species and services along with it.

2. *Condition of natural ecosystems.* Ecosystems can be stable in extent but degraded and can recover very effectively with the right actions and under the right conditions. Measurement of condition reflects ecosystem health and resilience and allows us to measure its recovery.
3. *Landscape condition.* This indicator provides an indication of the health of a landscape as a whole for certain levels of biodiversity and ecosystem services to be maintained within it. This is particularly important for 'mosaic' landscapes with a combination of natural, anthropogenic and productive areas where it is important to reduce further fragmentation, and increase connectivity and the ecological viability of ecosystem patches and the species that depend on them. It also provides vital information for landscape use planning.
4. *Extinction risk and species abundance.* This provides an indication of the importance of an area or landscape for avoiding extinction, from a local and global perspective and follow the state of populations that are either rare or still widespread but declining.
5. *Natural processes.* Integrity and functionality of natural processes, such as carbon storage, hydrological cycles, sedimentation but also wildlife migration, capture more holistically the interactive dimension between various elements of nature in supporting systemic functionality and resilience.

Beyond universal metrics, a set of case-specific metrics are also needed to ensure that enough attention is paid to context-specific priorities. These could be highly threatened species or common species declining rapidly at a local scale, threatened ecosystems or the proportion of semi-natural habitat in a landscape dominated by agriculture or other intensive human uses. In the spirit of striking the right balance between credibility and practicality, this two-tier approach of universal and case specific metrics provides a consistent and comparable universal baseline of biodiversity measurement, but with the flexibility to address more sensitive or critical biodiversity features when needed.

There are also two data types, *Individual* – for example, which refer to data collecting at an individual site, and *Contextual* – recognising the need for all actors in a land/seascape to contribute to a nature-positive outcome. The value of applying both individual and contextual metrics is that they allow individual actors to identify their specific impacts, dependencies and outcomes but also bridge the gap between site-specific data and broader landscape-level information, encourage collaboration and engagement between various stakeholders at the landscape or jurisdictional level and assess the change in the state of nature at a more meaningful scale.

The biggest challenge for companies in measuring their footprint and nature-positive outcomes relates to their value chains, both upstream (e.g. sourcing of raw materials, intermediate products, machinery) and downstream (e.g. distribution of products, product use and end-of-life). For investors, it is about identifying which companies or assets in their investment portfolio contribute to nature loss, and how they assess and address their risks and impacts. In both instances the focus is on the material, compliance, legal and reputational risks posed by their impacts on nature. In adopting the notion of '*double materiality*', corporates are looking at the impacts of investments and business decisions on nature and biodiversity, as well as the impacts nature loss generates on their productivity, profitability, compliance and reputation.

Biodiversity decline or recovery can most accurately be monitored at the local level. The lack of *traceability* to local sites represents a serious obstacle to measuring impacts, developing adequate responses and assessing nature-positive outcomes. Companies or financial institutions should engage value chain partners or companies in their investment portfolio. Users with partial traceability to a particular sourcing region, such as a defined landscape or sub-national jurisdiction, may use statistical methods to guide initial priority actions.

Companies should improve value chain traceability and start measuring and reporting on the proportion of their value chain which is traceable with the target of increasing traceability over time. As a starting point, companies should focus on segments of their value chain where the potential impacts and dependencies on the state of nature could be most material. The only way to progress is to engage and support suppliers and customers to collect their location-specific data. This is where starting to use the simpler entry level of the State of Nature Metrics framework can help smaller providers along the supply chain to engage and begin to assess pressures, responses and outcomes in terms of changes to the state of nature in their operations.[1]

– Who can claim what? Opportunities, risks and complexities around claims of attributions and contributions

Once universal and comparable State of Nature Metrics to credibly measure net-positive (Nature Positive) outcomes are agreed, the inevitable question is whether an entity can *claim attribution of* or *contribution to* the outcomes. What can an entity legitimately claim?

Linking actions to nature-positive outcomes is complex for several reasons. Firstly, the degradation or recovery of nature in a site and even more so in a land/seascape is often impacted by multiple variables and entities at once, even by external factors that are not directly or uniquely attributable to the entity itself such as climate change, forest fires, upstream pollution, etc. Secondly, nature involves so many elements that an entity could deliver a net-positive outcome on one element (e.g. forest cover) but not on another (e.g. a threatened species population in the same forest). For a company to talk about 'nature-positive pears' but not 'nature-negative-apples' would clearly be an example of the most classic greenwashing behaviour. However, as Nature Positive was designed as a measurable goal in order to allow for stronger accountability, it is also important that successful efforts in halting and reversing biodiversity loss are recognised and rewarded.

After making investments, changing practices and even business model to deliver nature-positive outcomes, entities would legitimately feel that they should be recognised for their actions. Let's take the example of a farm which has embraced regenerative practices, such as drastically reducing the use of chemical inputs and land use emissions, avoiding conversion of natural habitats and setting aside a proportion of the land for natural regeneration. If this farm demonstrates, through credible State of Nature Metrics, an increase of above and below ground biodiversity against the agreed baseline, it would have a legitimate expectation to be recognised as a farm which is delivering a measurable and genuine nature-positive outcome by taking effective corrective measures which address its primary *in situ* impacts, as well as downstream and indirect ones such as emissions and runoff of fertilisers, herbicides and pesticides. At the other end of the spectrum, for a company with a complex value chain and multiple activities, products and suppliers, it is much more difficult to deliver measurable nature-positive outcomes across all its operations or supply chains. In these situations, nature-positive outcomes are likely to be more specifically defined with the expectation and commitment to increase coverage and scope over time.

Every action that helps improve the state of an ecosystem, landscape or species is welcome and important. But the Nature Positive goal refers to a measurable net-positive improvement of the state of nature based on a meaningful set of ecosystem, landscape and species indicators and metrics. This is what sets apart the 'pre-' from the 'post'–Nature Positive goal 'era'. The Nature Positive goal sets the new ambition, a net-positive improvement in the state of nature, that every entity should aspire to contribute to and be measured against it. The Nature Positive goal is a global societal goal, but to be achieved it requires all entities to contribute to it. The key question is: what constitutes a 'sufficient contribution' to ensure that entities avoid merely doing 'something' instead of taking the necessary actions to halt and reverse nature loss?

For most companies, the site level is the most practical way to assess negative impacts on nature, take actions to address those impacts and deliver nature-positive outcomes. These are both sites of their own operations and of the entities in their value chain. This is an initial but essential level because this is where impacts materialise and corrective actions can be taken sooner and most directly, and where measurements are most practical. This is the level where each entity must take direct responsibility, and for which it is directly accountable.

However, true nature-positive outcomes are mostly meaningful on a larger scale, for example at land/seascape level. Improvements of the state of ecosystems and species, and even more so for natural processes (e.g. hydrological cycle, sedimentation, carbon storage, wildlife migrations, etc.), are usually more effective and resilient at a large-enough scale to ensure viable species populations, ecological functions and the resilience of ecosystems and processes. Coordinated efforts at the landscape level is key, but efforts at site level are the ones that contribute to landscape level outcomes. It is not an 'either-or'. This is to say that both scales – site and landscape – are critical and interconnected.

Operational sites are easier to identify, and once the traceability challenge is addressed, so are the sites where value chain actors operate. However, the land/seascape approach adds both value and complexity. Defining a 'land/seascape' is inherently subjective, depending on ecological, geographical, geopolitical or socio-economic criteria and the specific context. A '*land/seascape approach*' is a method of managing an area that balances 'protection and production' in a coordinated way among stakeholders, ensuring long-term sustainability.

The effectiveness of this approach relies heavily on voluntary or regulated land/sea use plans, which typically require committed local governance to play a central role in convening, endorsing and enforcing these plans. However, challenges such as a lack of capacity, experience, vision, political leadership and issues like corruption and political opportunism often present significant obstacles to implementing an effective 'land/seascape approach'. These challenges are common across many regions. The difficulty in reaching agreement on the definition and boundaries of a land/seascape has led to the promotion of *jurisdictional approaches*, where boundaries and governing authorities are clearly defined. The Align project discussion paper on corporate nature-positive commitments offers a comprehensive set of recommendations to address the challenges associated with the land/seascape approach [30].

Stakeholders have emphasised the need for clear and consistent guidance on establishing credible links between actors' efforts and nature-positive outcomes, as well as on defining what constitutes a legitimate claim. Ultimately, every outcome contributes to the global Nature Positive goal, but achieving the ambitious global target of halting and reversing biodiversity loss requires each individual actor or group of actors to step up and take responsibility in delivering the necessary contribution in terms of net-positive (Nature Positive) outcomes. The global nature of the Nature Positive goal cannot serve as an excuse for individual entities not to deliver their fair share in relation to their impacts. Since biodiversity is inherently local, net-positive outcomes should primarily be delivered at the local level – where impacts occur – or within equivalent ecosystems or land/seascapes. This requires each individual actor to do their part in the sites and land/seascapes where they generate direct or indirect negative impacts.

Alignment on standardised approaches to measure nature-positive outcomes, along with clear principles and boundaries for claims of attribution of and contribution to nature-positive outcomes, is both critical and urgent. This is essential to drive action, improve accountability and ensure fair recognition. Initiatives such as the NPI State of Nature Metrics and the Align project on corporate claims aim at fostering the discussion and reaching the necessary consensus to advance this agenda.[2]

In addition to agreeing on standardised indicators and metrics to measure the state of nature, we may consider determining a '*biodiversity budget*' similar to the carbon budget used to assess how much CO_2 equivalent emissions are still permitted in order to remain below 1.5 °C of average global temperature increase. Such a budget will have to reflect the localness and diversity of nature and could be broken down by ecosystem typologies. The comparison between the use of biodiversity and nature's ability to replenish it can be a measure of a biodiversity budget. That budget today is hugely negative. Additionally, the planetary boundaries science applied to the biosphere has shed some clarity on what we should do to limit our impacts on the biosphere, beyond which self-reinforcing or cross-cutting negative feedbacks will start driving irreversible decline and changing the state of ecosystems and the services they provide. As these limits are being specified for various ecosystems and standardised State of Nature Metrics are being agreed, would the calculation of a multi-scale 'biodiversity budget' be possible and useful to drive our efforts in avoiding local and global tipping points and deliver nature-positive outcomes?

4.5   Now we have a global goal and a global biodiversity plan; it's time for action

We know what's happening to nature, we understand the risks and consequences and we know what the solutions are. We needed a plan that brings all this together and unites the world in dealing with the existential threat of nature loss. A plan that is agreed globally and implemented locally. A Paris Agreement–style plan for nature that starts with a clear, time-bound and measurable global goal. This was the rationale behind the call to adopt a Nature Positive Global Goal.

We have begun the transition towards a low-carbon economy. Not as quickly as is needed, but we *are* transitioning, inspired and driven by science, as well as an increasingly strong moral and business case and a shared mission on climate change: to become a carbon neutral society by mid-century and to stay as close as possible to below the 1.5°C average temperature increase. This has resulted in a clear and measurable Global Goal of net-zero emissions by 2050. A Global Goal that allows governments and companies to develop plans and science-based targets, and which injects a new level of accountability, showing who is on track and who is lagging. Today, this is our global compass to fight climate change and hold each other accountable for achieving the goal. However, reducing emissions is one crucial dimension of the challenge to avoid climate change. The other and equally crucial dimension is preserving and even enhancing nature's ability to sequester and store CO_2 from the atmosphere.

For this and many other reasons, we needed similar clarity for nature and biodiversity, the other side of today's ecological crisis. After many years of laudable but obviously inadequate action to protect, sustainably use and restore nature, as demonstrated by accelerated species loss, ecosystem destruction and the weakening of natural processes, the world adopted the Kunming-Montreal Global Biodiversity Framework (GBF) in December 2022 [31].

It is undoubtedly the most ambitious plan on nature the world has ever adopted. Firstly, the GBF sets, for the first time, a clear and measurable global goal for nature. This is expressed in its mission to "*halt and reverse biodiversity loss by 2030*". Halt the decline and stimulate the recovery of nature, a net-positive, nature-positive goal because net zero for nature is not enough. Contrary to the long recovery times associated with climate change, living nature, in most cases, can bounce back very quickly if given a chance, with immense benefits for people through the many contributions nature offers us, including climate change mitigation. The 2030 mission is part of a broader vision for a world living in harmony with nature. By 2050, biodiversity will be valued, conserved, restored and wisely utilised, ensuring the maintenance of ecosystem services, a healthy planet and essential benefits for all people. This vision is supported by four 2050 goals that also inspire the 2030 targets: a) protect and restore nature; b) prosper with nature; c) share benefits fairly; d) invest and collaborate.

The GBF is also the most measurable biodiversity plan we have ever agreed to. In fact it includes, for the first time, several measurable targets aligned to deliver this new ambition, including: reduce the loss of intact areas to near zero (target 1); restore 30% of degraded ecosystems (target 2); protect 30% of the natural ecosystems remaining on land, ocean and freshwater (target 3); eliminate, phase out or reform half a trillion dollars of harmful subsidies for nature (target 18); and reach an overall $200 billion per year for biodiversity conservation from all sources, more than doubling current domestic and international investments (target 19). Target measurability is key. The only target that was nearly achieved from the previous set of biodiversity goals (the UN CBD Aichi Targets), agreed upon a decade earlier, was the sole truly measurable one: achieving 15% terrestrial and 10% marine protected area coverage.

The GBF is by no means a perfect agreement. One major weakness is that it calls for the reform of our unsustainable production and consumption model, and mentions some key economic sectors like agriculture, aquaculture, fisheries and forestry (target 10) yet falls short of providing precise transition targets and overlooks others like infrastructure and extractives, manufacturing and tourism. While this is a major gap in the framework, it is clear that there is no way to achieve the mission to "halt and reverse biodiversity loss by 2030" without reforming the way we produce and consume, and without mainstreaming biodiversity into the rules and behaviours that regulate our society and economy. Despite these weaknesses, the Kunming-Montreal GBF remains the most ambitious and targeted biodiversity plan we have ever had. It is our best chance to move towards a nature-positive future.

4.6 The transition to a Nature-Positive economy: how business and finance can contribute to the Nature Positive goal

Since the launch of the Nature Positive goal in 2020, there has been growing momentum for and significant uptake of the nature-positive vision by businesses and investors. Great impetus came from the adoption of the Kunming-Montreal Global Biodiversity Framework and its mission to "halt and reverse biodiversity loss by 2030", consistent with the definition of the Nature Positive Global Goal. Many companies, consulting firms and assurance providers increasingly refer to Nature Positive as the reference corporate global goal for nature. The *Financial Times*, on 21 October 2024, published a full-page article, "The call of 'nature positive'", publishing the same article online with the headline: "The new corporate green goal: being nature positive" [32].

However, the Nature Positive Global Goal has emerged much more recently than the carbon-neutral goal included in the Paris Agreement. The result is that biodiversity is still a relatively new area of focus for many companies and investors. It is also perceived as complex, and as something that generates additional burden on companies that are already busy assessing and reporting on their climate obligations. The reality, of course, is that the climate and nature goals are artificially separated by conventions, regulation and standards while in fact they are mutually supportive and interdependent. One cannot be achieved without the other. Moreover, most businesses have typically approached biodiversity from a philanthropic angle and only very recently as part of their overarching Corporate Social Responsibility (CSR) and Environmental, Social and Governance (ESG) agendas.

Nature represents a *double materiality* issue in terms of impacts and dependencies for companies and investors. In fact, every company depends on and has negative impacts on nature through their direct operations or across their value chains. The tension between dependency and negative impacts leads to physical risks (operational and financial risks related to impacts on the quality and availability of nature's services and resources), regulatory, compliance and liability risks (operational limitations and transitions, costs of sanctions, legal suits) and reputational and market risks (consumer and investor expectations and preferences). It goes without saying that unaddressed risks can ultimately lead to financial losses. Moreover, impacts generated by one company or sector can increase risk and costs to others, hence the importance of the pre-competitive, level playing field alignment provided by the Nature Positive Global Goal. To derisk their operations and supply chains and to fulfil their ESG commitments, companies must identify and address their impacts on nature. That is through the adoption of a '*nature-positive strategy*'.

A nature-positive strategy is, for many companies, still very much a black box, partly because of the fragmentation of frameworks, standards and metrics available to assess and address impacts and dependencies, and primarily because of the lack of understanding and experience in dealing with biodiversity. Consequently, so far only a few companies have developed proper nature-positive strategies and plans. The 2022 Corporate Sustainability Assessment of S&P Global found that less than 20% of Fortune 500 companies had published biodiversity commitments, and less than half of these included an agreed timeframe [33.]

According to a 2023 McKinsey survey, global companies are paying more attention to nature, but carbon remains their primary focus. While four out of five companies in the Fortune Global 500 have set carbon-reduction targets, one in five companies track some dimensions of their impacts on nature, but only 6% have actually set targets on biodiversity and 13% specifically on forests [34].

Planet Tracker's *Nature Scorecard* [35] assesses companies' participation in target setting, disclosure and action-orientated initiatives as an indicator of their commitment to the nature agenda. The results show a low level of voluntary engagement, and the report recommends stopping what it calls 'bio-crastination'

[36]. The reality is that for many companies at the early stage of their sustainability journey, nature is not a priority, nor is it perceived as a potential material or transition risk. Many companies' and investors' poor understanding of biodiversity, coupled with the lack of a clear global guidance and agreed standardised metrics to assess the effectiveness of actions on nature, has generated a lack of confidence in developing plans on biodiversity, and has even led to *'greenhushing'* – the decision by companies not to publicly communicate their sustainability plans and efforts for fear of being criticised or accused of greenwashing.

When it comes to the response of the private sector to the nature-positive challenge and opportunity, the picture is still inconsistent and even contradictory. This is not surprising nor necessarily an indication of lack of interest or commitment but is perhaps emblematic of an early stage of awareness and transition. For example, on the one hand, according to Environmental Finance Data [37], sustainable bonds issued in 2023 were worth $150 billion. Notably, the proportion of bonds with 'terrestrial and aquatic biodiversity conservation use of proceeds' rose from 9% in 2022 to 37% in 2023. On the other hand, Planet Tracker recently analysed 26,500 votes cast for biodiversity proposals in sustainability and ESG funds and found that only 38% were in favour. Votes against the proposals were due to the perception that they were overly prescriptive or didn't demonstrate sufficient benefit to shareholders [38].

Although it is apparent that for many companies and investors biodiversity is a relatively new subject, we are seeing significant upskilling in the private sector, leading to a better understanding of the issue and its relevance. From an initial quasi-philanthropic approach to conserving nature, businesses and investors have started to look more holistically at their impacts and dependencies on ecosystem services and the negative impacts the decline of those services could have on the profitability and even existence of their operations.

Global accords like the Kunming-Montreal Global Biodiversity Framework, the rise of voluntary standard setting and reporting frameworks, as well as the first regulations on mandatory disclosure, are sending clear signals to the markets about the relevance of the biodiversity agenda. This should stimulate companies and investors to increase their preparedness for the arrival of regulated mandatory requirements on biodiversity such as the recently approved European Sustainability Reporting Standards (ESRS) under the EU Corporate Sustainability Reporting Directive. Consumer demands and expectations have also increased pressure on both regulators and the private sector [39].

Corporates have long been under the scrutiny of civil society, but today's measurable net-zero emissions and net-positive biodiversity goals are increasing the pressure and clarity on both ambition and delivery. Consequently, an increasing number of public and private investors are asking themselves how to identify companies and activities that have a negative impact on nature, how they are addressing their impacts, how to derisk their investment portfolio from nature-related risks and how to measure them.

Some investors are taking a more active role. In the Nature Action 100 Initiative [40], investors are proactively assessing 100 companies considered to be influential and transformational in their respective sectors. The Principles for Responsible Investment's Spring initiative convenes investors to use their influence to halt and reverse global biodiversity loss by 2030 with an initial focus on forest loss [41]. In the Share Action's Pesticides Working Group [42], investors focus on six global agro-chemical companies' performance on nature. To undertake effective nature-positive transitions, business leaders and investors need clear, high-integrity and practical guidance. Corporate CEOs and boards of directors often lament the lack of aligned guidance on nature. This is also a concern shared by auditing firms, as they are increasingly asked by companies to provide '*nature-positive assurance*' against their nature-related commitments and outcomes. The fragmentation of standards, methodologies and metrics undermine action, as well as comparability and accountability. As mentioned in the previous chapter, reaching agreement on a set of credible and practical metrics – and associated guidance, to measure and demonstrate genuine nature-positive outcomes is critical.

There are several high-quality methodologies, standards and reporting frameworks available or in development that companies and investors can use to drive their thinking on how to assess and disclose impacts, risks and dependencies, and develop a nature-positive strategy [43–47]. To improve global alignment, the main frameworks including the Global Reporting Initiative (GRI) [48], the European Financial Reporting Advisory Group (EFRAG) [49], the Taskforce on Nature-related Financial Disclosures [50], Science Based Targets Network [51] and others have improved coherence and interoperability in their reporting frameworks, something which was strongly needed to avoid confusion, duplication and the burden of excessive reporting. The International Organisation for Standardization (ISO) is in the process of developing the Biodiversity standards and the International Sustainability Standards Board (ISSB) [52], aims to develop standards for comprehensive sustainability disclosures of investors and the financial markets.

There is also excellent guidance on how to develop a nature-positive transition plan from business platforms like the WEF and the WBCSD or environmental NGOs such as the International Union for Conservation of Nature (IUCN), the World Wide Fund for Nature (WWF), The Nature Conservancy (TNC), Conservation International, Fauna and Flora and the Wildlife Conservation Society amongst others. Global and specialised consulting firms are also growing their capacity in supporting the private sector on biodiversity and have published excellent guidance documents.

As previously mentioned, one significant gap that remains is the lack of agreement and clear guidance on what metrics to use to measure and demonstrate genuine nature-positive (net-positive biodiversity) outcomes. In fact, it is only by effectively measuring net-positive biodiversity outcomes that a company can demonstrate that it has taken the necessary steps to contribute to halting and reversing biodiversity loss in line with the Kunming-Montreal GBF mission and the Nature Positive Global Goal. The NPI is convening all main global environmental NGOs, business and finance sustainability platforms, sustainability standard and target-setting frameworks, consulting and assurance providers and representatives from academia, local governance and Indigenous knowledge networks to agree on a minimum set of metrics and related guidance on how to measure genuine and credible nature-positive outcomes. While reaching a consensus on the best metrics to measure nature-positive outcomes is in progress, it should not prevent a company from beginning its 'footprinting' assessment process and taking actions on their known and most material negative impacts.

The Nature Positive goal is ambitious and disruptive. It challenges 'business-as-usual' and requires a deep transformation in the way we run our economy and the way we live our lives. It forces us to consider a new way of valuing nature vis-à-vis our development agenda, decoupling economic development from nature destruction, the paradigm that has driven most of our nature-negative economy and society until today. It means doing business, making profits and operating investment portfolios that embed nature at the heart of decision-making to benefit all of humanity and life on Earth. It means new or repurposed investments. It means a different way to measure and report business and investment performance.

– Corporate Nature-Positive strategies

The most transformational dimension of the Nature Positive goal is that it calls for all stakeholders to measure their performance against a common and shared global ambition, and to meaningfully contribute to it. The time of 'doing something good' for nature and appeasing our own conscience is gone. Today, everyone's goal is to do what's necessary to halt and reverse nature loss and build a truly nature-positive economy and society. Companies must ensure that their nature-positive strategy is fit for purpose and can deliver or contribute to the delivery of a net-positive biodiversity result, an absolute improvement in the state of nature.

A nature-positive strategy is about committing to the right level of ambition, designing an effective set of targets able to address the company's impacts on the company's operations and across its entire value chain, put in place adequate accountability mechanisms, allocate the necessary resources and ultimately demonstrating net-positive biodiversity outcomes against the defined baseline.

The nature-positive journey of an organisation should include: a) stating the ambition through a *public commitment* to contribute to the Nature Positive Global Goal and to take all the necessary steps to contribute to it by delivering nature-positive outcomes; b) undertaking *immediate actions* on avoiding or reducing the known and most relevant impacts/pressures with the potential for the greatest outcomes; c) carrying out the *assessment* of impacts/pressures in the operations and across the value chain and increasing coverage and completeness over time, as well as opportunities for risk mitigation and value creation; d) developing a *nature-positive strategy* applying the mitigation hierarchy with time bound science-based targets and an implementation plan and increasing coverage and completeness over time; e) developing an *implementation and monitoring* framework to measure the outcome of the responses through credible pressure and state of nature (nature-positive) metrics. All these steps should be supported and enabled by additional actions such as a) investing in *training and incentives* for relevant staff at all levels in the company; b) ensuring *management responsibility* and *board oversight* for implementation and reporting; c) regularly *disclosing and reporting* on both positive and negative impacts; and d) ensuring the *engagement of stakeholders* from the start and through a genuine inclusive and participatory approach, transparency and strategic partnerships.

Above and beyond the interventions taken within the company's operations or value chain, e) *supporting systemic transformation* of norms and regulation at sectoral or political level can greatly help to support the overall Nature Positive goal as well as creating shared ambition and common rules. Finally, f) a nature-positive strategy should be *integrated* with and in support of strategies on climate and other SDGs to maximise synergies and co-benefits across the social and environmental agendas.

The achievement of a Nature Positive Global Goal depends on the implementation of a number of key targets, to which each stakeholder's nature-positive strategy should contribute. We know what needs to be done, and this is captured in the Kunming-Montreal Global Biodiversity Framework. Firstly, we must *protect* much more (the GBF target is at least 30% by 2030; scientific research points at a target of 50% or more) of the world's ecosystems, prioritising the most 'intact' and functional. With just over 17% and 8% of land and ocean under variable degree of protection, the target is to at least respectively double and more than triple these figures by the end of the decade.

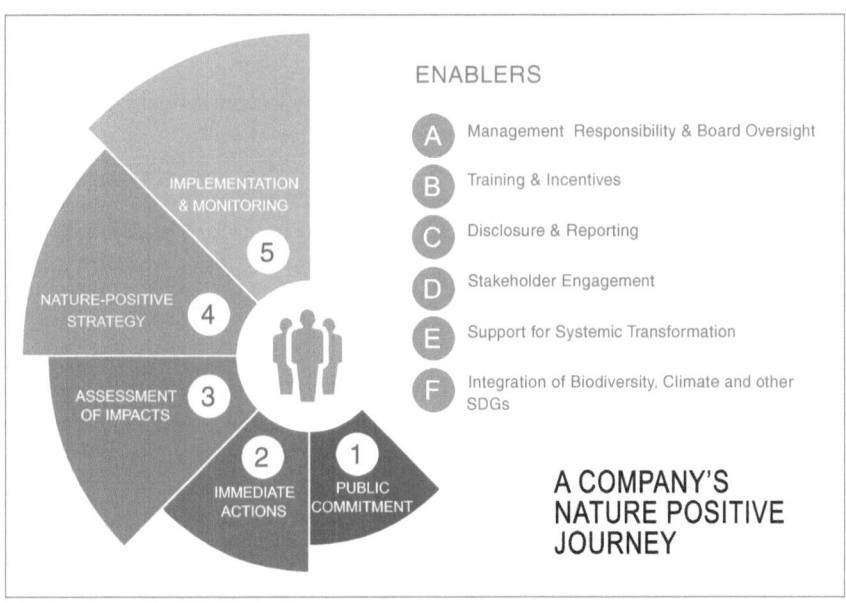

Figure 4.2 An organisation's nature-positive journey.

Source: Marco Lambertini

This requires us to accept the necessary trade-offs that come with sparing large areas of land, ocean and fresh water from traditional destructive uses and short-term profits in exchange for greater environmental benefits that will last far into the future and support the welfare and development of present and future human generations. This is something that doesn't come naturally to our short-term approach to decision-making. For example, too often businesses, investors, politicians and even members of the public still see the creation of new protected areas as a missed economic opportunity when compared to mining, logging, conversion to agriculture or fishing, rather than considering it a necessary investment to maintain natural infrastructure, capital and services that underpin the stability and productivity of our entire economy. Achieving this mindset shift is at the heart of succeeding in achieving this target.

Secondly, we should *restore* as much as we can (the GBF target is at least 30% by 2030) of the ecosystems lost or degraded. We know that many ecosystems are locally on the verge of collapse, leading to a change in their state and an alteration of the services they provide. While halting this loss must be our priority, we should also invest in restoring what's feasible. Many wild habitats and species have an intrinsic ability to bounce back, and to do so quickly, under the right conditions. However, like in the case of protected areas, restoration efforts require a conscious land use planning decision of 'giving back' these areas to nature in exchange for greater present and long-term existential services.

It is important to underscore that nature restoration as well as protection is not necessarily about excluding human activities but ensuring their compatibility and sustainability whenever and wherever possible. For example, emerging agro-ecological, agro-forestry and regenerative agriculture practices involve bringing back certain natural features and restoring the presence of biodiversity in productive landscapes without changing their land use, but in fact increasing resilience of production, livelihoods and local economies. A terrestrial protected area may include core protected zones or enforce strict visitation regulations to safeguard sensitive, threatened species, such as mountain gorillas in Volcanoes National Park in Rwanda or Virunga National Park in the DRC. Similarly, a marine protected area might implement a no-take policy to conserve fish stocks while permitting ecotourism activities.

Thirdly, and this is the most complex, we should *transition* to nature-positive practices and outcomes in the key economic sectors which today are the main drivers of nature loss, particularly agriculture, fishing, forestry, extractives, energy, manufacturing and infrastructure, amongst others. Just as the goal of net-zero emissions has disrupted the energy sector, accelerating the advent of renewable energy, in this case the disruption of a net-positive biodiversity goal lies in the transformation of nature-negative practices and technologies into nature-positive ones.

And finally, perhaps the most disruptive but also crucial and impactful element of all, we must *redirect* private and public financial flows to support these nature-positive transitions, through regulatory norms that incentivise nature-positive practices and penalise nature-negative ones at the national level, as well as in the context of international trade.

These broad categories of actions should be reflected, as appropriate, in each company's nature-positive strategy. While the Nature Positive goal was designed to create stronger accountability around a common ambition, companies and investors can still use their nature-positive strategies to achieve market differentiation and gain a competitive advantage. While contributing to a net-positive biodiversity outcome by 2030, based on a 2020 baseline, is expected of all actors, there is no upper limit to how much Nature Positive an outcome can be by 2030. The 'dot' on the nature-positive curve in 2030 is just above the 2020 baseline, not to signal the maximum expected outcome but rather the minimum goal. The ambition level of companies in their nature-positive strategies and their effectiveness in delivering nature-positive outcomes will become crucial factors in determining their recognition and value. This will influence their exposure to risks and costs, affecting their credibility, investors' confidence, societal recognition and ultimately their business value.

Achieving a net-positive biodiversity outcome requires corporates to adopt a disciplined approach and follow the *mitigation hierarchy* applied to impacts from direct operations as well as those which occur along the value chain. The mitigation hierarchy has been used as the guiding approach to effectively address corporate environmental impacts [53, 54].

As mentioned before, the fundamental steps of the mitigation hierarchy are, first of all, to *avoid* all avoidable harm to nature (such as re-routing infrastructure projects away from protected or high-value ecological areas). Secondly, to *mitigate* and minimise unavoidable impacts on nature (such as reducing disruption to natural habitats by limiting the number of access roads to a building site or building adequate wildlife crossings along a linear infrastructure). Thirdly, when possible, to *restore* loss of extent, condition and functionality of nature (like restoring *in situ* areas impacted by construction operations or building adequate wildlife crossings along a linear infrastructure to restore corridors and allow wildlife mobility). Fourth, to *compensate* for residual unavoidable impacts by creating positive actions that lead to ecologically comparable, or 'like-for-like' outcomes, resulting first in no net loss and subsequently in net-positive gains of nature (such as investing in conserving or restoring an area with the same ecosystem typology). Companies can explore options to compensate through 'insetting' or 'offsetting', by first improving the state of nature in landscapes/seascapes under control or influence. Lastly, there is an opportunity to *invest* in further conservation or restoration '*ex situ*' to achieve a greater net-positive outcome.

More recent variations of the original mitigation hierarchy framework described earlier are the 'conservation hierarchy' which includes actions to address historical, systemic and non-attributable biodiversity loss and the Science-Based Target Network's AR3T approach which adds the act of 'transformative change' at the systemic level [56, 57].

A common message contained in all these approaches is that companies don't have to wait to complete their entire assessment process but can and must start taking early actions on the avoidance and mitigation of known and most material impacts. This should not, however, mean they halt the development of their full

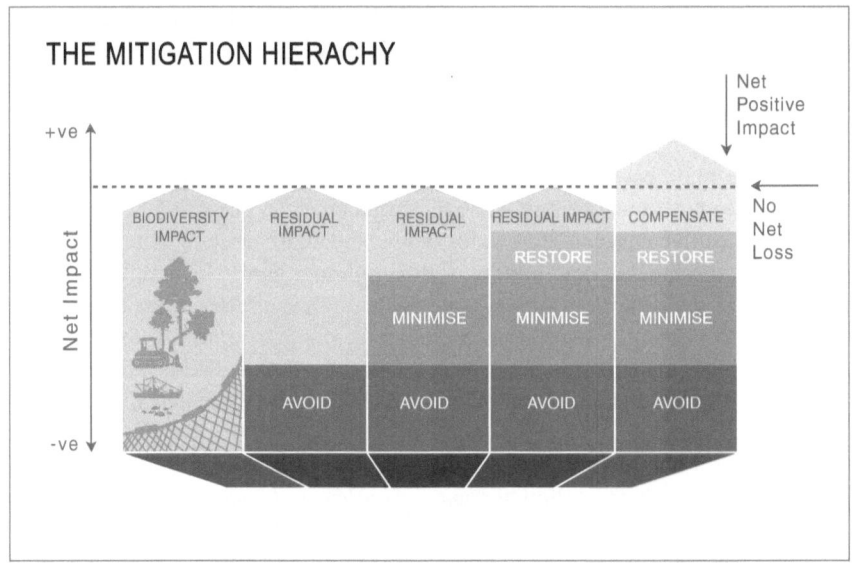

Figure 4.3 The mitigation hierarchy.

Source: adapted from The Biodiversity Consultancy [55]

strategy and target-setting process. On the contrary, by taking early steps, they could gain valuable experience and build confidence in dealing with more complex issues later in the process. A holistic and high-integrity approach to developing an effective and credible nature-positive strategy should also include clear measurable and time-bound targets alongside the mitigation hierarchy process.

While the definition of the Nature Positive goal is clear and increasingly understood, the understanding of how to develop and implement an effective nature-positive strategy is still emerging. This includes developing clearer and more comprehensive implementation guidance for different sectors, on footprinting methodologies, State of Nature Metrics, how to deal with value chains, landscape and jurisdictional scale and developing principles and guidance on contributions, attribution and claims. The concept of a '*biodiversity protocol*' should be explored, similar to the Greenhouse Gas (GHG) Protocol and the role it plays ensuring integrity in GHG measurement. As I have repeatedly said, this doesn't mean that companies and investors should wait until everything is crystal clear and codified. As the nature crisis accelerates, it is imperative that actions to reduce negative impacts on nature are taken now. This also helps to build the experience and confidence needed to address implementation challenges further down the road.

There are a few key words that characterise the approach to developing and implementing a corporate nature-positive strategy. It is probably superfluous to say, but still important to stress, that the first keyword is '*high integrity*'. This refers both to genuinely adhere to the overall ambition of the strategy which should aim to deliver net-positive biodiversity (nature-positive) outcomes and to the thoroughness of the approach in designing and implementing the strategy. The mitigation hierarchy described earlier is a well-established and recognised approach to ensuring that all necessary steps are taken, and that they are taken in the right sequence. High integrity must apply to the process of assessing the company's footprint, adopting avoidance, mitigation and restoration measures and, finally, compensating for the absolute minimum number of unavoidable impacts. Applying a high-integrity lens to all these steps is key to delivering a genuine net-positive outcome, being recognised for it and avoiding greenwashing.

The second word is '*transparency*'. Much focus has been put on disclosure as an indicator of a strong commitment and openness to scrutiny. It goes without saying that in the context of a nature-positive strategy, disclosure should drive action, not just reporting. Transparency is required in disclosing methodologies, data and reporting on progress.

Although a net-positive outcome is the result of the balance of negative and positive impacts, the TNFD framework has rightly stressed the need for reporting separately on both to gain clarity on the remaining negative impacts of the company or investment portfolio. Inevitably, a nature-positive transition is a journey, and for many companies a journey they have never embarked on before. So, while integrity and transparency are key to the outcome, equally important are the words '*commitment*' to the Nature Positive goal and '*confidence*' in the value proposition of a nature-positive transition. Firstly, companies that improve their efficiency in using resources such as raw materials, energy and water will realise operating cost savings as well as contributing to the climate and nature agendas. Secondly, companies that lead on biodiversity will see increasingly better access to capital, incentives and benefit from emerging 'green' fiscal and trade policies. As investors are increasingly integrating ESG performance into their decision-making, this will also be reflected in market valuations. Companies with strong social and environmental credentials will be rated as more resilient to climatic and ecological disruptions, regulation requirements and consumers' expectations. Alongside social indicators, investors are paying increasing attention to reducing impacts on water, climate and biodiversity, and biodiversity is becoming an area of increasing focus.

The last thing we need is for companies and investors to lack the confidence or feel anxiety when committing to a Nature Positive goal. Early adopters committing, acting and initiating the journey towards a nature-positive contribution will be recognised and rewarded. Sectorial approaches are also to be encouraged as they have the advantage of creating common approaches and a level playing field within the sector, but also provide exposure and recognition, inject confidence and allow companies with similar needs, challenges and opportunities to learn from each other. While fully embracing the Nature Positive ambition, companies should not be shy when it comes to talking about the challenges they face.

Mainstreaming biodiversity in our society and economy is considered the foundational element as well as the main outcome of a transition to a nature-positive future. It fundamentally aims to integrate socio-economic development and environmental protection at the policy, planning, finance and implementation level, breaking down siloes and avoiding negative trade-offs, embracing a *whole-of-government* approach. Corporations face the same challenge. Like in governments where decisions on environmental issues are normally relegated to the Ministry of Environment and then often countered by economic ministries, in companies they are mainly dealt with by CSR and sustainability divisions. While technical capacity on biodiversity needs to further develop in many corporations, it is equally crucial that nature-positive strategies are the result of a *whole-of-company* commitment starting with the Chairperson, CEO and the CFO, risk professionals, operations units right down to the shareholders and the various stakeholders along value chains. However, the challenge extends beyond capturing top management's attention – it requires building the necessary expertise at these levels to effectively oversee nature strategies and develop a system of incentives related to nature-positive performance. This is crucial to ensure that corporate nature-positive strategies move beyond commitments and are supported with adequate resources for implementation.

For most sectors, a nature-positive transition demands significant changes in how they operate. Yet, a study by the Boston Consulting Group of over 500 sustainability initiatives worldwide revealed that only one in 15 led to a transformation of the company's business model. This is not surprising and indicates that we are at the start of the process of transition [58]. For that to happen it is key that the drive comes from the top governance and management of the organisation.

When it comes to developing a corporate nature-positive strategy, for many companies the main challenge is in understanding and addressing its supply chain impacts, in other words the 'traceability' of its impacts and their locations. While impacts in the company's direct operations are relatively easier to identify and quantify, assessing impacts across complex and global supply chains is a significant challenge for many companies, for example, trading products, consumer goods or commodities. But for many companies most impacts occur outside their direct operations, therefore the scope of their nature-positive strategy must extend to their value chain to truly contribute to halt and reverse biodiversity loss.

Low traceability or a lack of traceability of the impacts through suppliers and sourcing locations is a major obstacle to developing and delivering credible nature-positive strategies and are key elements companies must address over time. Knowing the location of impacts along the value chain and being able to trace the source of procurements along the supply chain is key to developing informed and effective nature-positive strategies. A further challenge is related to the fact that a company's effort could be undermined through the actions of others operating in the same land/seascape or supply chain. Companies seeking to deliver measurable nature-positive outcomes need to act at their operation sites but also *vertically* along their value chain and *horizontally* across the other stakeholders operating in the same landscape or jurisdiction, from the same or different sectors. It is about mainstreaming biodiversity across the company and suppliers. However, these complexities should not deter a company from committing to and embarking on the process of developing a nature-positive strategy based on the current knowledge of its impacts and to become more accurate over time.

An important additional step contained within the 'T' of 'transform' of the ACT-D framework [59] is for corporates to contribute to the systemic change required at the sectorial or the broader level (subnational, national or global) to deliver the Nature Positive goal. This is about actively supporting the transformation of systems (fiscal, financial, regulatory) that today lead to nature loss. An example of this is when companies advocate and support legislative changes, the redirection of environmentally harmful subsidies and the establishment of nature-positive incentives. This is a delicate topic as over the years and to these days corporate lobbying has developed a track record of opposing progressive environmental regulation and redirection of investments.

According to a Corporate Europe Observatory report published in 2017 [60], there were an estimated 25,000 lobbyists operating in Brussels on behalf of lobby groups and representing a multitude of sectors and corporations . . . and most do not advocate for progressive environmental legislation! The delay and weakening of the EU anti-deforestation law, or the halting of the Australian Nature Positive bill are recent results of industry opposition to environmental reforms.

While the number of lobbyists has probably increased since then, there are also signs that things are changing. More and more progressive companies have joined sustainable business coalitions that publicly support and advocate for ambitious environmental frameworks and rules to level the playing field, such as the UN CBD Global Biodiversity Framework. The High Ambition Business Coalition to End Plastic Pollution has advocated for global rules and standards on plastics, including reduction targets. Companies with business models based on nature-negative practices, which have been highly profitable up until now, are more resistant to change. However, as evidence of the material impact of nature loss on the sustainability and profitability of the company grows, shareholders and investors will likely be persuaded to support the transition and companies be persuaded of the benefits coming from nature-positive business models. The leadership role of company CEOs and Board Chairs in promoting this new culture is key.

If Rome wasn't built in one day, neither will a nature-positive world. Achieving a genuine nature-positive outcome requires a company or investor to commit to the nature-positive ambition and embark on a journey of understanding their impacts and dependencies. Then it involves setting and implementing targets for each impact, including those within their direct operations as well as value chain. This can be complex and should start with a process of internal promotion, adoption and socialisation that ensures commitment at all levels within the company.

Companies and the investors who fund them hold the key to building a nature-positive future. Just as it is for society and for us as individuals, the business case for any company to embrace a nature-positive transition is terrifyingly simple: it is about derisking their present and their future, embracing opportunities for new value creation and, ultimately . . . ensuring the survival of their business.

4.7   Markets can't do it alone: the critical role of regulation

The way we holistically and economically value nature is the foundation of a nature-positive future. Achieving a nature-positive future where economic development is not at the expense of nature requires a collective effort and the overcoming of established attitudes, policies, practices and vested interests.

This shift, this new cultural and economic paradigm around nature and her resources, cannot just be left to markets alone. In the absence of regulatory compliance or strong stakeholder pressure, there is no market impetus to invest in actions that avoid impacts on nature. There is also little incentive to invest in halting and reversing the decline of resources and services that are largely free, especially when the negative externalities are not financially accounted for. Growing awareness of nature's decline and our dependency on it is raising concerns among progressive and forward-thinking businesses and investors. They are beginning to support regulations that establish a level playing field through common standards and reporting norms.

Nature degradation is the very result of a failure of our current distorted market-based economic system. Voluntary or market-based efforts, from sustainable certification schemes to carbon offsets, have shown repeatedly how markets are not capable of transforming or scaling on their own. While voluntary frameworks play a key role in raising awareness, creating a vanguard corporate and consumer movement, testing practical approaches and generating confidence in new sustainable models, regulation is key to stimulating a large-scale nature-positive transition.

We need policies and regulations that can drive market transformation at the necessary speed and scale, offering direction and incentives for sustainable practices and appropriately pricing nature loss to foster the innovation required for low-footprint and nature-positive business models.

However, not all policies are created equal [61]. There is a big difference when policies are enshrined in law or expressed as non-statutory instruments variously described such as plans, strategies, guidelines, incentives and so on. We tend to call all of them 'policies', but their ability to be enforced from a legal perspective and drive transformative change such as nature-positive outcomes varies enormously. After the EU Habitats Directive came into force [62], governments were being taken to the European Court of Justice for approving development projects which were affecting the status of a threatened species or habitats. Projects were stopped and fines imposed. That was possible because EU Directives are binding on member states. On the contrary, 'non-statutory policy' instruments, which do not go through formal legislative processes, serve as flexible tools. While they avoid the lengthy and often politicised legislative process, they primarily offer guidance for decision-makers and do not always result in judicial review (where a court identifies procedural errors in decision-making). Any advocacy plan or campaign to instigate the adoption of new policies requires a careful consideration of what instrument to advocate for based on a technical and political feasibility assessment and on the effectiveness of enforcement.

When considering the role of regulation in driving and supporting nature-positive outcomes, one must focus on what perhaps is the most perverse manifestation of the 'development paradox' and our cultural approach of taking nature for granted. This is what we could call the *'subsidies paradox'*. The figures speak for themselves. Governments in 2022 spent at least $1.7 trillion of taxpayer money on environmentally harmful subsidies, equivalent to around 2% of global GDP. In addition to this, another estimated $5 trillion in private funding went to support nature-negative practices, particularly fossil fuel extraction, mining, commercial fishing and industrial agriculture. In comparison, less than $200 billion is spent annually on nature-based solutions through conservation and restoration, and not necessarily very effectively [63]. While these figures already reflect the huge imbalance between nature-negative and nature-positive financial flows, it is very likely that the nature-negative calculations are underestimated because of the difficulty in analysing the large volume of private finance, while the nature-positive figures may be overestimated, given the difficulty of avoiding double-counting in public spending, not to mention questions around effectiveness. Finance that indirectly affects nature is not part of these figures, and it certainly represents a much larger amount than the one associated with direct impacts, involving a much broader set of sectors.

Updated figures, which include subsidies for non-energy mining and plastics production and reflect an increase in fossil fuel subsidies, estimate the total of environmentally harmful subsidies in 2024 to be at least $2.6 trillion. This is equivalent to 2.5% of global GDP and represents a $570 billion increase when adjusted for inflation. This total includes over $1 trillion in subsidies for fossil fuels, $600 billion for intensive agriculture, $390 billion for water engineering projects, $180 billion and $175 billion for unsustainable forestry and destructive infrastructure respectively, $55 billion for unsustainable fishing, $40 billion for mining and an extraordinary $30 billion for plastics! [64] The result is that nearly 7% of global GDP is dedicated to wrecking the natural world. By comparison, in 2023, global military expenditure reached an all-time high of $2.24 trillion, or 2.3% of global GDP [65]

We are spending at least three times as much money in waging a *war on nature* than we are on conflicts between humans, as if nature was our worst enemy. A silent, often invisible yet catastrophic war on nature is being waged every day, day after day. Damages and casualties are not counted. As Carlos Manuel Rodriguez, the former Costa Rican Environment Minister and current CEO of the Global Environment Facility, said in an interview: *"There is not a single country today that invests more in protecting nature than it spends on activities that destroy it. This must change"* [69].

The EU, in parallel with a very ambitious and progressive legislative package called the Green Deal, is still issuing over 350 billion euros a year in subsidies to the agriculture sector, largely supporting nature-negative industrial farming practices. Currently, 75% of the payments to farmers under the EU Common Agricultural policy are benefiting disproportionally the large and intensive farms instead

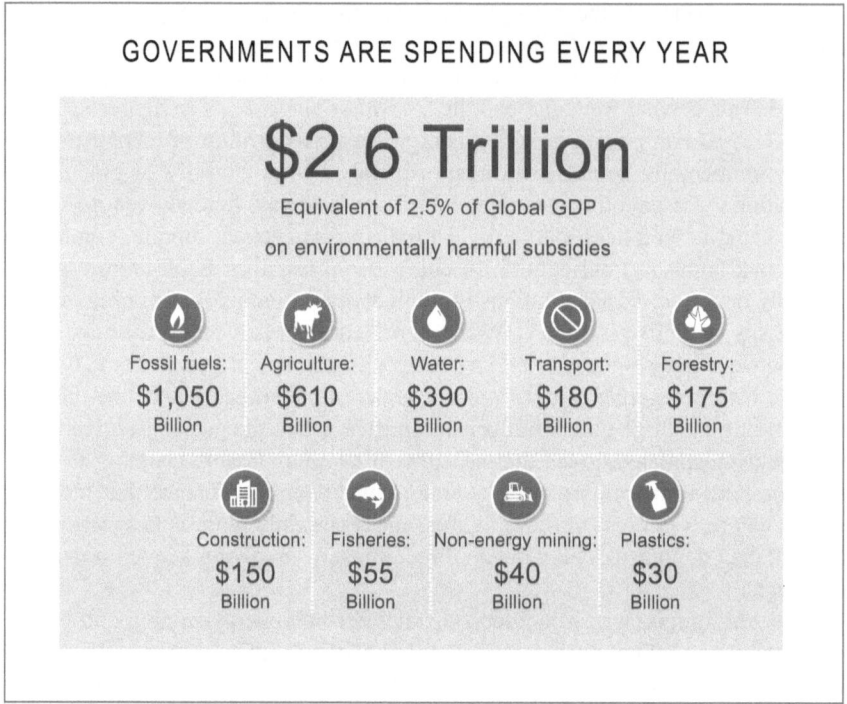

Figure 4.4 Annual government spending in environmentally harmful subsidies.

Source: Adapted from D. Koplow, R. Steenblik 2024 [66]

of focussing on the needs of the smallholders and the transition to sustainable practices. This is a blatant policy contradiction within the very same institution. The same happens in most countries around the world.

Fiscal policies, the way governments apply taxes and spend tax revenues, are a powerful regulatory tool to drive and support nature-positive transitions and innovations through mechanisms of incentives and disincentives. The Kunming-Montreal Global Biodiversity Framework specifically calls for the elimination or redirection of $500 billion per year of harmful environmental public subsidies by 2030 [70].

The subsidies paradox is the result of a siloed approach to development, whereby economic and development goals are not integrated with carbon-neutral and nature-positive ones. Instead of maximising synergies between development and the environment, trade-offs are often overlooked, leading to ineffective management. The biggest challenge is for government ministries and regulatory bodies to align on delivering integrated socio-economic and environmental goals, in other words the implementation of the Sustainable Development Goal framework. Governments and their central banks and financial regulators have the power, as well

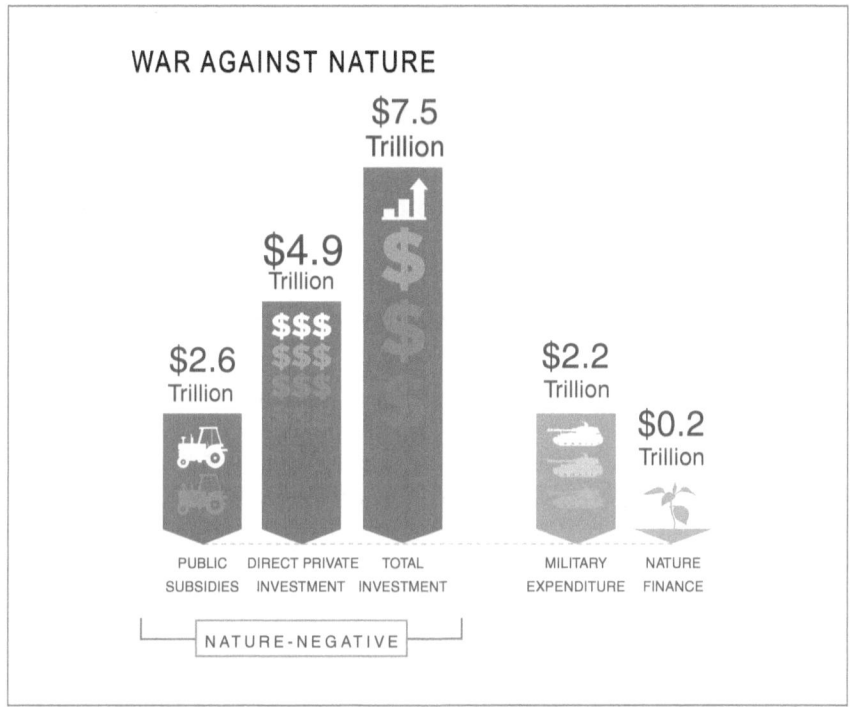

Figure 4.5 Financing our global war on nature.

Source: Adapted from UNEP, 2023; D. Koplow, R. Steenblik 2024 [67, 68]

as an opportunity, to incentivise sectoral transitions by redirecting financial flows towards sustainable practices and innovations [71].

In 2016, Colombia implemented a carbon tax, some of the proceeds of which went towards a fund to protect biodiversity. This is an example of how regulated markets can mobilise finance for nature as well as reduce the impacts in the first place. The EU Emissions Trading System (EU ETS), the world's first major carbon market launched in 2005 and the largest, was for years considered highly ineffective as the price per tonne of emissions was too low. Since permit prices have risen, it has started to contribute to the acceleration of the phase-out of coal from the EU energy market.

One successful approach to redirecting subsidies has been to support landowners, including small-scale landowners, with Payments for Ecosystem Services (PES). Costa Rica is one of the countries that has pioneered these successful schemes. The Central American country had a high deforestation rate throughout the 1980s, with forest cover dropping from over 50% of total land area in 1950 to 24.4% by 1985. In 1996, the government introduced a PES programme incentivising landowners to increase native forest cover. Today, natural forest has recovered and once more covers over half the country [72].

Even though regulation and financial flows are going to be the fundamental drivers of a nature-positive transition, beyond laws and money, we need to embrace the nature-positive transition culturally, morally, emotionally and politically. It's crucial to understand the risks and costs of failure and the benefits and opportunities of success. Only through a conscious and determined commitment will we be able to generate the will and confidence to embrace a truly nature-positive transition. This requires overcoming powerful biological, psychological and socio-political barriers.

4.8   Resisting the resistance: blockages and barriers to embracing the 'Great Transition'

"The old world is dying. The new one is struggling to emerge. And in this time of light and shade, monsters are born" [73]. Italian philosopher Antonio Gramsci wrote these strong and dark words to his mother between 1929 and 1935 while imprisoned by the Fascist regime. It was the time when the Nazis and Fascist movements (the 'monsters') were emerging, across a Europe gripped by a deep socio-economic crisis while America was facing the Great Depression – a perfect storm of conditions that led to the eruption of the Second World War. I see a striking parallel with today's world, where the old and failing economic model is refusing to give way to the new, more sustainable and equitable future that is struggling to emerge – and where today's 'monsters' are the vested interests and their political executors who attempt to obstruct by every possible means the transition to a better, safer and more just future for all. In short, a nature-positive future.

A heads-up. This chapter deals with a delicate and controversial subject, and it is in fact intended to provoke thoughts, reflections and discussion.

What would you prefer, a world polluted by toxic chemicals, devastated by drought, floods and forest fires, deprived of the beauty and services of nature, threatened by outbreaks of disease which cost millions of lives and trillions of dollars, or a stable, safer world full of nature and at a fraction of the cost?

It sounds like a stupid question. It's a no-brainer. Particularly as we are the most environmentally aware generation ever. According to a recent IPSOS survey across G20 countries, 59% of people interviewed said they are very worried about the state of nature today, almost two-thirds believe we are close to climate and nature tipping points and over half feel personally exposed to environmental risks [74]. So why do so many leaders and individuals lack the clarity and resolve that is so desperately needed? Why are we repeatedly taking one step forward and a half step back? Why are we delaying the much-needed transition and getting dangerously close to points of no return?

The way humans are: attitudes to danger and change

I believe that a big part of the answer to these questions lies in the way evolution has shaped our reaction to *limits*, *danger* and *change*. To better understand the deep-rooted challenges of a nature-positive transition, as well as other major socio-economic shifts like decarbonisation and equality, I'll turn to evolutionary biology and psychology. This will help identify the biological and psychological roots of the barriers that hinder us from embracing what is both a logical and the only transition towards a safer and more equitable future. We first need to acknowledge that our predominant focus on survival consciously or subconsciously influences a lot of our responses and decisions, leading us to prioritise our short-term self-interests.

Yes, from an overall evolutionary standpoint, our main goal is 'simply' *survival*. This is not unique to humans but is codified in all living forms, passed on in our genes for a countless number of generations to ensure the life of individuals and the existence of the species. Surviving in nature is hard. Struggling for survival is how we have spent most of our history as a species and how evolution has forged us to succeed. The survival of individuals is the fundamental condition through which Life, this rare and most extraordinary biophysical condition encountered on our planet and perhaps in a few other places, can continue and evolve. Only through securing our survival and welfare can we contribute to the next life forms that will succeed us – our offsprings in the short term, and perhaps other species in the long run. This is intuitive and obvious, but we rarely consciously think in these terms. As individuals we have been selected based on our ability to survive and our ingenuity to find bespoke solutions to specific challenges. Furthermore, like many other social animals, but in ever more sophisticated ways, we have developed norms that help us to feel secure, live and thrive together where the survival of the group enhances that of the individuals and vice versa. Consequently, fear and sense of security, avoiding danger and achieving well-being are powerful, perhaps the most powerful evolutionary drivers. Accordingly, threats and rewards, losses and gains are the most powerful drivers of the way we make decisions. They are at play in our individual, family and community life, in the workplace and in society. It all goes back to the 'selfish gene' or the *self-preservation instinct* that dominates our existence and that of all life on Earth.

The most fundamental condition for our survival – shared by all living forms – is access to vital resources like food and water. The *limitation of* and *competition for resources* has shaped extraordinary physical and behavioural adaptations across the natural world. Human beings, from ancient to modern times, have always been confronted with the challenge of securing resources and by resource scarcity.

One of our initial main survival strategies when dealing with limited resources has been to migrate to new richer and safer territories. This is what triggered many of the great migrations throughout our species' history. Even in the most stable societies in modern times you may argue that we are still resorting to the same strategy. However, instead of moving ourselves *en masse*, today we use trucks, ships and planes to migrate massive quantities of resources from distant territories to us. This has created an even greater disconnect between us and the resources we use. In addition to ignoring their origin and the modalities in which they are procured, some of these resources are today so processed and manipulated that they are unrecognisable from their original form. This affects our perception of these resources, makes us unaware of their decline and therefore less inclined to use them sustainably.

After the first great migrations of Homo sapiens across the continents, leaving a wave of extinctions of other life forms including other Homo species in our wake, we became more settled and learned how to exploit resources sustainably. Failing to do so would have simply resulted in starvation, disease and more conflicts. This culture of sustainable use is still rooted in several Indigenous populations today in direct contact with and with a direct dependence on wild nature. But with the advent of agriculture, animal farming and great technologies, we have again gradually lost this balanced relationship with nature and our knowledge of living sustainably. After a century of exponential socio-economic progress pushing our economy and society out of balance with the natural world, we are today beginning to discover the importance of planetary boundaries and the consequences of overshooting them.

Danger is the most powerful evolutionary driver because it is what most directly threatens our survival. When it comes to our reaction to danger, it is when confronted with an obvious and imminent threat that we respond most decisively. When confronted by choices, we will choose what we believe will secure our safety, reward us the most and enhance our well-being. While this innate drive also likely underpins our greedy and unsustainable economic development model, the growing awareness of the threats posed by the climate and nature crises should ideally motivate us to embrace change, once more for the sake of our survival. Our greatest chance to survive the 'Great Threat' of the Anthropocene is to emotionally and rationally internalise that climate change and nature loss are threats, and that our safe choice is sustainability. However, deep psychological barriers hinder this shift.

Firstly, behavioural psychology studies show that we (as well as other animals) tend to manage threats by mainly avoiding them. The uncertainty implicit to *change* is somehow perceived as a threat. So, the general reaction is to avoid change. When it then comes to dealing with change, we hate loss more than we appreciate gain. In other words, we treasure what we have, even if it's not satisfactory, more than what we could have but haven't yet experienced. It is another way to avoid the potential danger of being less secure, worse off. We are more likely to embrace change when we experience or clearly envisage a benefit.

Avoiding danger and risks are 'safe choices' in evolutionary terms, but our fundamental risk-averse and change-averse nature are today powerful obstacles to the nature-positive transition. The seductive nature of 'business-as-usual' is that we know it. 'Better the devil you know', as the saying goes.

As one of my dearest mentors, the brilliant late Italian ethologist, Professor Danilo Mainardi, used to say that during our development journey, our culture and socio-economic structure have gotten us increasingly out of sync with the planet's ecological rules and we have become *maladapted* to living in balance with the natural world. And I would add that as our nature is biological before cultural and our dependence is ecological, we cannot illude ourselves into thinking that this unbalance could last for much longer. Being maladapted to nature is not a safe position to be in.

Secondly, humans are a reactive species, brilliant in responding to immediate threats and solving pressing problems, but poor at dealing with issues perceived as distant risks. Our inclination towards short-term choices is an important factor, with huge implications when facing issues often still perceived as longer-term, like the climate and nature crises. Our short-term focus often makes us reluctant to make even small sacrifices, even by giving up or reducing non-essential comforts that we enjoy, in exchange for much greater long-term benefits. Diet shifts are a typical example. Despite the growing awareness of the impacts that eating farmed meat has on the climate, biodiversity and human health, our eating habits are changing extremely slowly. I keep attending sustainability conferences, even those focussed on climate, biodiversity and food, where the food served is far from sustainable. Taste and cultural identity are still the main reasons to keep eating meat or even not to reduce it, despite knowing the direct and indirect harm it causes to us, let alone to biodiversity and farmed animals themselves.

We tend to excel at making decisions that address single issues and offer the most immediate and direct solution. However, considering future consequences or threats which do not imminently affect us or anyone in our immediate social vicinity is far more difficult for us. Furthermore, for a threat to be better recognised it needs to be detected by our senses. Intellectually internalising an abstract threat requires a much bigger effort and deeper understanding of the issue at hand. We may react promptly to the threat of a flood or a typhoon, but we are not so inclined to address systemic climate change. Even as climate change intensifies, we still witness this attitude repeatedly, all over the world. In 2024, the King of Spain was called a murderer and had mud hurled at him by an angry crowd in Valencia, who accused authorities for not alerting the public quickly enough and preventing the damage inflicted by floods that tragically killed hundreds of people, but I am not sure how many have accused the oil or agriculture sectors or have even reflected on our own individual energy-intense lifestyles, the root causes of extreme weather events.

In the latest EU and US elections, the majority of people voted for candidates explicitly playing down the existence of human-induced climate change and even advocating for a revival of coal, oil and gas. And all this against the backdrop of an obvious intensification of extreme weather events, which devastate lives and livelihoods and already cost the economy hundreds of billions of dollars per year – a figure that continues to rise. The loss of biodiversity, compared to climate change and its extreme weather events, is even less perceptible and in fact practically invisible to most of us – for now hiding the threat it truly represents.

Perceptions are more important than reality

On a human timescale, biodiversity loss is perceived as both '*temporally distant*' – far off in terms of time – and as progressing slowly compared to the immediate threats we are evolutionarily wired to detect. This contributes not only to a lack of appreciation for the loss of nature but also the consequences of that loss. Indirect and long-term consequences are objectively difficult to conceptualise, and so are unintended consequences of certain actions. Moreover, the so-called '*shifting baseline syndrome*' is a result of another kind of temporal bias. The term refers to each generation losing track of and forgetting the scale of nature and biodiversity decline, therefore creating a new baseline for themselves [75]. This leads to the continued but undetected decline of biodiversity going unnoticed. In North America, there are currently around 600,000 buffalo (bison), while there are about 450,000 elephants in Africa. The numbers seem significant until we learn that just two centuries ago, the bison population was 30 million and that less than 50 years ago, there were an estimated 1.3 million elephants in Africa. We are reassured by seeing some bees, butterflies and birds in the countryside, but we have no immediate sense of how many were there just a few decades ago. In fact, in the European Union, 15 species of grassland butterflies have suffered a 29% decline between 1991 and 2020 [76], and North America alone has lost 2.9 billion birds, and overall, almost three-quarters of vertebrate populations have declined in the last 50 years [77, 78]. The same shifting baseline bias applies to the decline of habitats and ecosystems. To appreciate this decline, we need to resort to scientific data and trends, which are not often accessible or comprehensible to the public.

Nature loss is also perceived as *physically distant* from us. An increasing number of people live in man-made or highly modified environments where biodiversity is already less abundant and its decline less apparent. Most biodiversity is not that visible either, like fish in the ocean and rivers, worms in the soil or birds in the forest. Their decline is, for most, out of sight and out of mind.

Finally, when a threat is perceived so grave and so big it is almost intractable, and the reaction is even more extreme and diverse, largely influenced by individual differences in culture and values. With reference to the environmental crisis, we see responses ranging from *eco-anxiety*, *eco-depression* and *eco-anger* in the most aware and sensitive social groups, to *denialism* and *biased optimism* in the groups less responsive or those more directly affected by the transition. This range of responses is particularly visible in today's society confronted by complex, divisive and threatening issues. In a polarised society, people may adopt a position that in time becomes an *ideology* or identity even when it is at odds with the most recognised scientific evidence or pure common sense.

A person's hard line defensive stance can lead to denialism and become very difficult to change. This is something we witnessed during the COVID-19 pandemic, and the same psychological forces are at play regarding climate change and nature loss. The reactions are so impulsive that even the most rigorous and best communicated scientific evidence can still be met with denial and lack of trust. This is reinforced by the psychological mechanism of creating an '*identity*' often associated with a movement, school of thought or even political party, becoming an expression of '*cultural tribalism*' in which people find comfort at times of uncertainty and insecurity. Understanding human behaviour and the socio-psychological mechanisms that either prevent or enable diverse groups to come together and coordinate is key in addressing global-scale challenges such as climate change and nature loss. We must find a way to forge a '*global identity*' which extends to a '*global tribe*' to address 'common global challenges' together.

In less than two weeks between September and October 2024, hurricanes Helene and Milton, two of the fiercest hurricanes ever reported, battered Florida with huge economic losses and tragic impacts on lives and livelihoods. Following a deluge of misinformation and political manipulations, one of the most incredible conspiracy theories developed, stating that the hurricanes were 'geo-engineered' by the US government to intimidate and affect a specific sector of the electorate. A month later, the US electorate voted for a climate change denialist to be their next president.

Psychologists who are experts in conspiracy theories explain that, when faced with uncontrollable events like ever more frequent and destructive mega-hurricanes, people tend to embrace "conspiracy theories that preserve their sense of safety and help regain a sense of control" [79]. The 'identity' theory may have almost certainly also played a role.

The Norwegian psychologist Per Espen Stoknes identifies five psychological barriers which prevent a response to climate change [80], and which can also be, to a large extent, applied to biodiversity loss. These are distance, doom, dissonance, denial and identity. Here are my slightly revised definitions of these which can be applied to the nature crisis. '*Distance*' includes perceiving the issue as distant in time, in space, in sensorial perception and in social terms. '*Doom*' is a pessimistic perspective of the issue leading either to cynicism, inaction or simply accepting the ineluctability of the issue (this can lead to depression, anxiety or anger that we are particularly witnessing in young generations triggered by a deep sense of frustration and injustice). '*Dissonance*' occurs when our actions do not align with our beliefs, and when we don't have the resolve to act coherently but we rather find excuses to self-absolve ourselves, perhaps through resorting to 'single action bias' – doing something and convincing ourselves that it is enough when it obviously isn't. '*Identity*' is our decision to adopt a view that conforms with a view or ideology that others share, most likely connected to a social group we are a part of or decide to associate with (these identities may be formed on different scales, such as social, political, ethnic, geographical and professional). Finally, '*denial*' can be the outcome of all the previous responses when we decide, consciously or not, to address the issue by actively denying it exists. When faced with seemingly uncontrollable threats like climate change, ocean acidification or mass species extinction, some people resort to denial or even conspiracy theories to retain their sense of security. It is the ultimate defense mechanism to avoid accepting an uncomfortable or unmanageable truth or a conscious tactic to protect vested interests.

I would also add '*biased optimism*', the attitude or hope that things will sort themselves or the solution will come from unlikely or miraculous external sources. When it comes to finding solutions to climate change or nature loss, many people put their hope in technology. There is no doubt that technological innovation plays a fundamental role in supporting and speeding up the carbon-neutral and nature-positive transition. However, it has been proven repeatedly that technology is only as good (or bad) as the system that uses it. An economic system that does not value nature will not develop and use technologies to protect nature or use sustainably her resources, but exploit and destroy it more efficiently. This is exactly what we

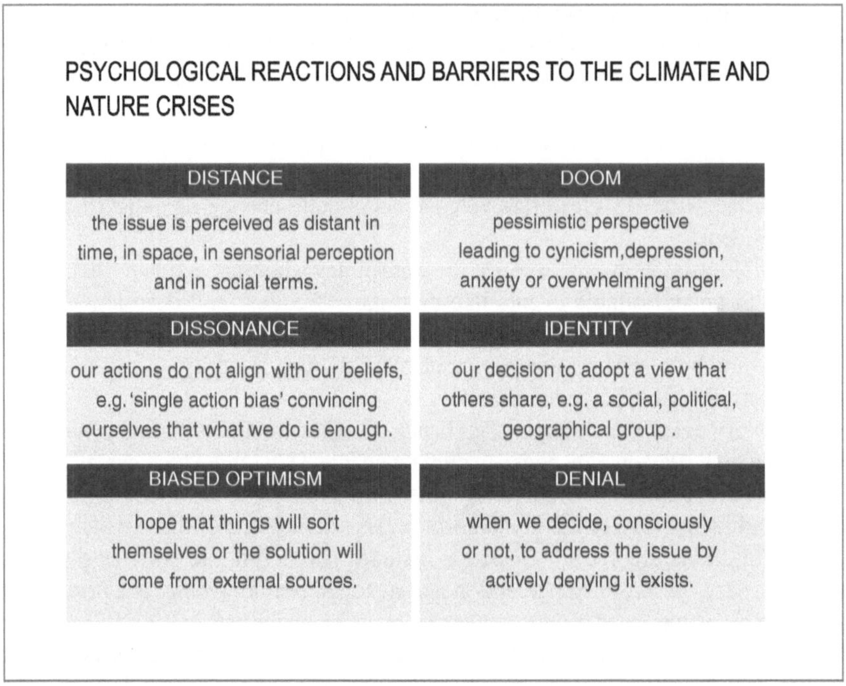

PSYCHOLOGICAL REACTIONS AND BARRIERS TO THE CLIMATE AND NATURE CRISES

DISTANCE	DOOM
the issue is perceived as distant in time, in space, in sensorial perception and in social terms.	pessimistic perspective leading to cynicism,depression, anxiety or overwhelming anger.
DISSONANCE	IDENTITY
our actions do not align with our beliefs, e.g. 'single action bias' convincing ourselves that what we do is enough.	our decision to adopt a view that others share, e.g. a social, political, geographical group .
BIASED OPTIMISM	DENIAL
hope that things will sort themselves or the solution will come from external sources.	when we decide, consciously or not, to address the issue by actively denying it exists.

Figure 4.6 The main psychological reactions and barriers to act on the climate and nature crises.

Source: modified from Stoknes, P.E., 2015. [80]

have seen happening until now. We've progressed from axes to chainsaws, manure to chemical fertilisers and pesticides and from fishing rods to vast oceanic lines, immense trawlers and sonar technology for fish detection. Technology can absolutely be an essential enabler of sustainability, but only when it is deployed within an agreed sustainable production and consumption framework, and with the clear purpose to operate within planetary boundaries. The shift in our perceptions and attitudes towards nature needs to happen first. Putting our hopes in future technologies that have been untested for both efficacy and externalities is very risky and could lead to 'biased optimism' that can divert focus from addressing current system failures. In fact, this diversion could even legitimise 'business-as-usual' or offer an alibi for delay and inaction.

The connection between human stress and behaviour is intuitive, but its psychological and neurological aspects are complex, as with most behaviours and emotional responses. This issue is increasingly important given rising stress levels worldwide. From a biochemical perspective, stress induces cellular inflammation, which activates a chain reaction involving interconnected inflammatory and neural processes, negatively impacting behaviours like decision-making. Researchers have examined stressors such as climate change, conflicts and the COVID-19 pandemic. External stressors can even trigger changes in the genome of cells to counter or adapt to the stressor [81]. This opens a fascinating field of research to investigate reactions to stressors on a biological and neurological as well as psychological and behavioural basis.

How to make 'Business Unusual' be seen as better and safer than 'Business-as-Usual'

Overwhelming evidence from psychological studies highlights the link between social and environmental responsibility and *altruistic and empathetic values*, suggesting that altruistic individuals are more likely to adopt sustainable behaviours. Social and environmental responsibility requires limiting individualistic interests and embracing altruistic behaviours. Research links oxytocin, a peptide hormone known to promote social bonding, love and emotionally positive feelings, with altruism [82]. Altruism and empathy have evolved to strengthen social bonds, enhancing individual and collective chances of survival and well-being. Interestingly but not surprisingly, oxytocin is not only present in humans and all mammals but throughout the animal kingdom. Fish, birds, reptiles and even ancient species like octopuses have their own version of oxytocin.

The boundaries of altruism and empathy are naturally more easily confined to socially close circles like family, friends and community and our own species. It becomes much harder to find evidence of altruistic sentiments and behaviours that apply towards socially distant categories such as future 'unknown' generations of people or non-human life forms. Exceptions are animals with whom we establish close and 'human-like' emotional connections, like our pets, or through inspirational and emotional interactions like taking a safari to see elephants or snorkelling with dolphins – or simply immersing ourselves in a nature documentary. We love our cat and dog, parrot or rabbit, but we happily eat pigs, cows and chickens that have mostly been industrially raised and endured terrible suffering. We mourn the loss of compatriots, but we are able to dehumanise people from other countries, races or faiths to justify discrimination and even wars or acts of terrorism.

Research also highlights that altruism affects social behaviours more easily than environmental ones. Researchers examined people's preferences for supporting social versus environmental charities and found that the release of oxytocin increased a tendency to favour social causes. This suggests that for environmentally sustainable behaviours to become more relevant, we need to internalise the essential role a healthy environment plays in human well-being.

Another fascinating dimension is that, whether due to our aversion to change or the 'single-action bias', we tend to favour increasing sustainable behaviours over reducing unsustainable ones. Neuroscientific research shows that brain regions linked to memory are particularly activated when we imagine increasing sustainable actions. Conversely, the idea of reducing unsustainable behaviours activates key regions in the brain responsible for inhibitory control. This mechanism may suppress memories of unsustainable actions, allowing for forward-thinking about sustainable alternatives [83]. This seems to explain the difficulty we have in halting unsustainable behaviours versus the more rapid adoption of sustainable ones, highlighting the importance of developing viable and accessible alternatives.

Transitions also face challenges of a related, but darker, nature – *'privileges'* and *'vested interests'*. Much like what's happening with the push for net-zero emissions in climate action, as society and the economy begin to grasp the scale of change, effort and investment needed for a nature-positive future, a populist and regressive rhetoric labelling these transitions as "anti-development" is resurfacing. The reality of course is that failing to transition to a sustainable net-zero and nature-positive model is the biggest risk of undermining future economic and, more broadly, human development prospects. But denying the need for change avoids the uncertainties of going through the transition and maintains what is perceived as the safer status quo.

Decades of investment in today's unsustainable economic model have built powerful sectors and benefited elites which are determined to perpetuate the status quo as long as possible to maximise returns on investment and retain their privileges. Well-paid, experienced and trained lobbyists, communicators and marketing specialists are hired to manipulate the public opinion, influence or even corrupt politicians and preserve the status quo. This is the dark side and 'two-sided' relationship of some corporates and sectors in relation to the sustainable agenda. Inspiring speeches at conferences and, behind the scenes, heavy lobbying to block change.

Every centre of political decision-making from Washington to Jakarta and Sao Paulo experiences the same situation. Powerful regressive efforts have derailed, delayed and watered down environmental legislation over the decades and continue to do so to this day. Only visionary political leadership, empowered by the backing of the general population, can make the difference in resisting these pressures, as we saw with the approval of the 'Forest Code' in Brazil, the payment for ecosystem services regulation in Costa Rica, the 'Green Deal' policies in the EU, the Inflation Prevention Act in the US, many environmental reforms in China, among others. Of course, thankfully, we are also seeing more corporate leaders walking the talk by setting ambitious commitments and working hard to deliver them and support common rules and ambitious regulation, as was the case of the business coalition that has supported the Paris Agreement, the Kunming-Montreal agreement and is actively advocating for a treaty on plastic pollution.

Every deep transition is filled with uncertainty and insecurity, but when this is combined with complex and significant systemic change, the hard choices required may lead many to search for comfort in 'business-as-usual.' For this reason, it is key that the transition is supported by a narrative that generates as much clarity as possible on the issue we are trying to address, the journey we need to embark on and the destination we want to reach. Whether on climate or nature, it is key to strike the right balance between a science-based, honest description of the seriousness of the crisis and its consequences, the necessary steps and trade-offs during the transition and, most importantly, the benefits that we will derive from it. This is the only way to convince our 'selfish gene' to embrace change. Ultimately, for the process to happen in practice, socio-economic instruments must be put in place to support society during the transition, particularly the most affected sectors, to inject confidence, trust and positivity.

The International Labour Organization (ILO) defines a just transition towards sustainability as one that promotes decent jobs for all involved and transitions to "environmentally sustainable economies in a way that is fair and inclusive to everyone concerned – workers, enterprises and communities" [84, 85].

To overcome some of the barriers we discussed earlier, adopting the right narrative is key. The narrative surrounding the transition must carefully balance the inherent tension between old and new, efforts and rewards. Firstly, to avoid complacency, excessive optimism and any room for delay or inaction, we should not refer to 'risks', but rather to 'threats.' A risk is something that may or may not materialise in the future, while the impacts of biodiversity loss are materialising as we speak and are sufficiently defined by science.

Secondly, we should not focus solely on risks and threats but also highlight the opportunities, imperatives and the many synergies between a nature-positive future and the most pressing socio-economic challenges society faces today and tomorrow. Connecting the dots between the loss of biodiversity and issues like climate stability, water and food security, human health and well-being is crucial, including the moral argument of intergenerational justice and building a safer future for our children. We must avoid both communicating biodiversity loss as an abstract, looming, intractable global issue but also injecting false optimism and hope as if someone else will come and fix the problem for us. Neither will lead to the resolve needed to achieve the system change we so desperately require. The nature-positive concept effectively captures the opportunity to translate personal and societal goals into action, highlighting the benefits that both we and future generations will gain from living in harmony with nature.

From its conception to the present day, the conservation of nature has been presented and perceived as a noble and altruistic cause, most notably as a way to protect species from extinction. Although few will disagree with these high moral imperatives, and the public support for nature conservation is growing as shown in poll after poll, for most it is still not a priority, not something that occupies their mind daily, nor something that influences how they vote in an election. Although awareness has grown and concern has deepened, for many, nature remains secondary to most other social and economic preoccupations. We spend as much money on protected areas globally as we do in barber shops [86].

I would say that if we want to proactively shift societal perceptions and priorities, we should put more emphasis on the pragmatic value of conserving nature, appealing to our innate selfish 'values' and self-preservation instincts. This doesn't mean we should speak of nature solely from an anthropocentric or utilitarian perspective. Instead, it involves emphasising the tangible everyday benefits nature provides and what we stand to lose if we forfeit her services. While I personally deeply cherish Earth's incredible diversity of life and I have been in love with wildlife since my childhood, the nature conservation narrative should not focus solely on wildlife – it should be about Life itself and human life in particular. Not in an anthropocentric, arrogant way, but recognising the holistic nature of the challenge. This is what is truly at stake in the current climate and ecological destabilisation crisis, Earth's diversity of life as we know it and the well-being of present and future human generations.

Seemingly powerful statements like 'half of global GDP depends on nature' or 'nature's services generate $150 trillion equivalent to our economy' are meaningless, abstract and far from the personal lives of most people, who are worried about low wages, rising energy bills and the cost of food, not to mention actual poverty and malnutrition. I believe that many people who vote for regressive populist political manifestos, which reject the green transition and instead support a business-as-usual future, are not necessarily in denial about issues like climate change and nature loss. They are simply more concerned about other pressing issues.

Leaders that have increased their country's emissions or rate of deforestation were not elected to do that. They were voted in because of the failures of their predecessors or because they seemed more likely to tackle issues of immediate concern, from crime or inflation to healthcare and illegal immigration. Over 85% of European Union citizens recently expressed their strong support for rules to curb deforestation. Yet, many of the leaders recently elected by these voters wasted no time in delaying and watering down the anti-deforestation law approved just a few months prior by a previous EU Commission and Parliament.

If asked, who would honestly say that they are in favour of deforestation? That they enjoy the extinction of species or water pollution? But if the question was 'are you more worried about deforestation, illegal immigration, the return of inflation or the war in Ukraine?', then the answer would be very different and would better reflect people's primary concerns, priorities . . . and who they vote for. Deforestation would certainly not rank near the top, but that doesn't mean people are not aware of it or don't care about it.

What is needed is a narrative that concretely explains how a life with less emissions, less pesticides, less deforestation and less plastic is a better everyday life for people, on an individual and household level. And we need a convincing plan on how to get there, explaining how any transition, any phase-out will be managed fairly and supporting the affected categories of people. It is about presenting concrete alternatives to practices, products, investments that are today nature-negative and are undermining our future, not enhancing it. It is about presenting '*business unusual*' as a better alternative to 'business as usual' – demonstrating how a more sustainable lifestyle and economy can outperform the lifestyle and economy we are accustomed to. By building confidence in a different and better future, and by showing a clear path to achieving a safe and just transition, we can overcome the barriers posed by our instinctive aversion to risk and change.

By recognising the powerful psychological and behavioural drivers deeply rooted in evolution, we can more effectively understand society's varied response to existential threats like the climate and nature crises. Our inherent self-interest can shift to altruistic and empathetic values and an 'eco-centric' worldview which sees people and nature as connected and interdependent, if this equates to a safer future for us and our children. The key is to internalise the link between sustainable behaviours and the improvement of survival and well-being for ourselves, our families and ultimately all of humanity. This seems to be the most powerful psychological trigger and our best chance to mobilise society at scale, leading us to a nature-positive future.

4.9   The 'Great Chance': speed, scale and positive tipping points

We are at a crucial inflection moment, where humanity is facing its deepest and most unprecedented *'Great Choice' and 'Great Chance'*: continue on the same nature-negative path, accelerate towards catastrophic planetary tipping points and put our wellbeing and existence at risk – or embrace a turning point by changing course towards a net zero, Nature Positive, safer and more equitable future.

The glass-half-full view of today's state of affairs is that awareness is at an all-time high and we are at the start of a potential holistic change process that targets the heart of our culture, value system and economic model. The transition to a carbon-neutral and a nature-positive future has started. The glass-half-empty view, however, is that we are already late to act on climate change and nature loss, even potentially too late to make it in time before certain planetary tipping points strike. The speed and scale of the transition will be the dividing line between success and failure. And speed and scale are what any nature-positive strategy should focus on. That's why it is very significant that the Global Biodiversity Framework Nature Positive mission of 'halt and reverse biodiversity loss by 2030' has set both the scale of the ambition and deadline by the end of this decade. Our 'Great Chance' is now; the race to a nature-positive future is on.

We must accelerate the transition. It's time to translate our new awareness into systems change. We know the problem and we understand its gravity. We know that it is linked to an unsustainable economic model and a distorted way to value nature's services and resources; the same system that has created an ever-widening socio-economic inequality gap. Despite the gravity of the situation, there is a potential silver lining. Our growing perception of the risk and the threat to our future may help trigger powerful cultural and behavioural shifts or *'positive social tipping points'*.

The same principles that apply to negative tipping points also apply to the emergence of positive ones, although the capacity for agency within certain social groups and the networked nature of our social structures means that a targeted and limited intervention can induce systemic change. History shows most social tipping points don't require a vast majority but rather a significant minority consensus to occur. In this case we refer to changes in economic and social patterns that trigger behavioural and socio-economic change at scale. Like for negative tipping points, positive ones are triggered by reinforcing feedback. These can be economies of scale or new social norms and values winning out over resistance and uncertainty, increasing socio-economic benefits and becoming more socially acceptable than old practices, accelerating uptake and exponential shifts [87].

Public regulation and fiscal policies, including incentives and disincentives, are crucial for accelerating private investment and driving market transformation. This, in turn, promotes technological advancements and innovation, creating positive experiences for consumers, investors and markets, and generating social acceptance and support for change. These feedback loops involving *finance, technology* and *behaviour* can trigger a positive tipping point leading to lasting systemic and behavioural change.

The awareness of the climate crisis led to the Paris climate agreement, signalling the need for our global economy to decarbonise, and has triggered massive investment in renewable energy generation technologies. Based on this global policy instrument and public concern, several governments have started putting in place domestic regulation, fiscal policies and incentive schemes to develop the renewable energy market. The EU, the US and China, despite being amongst the top global emitters, were amongst the most influential champions of the Paris Agreement and have played an important role in promoting the production of renewable energy.

China has been scaling up investment in both technology and production while using its massive domestic market to boost manufacturing of solar and wind energy on an unprecedented scale, crushing prices globally and paving the way for renewable energy to be affordable and even competitive with fossil fuels. Feed-in tariffs (FITs) and incentives implemented at the time by key economies like EU countries and the US provided incentives to reward renewable energy production and support trade of technology and products, contributing to the creation of a conducive global market. It is worth noting that, in line with the psychological mechanisms discussed in the previous chapter, attempts to regulate the 'bad' were not hugely successful, but efforts to incentivise the 'good' worked much better and accelerated the transition by driving down cost as scale increased. This is both a reflection of the impacts of the economic forces resisting the change, but also how markets and consumers favour testing new incentivised directions before abandoning old habits. That's how positive socio-economic tipping points emerge and scale.

At the time of writing this chapter, the International Energy Agency has estimated that in 2024, investment in renewable energy technology and infrastructure will hit the $2 trillion mark, double that of fossil fuels [88]. Some researchers argue that we are beginning to see positive tipping points approaching vis-à-vis the climate crisis in terms of renewable energy and electric vehicles. Non-fossil fuel-based ammonia for fertilisers and green hydrogen could be next. We must reach similar acceleration towards positive tipping points in regard to sustainable food production and consumption, perhaps the most challenging and crucial component of a local-to-global, nature-positive transition. New regenerative, soilless and hydroponic agriculture practices, as well as cultured meat technologies, are the most promising trends that should be encouraged.

Redirecting public subsidies for agriculture, animal farming and fishing away from current nature-negative, destructive practices will be key to supporting low-chemical and low-energy inputs, regenerative agriculture and sustainable fishing and aquaculture amongst other sectors. One positive tipping point would be the combination of the cultural acceptance and market uptake of 'alternative proteins' such as plant-based substitutes or cultured meat. While the growth in non-meat-based proteins is significant, it is not yet driving a reduction in meat consumption.

But with more than 25% of Europeans now considering themselves flexitarian, vegetarian or vegan, targeted interventions could help accelerate the transition.

We already have the technology to make cultured (cell-based) meat, which means growing cells outside the living organism. In practice, stems cells are extracted from the muscle of a living animal and cultivated in controlled conditions providing nutrients for them to multiply. Two issues have hampered the development of this new industry: the technology needed to produce it on a commercial scale and the high costs involved. However, new techniques promise an uptick in production of lab-cultured meat and could bring it down to the price of traditionally farmed organic meat, with a view to eventually match the price of the cheapest intensively industrially produced meat. For example, researchers have found a way to grow chicken meat cells at a density four times greater than with previous procedures [89].

Similar developments are happening for fish and other seafood. But the third hurdle for 'cultured meat' is 'culture' itself. This is where the unfortunate use of terms like 'synthetic meat' or 'lab-grown meat' have paved the way for propaganda and negative labels such as 'artificial or 'fake' meat.

Not only should cultured meat taste like conventional meat but it would also offer significant advantages. It should eliminate pesticides and antibiotic residues found in traditionally intensively farmed meat; it should also allow for control over fat and nutrient content and be entirely free from animal suffering. Despite these cultural challenges, the potential is huge, and the development of this new industry is probably already unstoppable. This is hopefully an example of an approaching 'environmentally positive' social tipping point.

A study from the Food and Land Use Coalition and Exeter University estimates that just a 20% market share of plant-based and alternative proteins by 2035 would save 400–800 million hectares of land being used for livestock and their fodder, equivalent to 7–15% of the world's farmland today. That would reduce pressure on forests, and the spared land could be used for natural ecosystem restoration, absorbing CO_2 and providing many other ecological services [90].

The nature-positive transition faces a main challenge upon which its success depends and that is confidence in a different future. Moving away from consolidated attitudes and practices does not come naturally. Evolution has taught us to trust what has been tested and to change behaviour or actions only in the face of an immediate need, benefit or threat. This manifests in extraordinarily simple behaviours. When we enter the room on the second day of a workshop the majority of us will return to the same seat we sat at on the day before. We tend to follow the same route going to work even though there may be new alternatives.

Our reluctance to change has created consolidated practices and behaviours. The problem is that many of these were developed at a time when natural resources were plentiful and Earth systems strong and stable. Mining, agriculture, fishing and forestry didn't have to be that efficient and sustainable. A significant amount of waste was acceptable, even economically beneficial, as it didn't lead to additional costs for collection and disposal. In the past, with a human population amply within the planet's limits, it wasn't necessary to pay much attention to externalities because our impacts were minimal compared to nature's levels of resilience. This is why, over the decades, vast quantities of mining tailings were dumped on land or into the sea, wood was left to rot after logging, polluted waters discharged in rivers and lakes and bycatch of dead sea creatures was discarded back into the ocean – practices that largely continue today.

We haven't been adapting our practices alongside the depletion of resources or degradation of natural systems, so much so that today the lifestyle and practices of many of us are unsustainable and do not fall within the boundaries of what the natural world can offer us. As the 'Great Decline' progresses, we have reached a challenging situation when, as ecosystemic health has reached crisis point, it is no longer just a question of incremental changes through the adoption of low-footprint practices but a need to transition to very different models, a truly nature-positive society and economy.

Understandably, the scale of the transition has been met with concern by some groups, such as coal miners protesting against the phasing out of the sector to make way for renewable energy, or by farmers unhappy with regulation aimed at reducing the use of pesticides and herbicides and the increase of land set aside to support regenerative agriculture. These are manifestations of the inevitable and partly legitimate anxiety generated by change and by the lack of clear, well-explained and supported transition plans. The focus must be on accompanying the transition by not only clearly communicating the destination, needs and advantages of embracing change but also putting in place the necessary fiscal policies, financial incentives and retraining opportunities needed for a successful and just transition.

4.10   Onwards!

"(Don't) Hope for the best (act!), (because you cannot) prepare for the worst". This is how I'd like to rephrase the famous saying to capture the necessary change in attitude and practice that characterizes humanity's 'Great Choice' before today's 'Great Threat' of planetary systems disruption. The choice of a nature-positive future.

I was born shortly after the world had entered the new era of the Anthropocene and right at the start of the 'Great Acceleration'. I have lived through its excitement and failures. As a teenager I was writing the monthly newsletter for members of my local NGO on a typewriter, using carbon copies to save time, and eagerly checking my mailbox for letters from my first girlfriend or papers ordered from the Milan Natural History Museum. Then came fax machines, early large and clumsy computers, huge mobile phones and a 1983 Steve Jobs speech predicting a future where people would spend more time in front of a computer than in their cars (or even more so on mobile phones, as we have discovered).

People seeing each other on a phone was something that featured in science fiction movies. I lived when the Nobel Prize–winning chemical DDT was sprayed in homes to kill mosquitoes, before it was banned for health and environmental reasons. I have seen the appearance of plastic bags and single-use bottles in shops and acrylic fibres in clothes, the advent of huge supermarkets and the explosion of food waste, the emergence of processed foods and fruits and vegetables imported from distant lands so they can be available at any time of the year. I saw oceanic and tropical fish species appear in the fish market of my Mediterranean seaside hometown and tropical timber in parquets and furniture, as well as roads filling with cars which have gradually doubled and then tripled in size.

I have seen all this and much more, and I am not even that old. This has taken place in just a few decades. A truly great acceleration, in every aspect of our lives. After the horrors and sacrifices of two successive World Wars, the future was something many could look forward to with a sense of hope and excitement. At the start of that new era, which would later be called the 'Anthropocene', nature was already suffering under human pressure, but this was nothing compared to what was to follow in the subsequent decades. Over the course of those decades, I have visited the most magnificent nature and witnessed her most reckless destruction. I have seen glaciers turn brown and coral reefs turn white. Nature was becoming a moral preoccupation for a growing number of people, but it could not compete with the powerful forces of the new 'God' of economic growth, wealth accumulation and rampant consumerism.

It is no coincidence that those were both the early days of modern environmentalism and of large-scale conversion schemes, with endless roads slicing through untouched forests, mega-dams blocking the flow of mighty rivers and huge monocultures of sugarcane, soy, oil palm, pulpwood and rubber trees replacing natural habitats. It was also a time when vast mining and logging operations began destroying tropical forests, and the emergence of new environmental threats like plastic, herbicides, pesticides and the tropical beef industry. As the impacts on nature grew and became more evident, more and more people started not only caring but also worrying. As is often the case, we start caring when something becomes rare and elusive.

The 'Great Acceleration' has also coincided with many positive developments that gave us hope for the environment and humanity's future. I have lived through the explosion of designations of parks and reserves everywhere in the world, the return of many species from the brink of extinction, the boom of nature tourism, the popularity of nature documentaries, the dramatic growth of the environmental and social justice movements and, more recently, the rise of renewable energy. I have met so many wonderful people who have a deep passion for nature and are uncompromisingly dedicated to her protection. I have had the privilege of being part of several of these exciting innovative and successful programmes that are making a difference on the ground as well as on a policy level. But I also had sufficient awareness to realise that all this progress wasn't enough. And so, we arrive at today's paradox, where an all-time high of public support for the environment is matched by an unabated increase in human pressure and degradation of the natural world. The paradox of a society split between neglect and anxiety, commitment and inaction, awareness and denial, demand for and opposition to change.

We are living in extraordinary times. Times of unprecedented risk but also of unparalleled opportunity for the future of humanity and life as we know it. Humanity finds itself at the most existential juncture and the most worrisome of times but also at a time of unequalled knowledge, awareness and chances. The world has never been more conscious of the problems and aware of the solutions and has never been more committed to addressing these challenges. Nature has never been higher on the political, corporate and societal agenda. More people are using social media or are taking to the streets to make their voices heard about the future they want.

The problem statement is crystal clear and extremely worrying. The consequences for humanity of a business-as-usual scenario are equally clear and terrifying. But these scenarios are not inevitable. Not yet. We have recently agreed on a Carbon Neutral and Nature Positive goal for our society and economy. That means a commitment to maintain the stability of the climate and the vitality and diversity of living organisms and ecosystems. Now the focus needs to be on making the transition happen at scale, in a way that is fast and fair, accessible and inspiring. Can we correct the course of unsustainable development before reaching climate and ecosystem tipping points? Can we learn how to live within planetary boundaries? Can we start valuing nature in its complexity and not just for what we can extract and trade? The answer to all these questions is only this: yes, we can. If we want to.

Transitions are complex and filled with anxiety. They clash with our widespread aversion to change and risk. The temptation to backtrack to business-as-usual is omnipresent. As imperfect as it is, to many people the status quo still feels safe. For many, abandoning a development model that has generated undeniable progress, despite being unequally distributed, is a choice filled with uncertainty, even for those who haven't benefited very much. Moreover, the elites who have benefited the most from the current system are not prepared to let it go. There is and will be resistance, whether it is based on genuine concerns, emotional panic, ignorance or selfishness, to protect economic interests and the privileges of the few versus the collective good. But we have the power of knowledge, technology and the 'power of the many' in the shape of a growing number of concerned, like-minded people around the world.

A safer and more sustainable future for people and nature remains in our grasp if humanity unites and acts together. In the last few years, the world has, against the odds, united around historic agreements on climate and biodiversity, despite geopolitical fractures, competition and posturing. Much of the business and financial sector has woken up to the reality of the risks posed by climate change and nature loss to their existence, let alone their profitability. This can inspire confidence and resolve and keep us on course to act decisively. Despite all this growing potential for change, today's reality is grim. Global heating and biodiversity loss, as well as socio-economic inequality, continue unabated. The challenge is significant, rooted in our flawed economic and development models, our forgotten dependency on nature and our deep-seeded aversion to change. However, the systems that govern our society and economy today are of our own making, and they are systems we can change. We need political and corporate leadership to listen to science and to the fast-growing section of society which demands action to avert ecological and climate catastrophe. More critically we need society to want to change.

We should count on the fact that, for the first time in the history of our civilisation, more and more of us are beginning to understand that we cannot continue to take nature for granted. We cannot continue to dominate the planet and exploit its resources wastefully and destructively without consequences. And we are already seeing these consequences: loss of life, livelihoods and economic assets, with things set to get worse unless we course-correct with great urgency.

Change is possible, and more importantly, change is necessary.

We owe it to the amazing diversity of life that makes this planet a 'living planet'. We owe it to the world's children and to the generations to come, the ones who currently have no voice and no seat at the table around which decisions are made but will suffer the consequences of those decisions.

In fact, change is inevitable. We can either embrace it and trigger the 'Great Transition' towards sustainability, or it will be forced upon us by the collapse of Earth systems. And it won't be fun. Even if the change required is significant and disruptive of today's business-as-usual development model, the transition to a nature-positive economy comes with great opportunities and against a counterfactual of increasing costs and disastrous socio-economic global destabilisation, affecting all corners of society, with no one excluded.

We are living in times of crisis. We must avoid the socio-political fractures and polarisations that often arise when society is under pressure, resist short-term vested interests, ignite our capacity to unite in the face of the greatest challenges, embrace resolve, work collectively and channel our survival instinct and 'selfish gene' towards the only safe direction: a nature-positive future. Many nations, people and communities have pulled together in times of great challenges throughout the course of human history. Today, we need to do it not as one community, not as one nation, but as *one humanity*. Together, a people-positive and nature-positive future will be possible.

Just seven decades ago most of us didn't see limits to the exploitation of the natural world. Our modern economic model is still dangerously based on this false perception. Postponing change by even a few decades may mean it is too late to course correct. The choices made by this generation's political and corporate leaders and, more importantly, by all of us will have wide-ranging consequences for generations to come. The future is not in the present, but the ecological, social, political, and economic choices that will shape it are. The future is, in fact, today.

> Fifty years into the future it will be too late to do what is possible now. We are in a 'sweet spot' in time where the decisions we make in the next ten years will determine the direction of the next 10,000.
>
> (Sylvia Earle, American marine biologist)

Achieving a nature-positive future requires a massive collective effort. The corporate and financial sectors, with their huge ecological footprint but also their ability to allocate capital, have a critical role to play in making nature-positive a reality. This is not just about risk-mitigation or regulatory obligations but part of a vision and strategy for new value creation. It will require every sector and every organisation, every government and every community, to recognise the true value of nature to our economy and society and to understand the need for transitioning to a new operating model, a new '*Nature Positive business-as-usual*' which is able to continue to generate economic value and human development but is decoupled from climate change and nature loss. A nature-positive business-as-usual that preserves the ecosystems, species and natural processes which have sustained our lives and fuelled our development since the start of our civilisation, so that the natural world may continue to support future generations to come.

A nature-positive transition is deep and not without challenges, but it represents change for a better future. A future of opportunities rather than risks. We know that today's business-as-usual model is not an option for much longer as the unavoidable socio-economic consequences of nature loss and climate change will, sooner than later, force a change and will most likely lead to great social disruption. It is better to ensure we are driving change in a way that is effective and equitable. It is time to realise that for humans to thrive, nature needs to thrive too. This has been the foundation of our development until now, but the approaching climatic and ecosystemic tipping points can change all that, probably in a matter of decades. This can't be dismissed as Doomsday rhetoric; this is science speaking.

The discussions around the nature-positive transition have led some people to question, due to our long exploitative relationship with nature and the current nature-negative economic model that is resistant to change, whether halting and reversing the decline of the world's natural systems is even possible. The scale of the cultural and systemic shifts required cannot be exaggerated. However, while this transition is undoubtedly challenging, the deep changes needed in our society and economy should not discourage us. The scale of nature loss and the risks this poses to our society and economy should be enough of an incentive to embark on this necessary transition, however challenging it may be.

The Nature Positive goal is undoubtedly *ambitious*, but it is *possible*. Firstly, because biodiversity has an extraordinary ability to regenerate and be resilient – if we act quickly enough to prevent ecosystems from reaching irreversible tipping points. And second, because of our growing understanding of the existential risks that further nature loss poses to our future. In fact, Nature Positive is not only possible but it is *necessary*; it is imperative to secure nature's contributions to and a safe future for humanity. A thriving and fair society and a prosperous and resilient economy depend on ecosystem health, stability and productivity. Nature is the foundation of everything. It is the foundation of Life (including human life). This is the rationale behind the need for becoming a nature-positive society.

We live in the Anthropocene, the era dominated by humans. This is a fact. But what Anthropocene do we choose? One that leads to climate and ecosystem disruption or one where human development is kept within planetary boundaries? A new chapter of the history of our civilisation will inevitably be written. It can be one where we struggle against a destabilised ecological and climatic world order, or one where we can learn to live in, with and from nature sustainably and harmoniously, and perhaps, most importantly, feeling part of nature. We are the ones holding the pen that will write that new chapter, one way or the other.

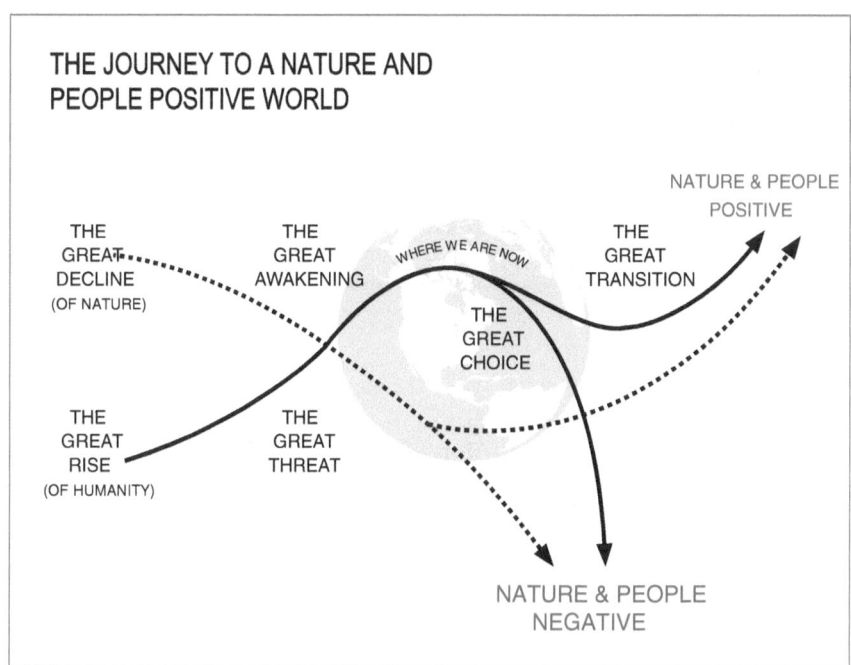

THE JOURNEY TO A NATURE AND
PEOPLE POSITIVE WORLD

Figure 4.7 The Great Moments of Humanity.

Source: Marco Lambertini

From the '*Great Rise*' of humanity, marked by unparallelled progress and development but matched by the '*Great Decline*' of the natural world, humanity has recently experienced a '*Great Awakening*' about the impacts and the risks of our unsustainable relationship with nature. This is leading a greater share of today's global population to acknowledge the '*Great Threat*' climate change and nature loss represent for all life on Earth, including present and future human generations. We are now at the most dramatic, exciting and consequential time ever experienced by our species. We are already living in a time of consequences that will become increasingly grave and irreversible. But consequences are the result of choices. And we still have a choice, the '*Great Choice*' to embark on the '*Great Transition*' towards sustainability and grab with resolve and optimism our '*Great Chance*' to turn a threat into an opportunity. The opportunity for a safer and more equitable future, a future that is People and Nature Positive at the same time.

Back to Gramsci's quote at the start of the previous chapter, our task is to accelerate the end of the 'old' – the current nature-negative model – and accelerate the transition to a 'new' nature-positive future, by demonstrating to society that it is the only possible future, and more importantly a better one for all. Because a people-positive future cannot exist without a nature-positive one.

Onwards!

Notes

1 At the time of printing this book the global consultation on State of Nature metrics convened by the Nature Positive Initiative has been concluded and the feedback incorporated. Following the piloting phase, the State of Nature Metrics framework is expected to be finalised in 2026 and will be published on www.naturepositive.org. Also, the initial State of Nature metrics framework is restricted to terrestrial biodiversity. Freshwater and marine state of nature metrics will follow a similar process.
2 After the finalisation of the universal State of Nature metrics, the Nature Positive Initiative has committed to also convene an engagement process to build consensus on guidance regarding claims of nature-positive outcomes.

References

1 Díaz, S., Settele, J., Brondízio, E.S., Ngo, H.T., Agard, J., Arneth, A., Balvanera, P., Brauman, K.A., Butchart, S.H. and Chan, K.M., 2019. 'Pervasive human-driven decline of life on Earth points to the need for transformative change', Science, 366(6471). Available at: https://doi.org/10.1126/science.aax3100
2 Moody's, 2021. ESG Solutions: Integrating Biodiversity Into a Risk Assessment Framework, 26 May. Available at: www.moodys.com/web/en/us/hosted-assets/esg-insights-bx6446-integrating-biodiversity-2jun2021.pdf
3 Nature Positive Initiative, 2023. The Definition of Nature Positive. Available at: https://4783129.fs1.hubspotusercontent-na1.net/hubfs/4783129/The%20Definition%20of%20Nature%20Positive.pdf
4 IPCC, 2018. 'An IPCC Special Report on the impacts of global warming of 1.5°C above pre-industrial levels and related global greenhouse gas emission pathways, in the context of strengthening the global response to the threat of climate change, sustainable development, and efforts to eradicate poverty', In V. Masson-Delmotte, P. Zhai, H.-O. Pörtner, D. Roberts, J. Skea, P.R. Shukla, A. Pirani, W. Moufouma-Okia, C. Péan, R. Pidcock, S. Connors, J.B.R. Matthews, Y. Chen, X. Zhou, M.I. Gomis, E. Lonnoy, T. Maycock, M. Tignor and T. Waterfield (eds.), Cambridge, UK and New York, NY, USA: Cambridge University Press, pp. 3–24. Available at: https://doi.org/10.1017/9781009157940.001
5 Net Zero Tracker, 2022. Net Zero Stocktake 2022: Assessing the Status and Trends of Net Zero Target Setting Across Countries, Sub-National Governments and Companies [online]. Available at: https://zerotracker.net/analysis/net-zero-stocktake-2022
6 MSCI, 2023. 'A periodic report on progress by the world's listed companies toward curbing climate risk', Net Zero Tracker. Available at: www.msci.com/documents/1296102/38217127/MSCI-NetZero-Tracker-July-2023.pdf
7 International Renewable Energy Agency (IREA), 2024. Clean Energy Market Monitor. Available at: https://iea.blob.core.windows.net/assets/d718c314-c916-47c9-a368-9f8bb38fd9d0/CleanEnergyMarketMonitorMarch2024.pdf
8 United Nations, 2015. Transforming the World: The 2030 Agenda for Sustainable Development. A/RES/70/1. Available at: https://sdgs.un.org/goals
9 Global Goal for Nature Group, 2020. 'A global goal for nature: Nature positive by 2030', Establishing a Goal for Nature-Positive Societies. Available at: www.naturepositive.org/app/uploads/2024/02/Global-Goal-Nature-Positive-2030-v11092020.pdf
10 Secretariat of the Convention on Biological Diversity, 2020. 'Global biodiversity outlook 5', Montreal. Available at: www.cbd.int/gbo/gbo5/publication/gbo-5-en.pdf
11 Nature Positive by 2030, 2021. Non-State Actors' Call for Governments to Strengthen the Post-2020 Global Biodiversity Framework: Secure an Equitable, Nature Positive, Net Zero Emissions World. Available at: www.naturepositive.org/app/uploads/2024/03/Non-State-Actor-Call-To-Action-2021.pdf

12 Leaders Pledge for Nature, 2022. The Race is On for a Nature-Positive World by 2030. Available at: www.leaderspledgefornature.org/wp-content/uploads/2022/01/LPN_ progress_180122.pdf

13 Nature Positive Initiative, 2023. The Definition of Nature Positive. Available at: https:// www.naturepositive.org/app/uploads/2024/02/The-Definition-of-Nature-Positive.pdf

14 Bull, J.W., Milner-Gulland, E.J., Addison, P.F.E., et al., 2020. 'Net positive outcomes for nature', Nature Ecology & Evolution, 4, pp. 4–7. Available at: https://doi.org/10.1038/ s41559-019-1022-z

15 Maron, M., Simmonds, J.S., Watson, J.E.M., et al., 2020. 'Global no net loss of natural ecosystems', Nature Ecology & Evolution, 4, pp. 46–49. Available at: https://doi. org/10.1038/s41559-019-1067-z

16 Langhammer, P.F., et al., 2024. 'The positive impact of conservation action', Science, 384, pp. 453–458. Available at: https://doi.org/10.1126/science.adj6598

17 Ledger, S.E.H., Rutherford, C.A., Benham, C., Burfield, I.J., Deinet, S., Eaton, M., Freeman, R., Gray, C., Herrando, S., Puleston, H., Scott-Gatty, K., Staneva, A. and McRae, L., 2022. 'Wildlife Comeback in Europe: Opportunities and challenges for species recovery', Zoological Society of London, BirdLife International and the European BirdLife International and European Bird Census Council. Available at: www.rewildingeurope.com/wp-content/uploads/publications/wildlife-comeback-in-europe-2022/ index.html

18 G20, Global Infrastructure Hub and Oxford Economics, 2018. Global Infrastructure Outlook: Infrastructure Investment Need in the Compact with African Countries. Available at: https://outlook.gihub.org/

19 IFC, 2012. Performance Standard 6: Biodiversity Conservation and Sustainable Management of Living Natural Resources. Available at: https://documents1.worldbank. org/curated/fr/898321491456820716/pdf/113846-WP-ENGLISH-PS6-Biodiversity-conservation-2012-PUBLIC.pdf

20 Ekstrom, J., Bennun, L. and Mitchell, R., n.d. 'A cross-sector guide for implementing the Mitigation Hierarchy', The Biodiversity Consultancy for the Cross Sector Biodiversity Initiative. Available at: https://www.thebiodiversityconsultancy.com/fileadmin/ user_upload/A_cross-sector_guide_for_implementing_the_Mitigation_Hierarchy.pdf

21 Global Goal For Nature Group, 2020. No-Net Loss/Net-Gain Meaning and Principles. Available at: www.naturepositive.org/app/uploads/2024/02/Apex-goal-Net-task-force_draft-October-20-1-1.docx

22 Bull, J.W., Milner-Gulland, E.J., Addison, P.F.E., et al., 2020. 'Net positive outcomes for nature', Nature Ecology & Evolution, 4, pp. 4–7. Available at: https://doi.org/10.1038/ s41559-019-1022-z

23 Allan, J.R., Possingham, H.P., Atkinson, S.C., Waldron, A., Di Marco, M., Adams, V.M., Butchart, S.H.M., Venter, O., Maron, M., Williams, B.A., Jones, K.R., Visconti, P., Wintle, B.A., Reside, A.E. and Watson, J.E.M., n.d. 'Conservation attention necessary across at least 44% of Earth's terrestrial area to safeguard biodiversity', Biorxiv.org Preprint. Available at: https://doi.org/10.1101/839977

24 Jones, K.R., et al., 2020. 'Area requirements to safeguard Earth's marine species', One Earth, 2, pp. 188–196.

25 Wilson, E.O., 2016. Half-Earth: Our Planet's Fight for Life. Liveright.

26 Dinerstein, E., Olson, D., Joshi, A., Vynne, C., Burgess, N.D., Wikramanayake, E., Hahn, N., Palminteri, S., Hedao, P., Noss, R., Hansen, M., Locke, H., Ellis, E.C., Jones, B., Barber, C.V., Hayes, R., Kormos, C., Martin, V., Crist, E., Sechrest, W., Price, L., Baillie, J.E.M., Weeden, D., Suckling, K., Davis, C., Sizer, N., Moore, R., Thau, D., Birch, T., Potapov, P., Turubanova, S., Tyukavina, A., de Souza, N., Pintea, L., Brito, J.C., Llewellyn, O.A., Miller, A.G., Patzelt, A., Ghazanfar, S.A., Timberlake, J., Klöser, H., Shennan-Farpón, Y., Kindt, R., Lillesø, J.-P.B., van Breugel, P., Graudal, L., Voge, M., Al-Shammari, K.F. and Saleem, M., 2017. 'An ecoregion-based approach to protecting

half the terrestrial realm', BioScience, 67(6). Available at: https://doi.org/10.1093/biosci/bix014

27 Nature Positive Initiative, 2023. The Definition of Nature Positive. Available at: https://www.naturepositive.org/app/uploads/2024/02/The-Definition-of-Nature-Positive.pdf

28 European Parliament and Council, 2022. Directive (EU) 2022/2464 of the European Parliament and of the Council of 14 December 2022 amending Regulation (EU) No 537/2014, Directive 2004/109/EC, Directive 2006/43/EC and Directive 2013/34/EU, as Regards Corporate Sustainability Reporting. Available at: https://eur-lex.europa.eu/legal-content/EN/TXT/?uri=CELEX:32022L2464

29 Locke, H., Ellis, E.C., Venter, O., Schuster, R., Ma, K., Shen, X., Woodley, S., Kingston, N., Bhola, N., Strassburg, B.N., Paulsch, A., Williams, B. and Watson, J.E.M., 2019. 'Three global conditions for biodiversity conservation and sustainable use: An implementation framework', National Science Review, 6(6), pp. 1080–1082. Available at: https://anthroecology.org/wp-content/uploads/2020/09/locke_2019.pdf

30 Arcadis, ICF, UNEP-WCMC, Capitals Coalition, WCMC Europe, 2024. 'Exploring measurement solutions for corporate Nature Positive Commitments: A discussion paper from the Align project', Align Project – Aligning Accounting Approaches for Nature.

31 UN Convention on Biological Diversity, 2022. The Kunming-Montreal Global Biodiversity Framework. Decision Adopted by the Conference of the Parties to the Convention on Biological Diversity. Available at: www.cbd.int/doc/decisions/cop-15/cop-15-dec-04-en.pdf

32 Savage, S., 2024. 'The big read', The Financial Times, 21 October. Available at: www.ft.com/content/4d12f8d1-c0df-4ab6-b374-741e9517448b

33 Gautam, N. and Rueedi, J., 2022. 'Nature is climbing the agenda, but corporate biodiversity commitments remain rare', Standard & Poor Global. Available at: www.spglobal.com/esg/csa/yearbook/articles/nature-is-climbing-the-agenda-but-corporate-biodiversity-commitments-remain-rare

34 Erben, I., Yang, D., Hopman, D., Jayaram, K., Katz, J. and Van Aken, T., 2023. 'Companies are broadening their commitments to nature beyond carbon', McKinsey & Company. Available at: www.mckinsey.com/industries/agriculture/how-we-help-clients/natural-capital-and-nature/our-insights/companies-are-broadening-their-commitments-to-nature-beyond-carbon

35 Planet Tracker, 2024. The Nature Scorecard. Available at: https://planet-tracker.org/the-nature-scorecard/

36 Planet Tracker, 2023. Bio-Crastination. Available at: https://planet-tracker.org/nature-scorecard/; https://planet-tracker.org/wp-content/uploads/2023/02/Biocrastination.pdf

37 Environmental Finance Data, 2023. Available at: https://efdata.org/

38 Baldock, C., Willis, J. and Kozlowski, N., 2023. 'Votes against nature', Planet Tracker. Available at: https://planet-tracker.org/wp-content/uploads/2023/05/Voting-against-Nature.pdf

39 The Economist Intelligence Unit, 2021. 'An Eco-wakening: Measuring global awareness, engagement and action for nature', WWF. Available at: https://www.worldwildlife.org/publications/an-eco-wakening-measuring-awareness-engagement-and-action-for-nature

40 Nature Action 100, n.d. Nature Action 100. Available at: www.natureaction100.org/

41 PRI, n.d. Spring, A PRI Stewardship Initiative for Nature. Available at: www.unpri.org/investment-tools/stewardship/spring

42 ShareAction, n.d. Biodiversity Initiatives. Available at: https://shareaction.org/investor-initiatives/biodiversity-initiatives

43 WBCSD, 2021. What Does Nature Positive Mean for Business? World Business Council for Sustainable Development. Available at: www.wbcsd.org/resources/what-does-nature-positive-mean-for-business/

44 Business for Nature, et al., 2022. How Business and Finance Can Contribute to a Nature Positive Future Now, October 2022. Available at: https://www.businessfornature.org/news/nature-positive-discussion-paper

45 Natural Capital Coalition (now Capitals Coalition), 2016. Natural Capital Protocol. Available at: https://capitalscoalition.org/capitals-approach/natural-capital-protocol
46 WEF, 2023. Nature-Positive Industry Sector Transitions. Available at: www.weforum.org/publications/industry-transitions-to-nature-positive-report-series/
47 TNFD, 2024. Discussion Paper on Nature Transition Plans. Available at: https://tnfd.global/nature-transition-plans/
48 GRI, 2024. GRI 101: Biodiversity. Available at: www.globalreporting.org/standards/standards-development/topic-standard-for-biodiversity/
49 EFRAG, 2023. ESRS E4 Biodiversity and Ecosystems. Available at: https://xbrl.efrag.org/e-esrs/esrs-set1-2023.html#d1e20709-3-1
50 TNFD, 2023. Guidance on Assessment of Nature-Related Issues: The LEAP Approach. Available at: https://tnfd.global/wp-content/uploads/2023/08/Guidance_on_the_identification_and_assessment_of_nature-related_Issues_The_TNFD_LEAP_approach_V1.1_October2023.pdf?v=1698403116
51 SBTN, 2023. Launch of World's First Science-based Targets for Nature. Available at: https://sciencebasedtargetsnetwork.org/news/news/launch-of-the-worlds-first-science-based-targets-for-nature-to-mobilize-businesses-to-address-nature-loss-climate-change-together/
52 IFRS, n.d. About the International Sustainability Standards Board. Available at: www.ifrs.org/groups/international-sustainability-standards-board
53 International Finance Corporation, 2012. Conservation and Sustainable Management of Living Natural Resources – Performance Standard 6. Available at: https://www.ifc.org/en/insights-reports/2012/ifc-performance-standard-6
54 Ekstrom, J., Bennun, L. and Mitchell, R., n.d. 'A cross-sector guide for implementing the Mitigation Hierarchy', The Biodiversity Consultancy for the Cross Sector Biodiversity Initiative. Available at: https://www.thebiodiversityconsultancy.com/fileadmin/user_upload/A_cross-sector_guide_for_implementing_the_Mitigation_Hierarchy.pdf
55 The Biodiversity Consultancy, n.d. Net Positive and the Mitigation Hierarchy. Available at: www.thebiodiversityconsultancy.com/services/site-level-advisory/mitigation-hierarchy/
56 SBTN, 2020. Initial Guidance for Business. Available at: https://sciencebasedtargetsnetwork.org/wp-content/uploads/2020/11/Science-BasedTargets-for-Nature-Initial-Guidance-for-Business.pdf
57 Business for Nature, 2023. The Nature Strategy Handbook: A Practical Guide for Businesses. Available at: https://nowfornature.org/read-the-handbook/
58 Young, D., Reeves, M. and Gerard, M., 2021. 'The secrets of sustainability front-runners', BCG. Available at: www.bcg.com/publications/2021/keys-to-being-a-leader-in-sustainable-business-model-innovation
59 Business For Nature, 2021. High-Level Business Actions on Nature. Available at: https://www.businessfornature.org/high-level-business-actions-on-nature
60 Lundy, D., 2017. 'Lobby Planet: A guide to the murky world of EU lobbying', The Corporate Europe Observatory. Available at: https://corporateeurope.org/sites/default/files/lp_brussels_report_v7-spreads-lo.pdf
61 Bell-James, J., Foster, R., Frohlich, M., Archibald, C., Benham, C., Evans, M., Fidelman, P., Morrison, T., Baggio, L.R., Billings, P. and Shumway, N., 2024. 'Not all conservation "policy" is created equally: When does a policy give rise to legally binding obligations?', Conservation Letters, p. e13054. Available at: https://conbio.onlinelibrary.wiley.com/doi/10.1111/conl.13054
62 EU Habitats Directive, 1992. Council Directive 92/43/EEC, 21 May 1992, on the conservation of natural habitats and of wild fauna and flora. OJ L 206, 22 July 1992, p. 7.
63 United Nations Environment Programme, 2023. 'State of Finance for nature: The big nature turnaround – Repurposing $7 trillion to combat nature loss', Nairobi. Available at: https://doi.org/10.59117/20.500.11822/44278

64 Koplow, D. and Steenblik, R., 2024. 'Protecting nature by reforming environmentally harmful subsidies: An update', Earthtrack. Available at: www.earthtrack.net/sites/default/files/documents/ehs_report_september-2024-update_final.pdf

65 Tian, N., Lopes da Silva, D., Liang, X. and Scarazzat, L., 2024. 'Trends in world military expenditure 2023', SIPRI. Available at: www.sipri.org/sites/default/files/2024-04/2404_fs_milex_2023.pdf

66 Koplow, D. and Steenblik, R., 2022. 'Protecting nature by reforming environmentally harmful subsidies: The role of business', Earthtrack. Available at: www.earthtrack.net/sites/default/files/documents/EHS_Reform_Background_Report_fin.pdf

67 United Nations Environment Programme, 2023. 'State of finance for nature: The big nature turnaround – Repurposing $7 trillion to combat nature loss', Nairobi. Available at: https://doi.org/10.59117/20.500.11822/44278

68 Koplow, D. and Steenblik, R., 2022. 'Protecting nature by reforming environmentally harmful subsidies: The role of business', Earthtrack. Available at: www.earthtrack.net/sites/default/files/documents/EHS_Reform_Background_Report_fin.pdf

69 Rodríguez, C.M., 2024. 'Time 100 Climate', Time Magazine. Available at: https://time.com/collection/time100-climate/

70 Convention on Biological Diversity, Kunming-Montreal Global Biodiversity Framework, Target 18: Reduce Harmful Incentives by at Least $500 Billion per Year, and Scale Up Positive Incentives for Biodiversity. Available at: www.cbd.int/gbf/targets/18

71 BIOFIN-UNDP, 2024. The Nature of Subsidies: A Step-by-Step Guide to Repurpose Subsidies Harmful to Biodiversity and Improve Their Impacts on People and Nature. Available at: www.biofin.org/sites/default/files/content/knowledge_products/The%20Nature%20of%20Subsidies%20%28Web%29.pdf

72 Quesado, A.U., 2024. 'Valuing nature: The case of tropical forests and Costa Rica', Philosophical Transactions of the Royal Society B: Biological Sciences. Available at: https://royalsocietypublishing.org/doi/10.1098/rstb.2022.0320

73 Giasi, F. (ed.), 2020. Antonio Gramsci. Lettere dal carcere. Einaudi.

74 Earth4All and the Global Commons Alliance, 2024. Global Commons Survey and Earth for All Survey. Available at: https://earth4all.life/global-survey-2024/

75 Pauly, D., 1995. 'Anecdotes and the shifting baseline syndrome of fisheries', Trends in Ecology and Evolution, 10, p. 430.

76 European Environmental Agency, 2024. Grassland Butterfly Index in Europe. Available at: www.eea.europa.eu/en/analysis/indicators/grassland-butterfly-index-in-europe-1

77 Rosenberg, K.V., Dokter, A.M., Blancher, P.J., Sauer, J.R., Smith, A.C., Smith, P.A., Stanton, J.C., Panjabi, A., Helft, L., Parr, M. and Marra, P.P., 2019. 'Decline of the North American avifauna', Science, 366, pp. 120–124. Available at: https://doi.org/10.1126/science.aaw1313

78 WWF and ZSL, 2024. The Living Planet Report 2024. Available at: https://livingplanet.panda.org/

79 Dinnick, I. and Jolley, D., 2024. 'Why might people believe in human-made hurricanes? Two conspiracy theory psychologists explain', The Conversation, 15 October. Available at: https://theconversation.com/why-might-people-believe-in-human-made-hurricanes-two-conspiracy-theory-psychologists-explain-241098

80 Stoknes, P.E., 2015. What We Think about When We Try Not To Think about Global Warming: Toward a New Psychology of Climate Action. Chelsea Green Publishing.

81 Vodovotz, Y., Arciero, J., Verschure, P.F.M.J. and Katz, D.A., 2024. 'Multiscale inflammatory map: Linking individual stress to societal dysfunction', Frontiers in Science, 1.

82 Marsh, N., Scheele, D., Gerhardt, H., Strang, S., Enax, L., Weber, B., Maier, W. and Hurlemann, R., 2015. 'The neuropeptide oxytocin induces a social altruism bias', Journal of Neuroscience, 35(47), pp. 15696–15701.

83 Brevers, D., Baeken, C., Maurage, P., et al., 2021. 'Brain mechanisms underlying prospective thinking of sustainable behaviours', Nature Sustainability, 4, pp. 433–439. Available at: https://doi.org/10.1038/s41893-020-00658-3

84 ILO, 2015. Guidelines for a Just Transition towards Environmentally Sustainable Economies and Societies for All. Available at: www.ilo.org/sites/default/files/wcmsp5/groups/public/%40ed_emp/%40emp_ent/documents/publication/wcms_432859.pdf

85 ILO, 2023. Achieving a Just Transition Towards Environmentally Sustainable Economies and Societies for All. Available at: www.ilo.org/sites/default/files/wcmsp5/groups/public/%40ed_norm/%40relconf/documents/meetingdocument/wcms_876568.pdf

86 Kenthley, 2024. 'The 2024 global market size & growth report – Barber shops', Kenthley Insights.

87 Winkelmann, R., Donges, J.F., Smith, E.K., Milkoreit, M., Eder, C., Heitzig, J., Katsanidou, A., Wiedermann, M., Wunderling, N. and Lenton, T.M., 2022. 'Social tipping processes towards climate action: A conceptual framework', Ecological Economics, 192. Available at: https://doi.org/10.1016/j.ecolecon.2021.107242

88 International Energy Agency, 2024. World Energy Investment. Available at: www.iea.org/reports/world-energy-investment-2024

89 Pasitka, L., Wissotsky, G., Ayyash, M., Yarza, N., Rosoff, G., Kaminker, R. and Nahmias, Y, 2024. 'Empirical economic analysis shows cost-effective continuous manufacturing of cultivated chicken using animal-free medium', Nature Food, 5, pp. 693–702. Available at: https://doi.org/10.1038/s43016-024-01022-w

90 Food and Land Use Coalition and University of Exeter, 2021. Positive Tipping Points for Food and Land Use Systems Transformation. Available at: www.foodandlanduse coalition.org/wp-content/uploads/2021/07/Positive-Tipping-Points-for-Food-and-Land-Use-Systems-Transformation.pdf

Part II

Transitioning to a safe and just future

5 The science of Nature Positive

Joseph Bull

If an alien scientist observed and categorized life on Earth, they may well note: "To a first approximation, all species are insects". This famous quote, attributed to the ecologist Robert May, captures brilliantly how important insects are to global biological diversity. They take myriad fantastic forms and – in both diversity and sheer biomass – are fundamental to the functioning of ecosystems. As a result, recent research which shows that there has been a dramatic decline in insect populations is sobering and has gathered widespread attention [1]. Examples include papers covering decades of changes in invertebrate abundance in Europe, which have revealed catastrophic declines during that time [2], neatly captured in the public imagination and by the press in discussions around the 'windscreen phenomenon' (the anecdotal observation that fewer insects are being killed on car windscreens than in the past).

This is a problem. It is exemplified by the billions of dollars that are spent transporting bees across the United States each year, in an effort to maintain pollination for numerous valuable crops [3]. But insects play a variety of roles beyond pollination; and so efforts to arrest the ongoing loss of insect populations and return them to their former glory are also an attempt to maintain the fundamental operating system underpinning the global economy.

And so, with insects, as is the case for much of wild nature, we need more than we currently have in the world today. But one of the challenges we have with setting global targets on nature is that we do not have a clear problem statement for nature in quite the same way that we do for, say, climate change. One the one hand, there are practical reasons to increase numbers of insects overall, but there are also plenty of insects which are problematic (such as agricultural pests, or those carrying disease).

But on the other hand, nature conservation is not just about practicalities, it is also about natural and cultural heritage; consequently, different natural features (e.g. mammals and birds) are often the focus of conservation research and practice [4]. Nonetheless, the world's governments have agreed to an overarching 2030 mission of 'halting and reversing' the decline in nature and beginning the long road to global nature recovery [5]. This is consistent with the idea of Nature Positive, which by definition can be considered to involve a mission

DOI: 10.4324/9781003474043-7

to "halt and reverse nature loss by 2030 on a 2020 baseline and achieve full recovery by 2050. In simpler terms, it means ensuring more nature in the world in 2030 than in 2020 and continued recovery after that" [6]. This overarching 'halt and reverse' goal is well-defined in technical terms in scientific literature, building upon many years of research concerning 'net outcomes' approaches to nature [7], as discussed in recent papers including that from my colleague at the University of Oxford, Professor Dame EJ Milner-Gulland. Importantly, the 'halt and reverse' goal does for nature what the headline Paris Agreement goals aim to do for climate change [8].

So – given that there are multiple reasons why we might want to conserve different aspects of wild nature – what do we include in the measurement of progress towards a nature-positive world? Trends in nature conservation are often measured in terms of 'biodiversity', that is, "the variability among living organisms from all sources . . . this includes diversity within species, between species and of ecosystems" [9]. There are hundreds of viable metrics that can be used to track changes in biodiversity, many of which are useful proxies and are complementary to each other, none of which can claim to be the most universal metric. Nature Positive in this sense could, for example, among many other things, require an overall net gain in the diversity, abundance, coverage, or functional roles of insect populations from where we are today – but that is ambitious, as it is a *long* way from where we are today.

Insects are not typically included in the most widely used biodiversity metrics – partly likely due to data and expertise gaps, although in the case of most species, they are likely not even taxonomically described. As an example, the International Union for Conservation of Nature Red List of Threatened Species [10], also known as the IUCN Red List, can be considered "the world's most comprehensive information source on the global extinction risk status of animal, fungus and plant species" [11]. Yet, despite 97% of all animals being invertebrates (like insects), they make up only 30% of all animal assessments on the Red List. The impacts of our current economic system on insect life, on which it depends, is symptomatic of much larger trends. Crucially, nature-positive goals – though requiring an absolute improvement in the state of nature – do not require that we stop all negative impacts on biodiversity completely; rather, they require that those impacts are avoided or minimised to the point at which any remaining negative impacts can be overcompensated for, through conservation gains. The application of a sequence of actions like this (i.e. avoid, minimise, restore, compensate) to biodiversity impacts is one form of what is more widely known as the 'mitigation hierarchy', and it is an absolutely foundational scientific and technical concept underpinning Nature Positive [12, 13].

It is important to be clear, then, about what is causing declines; for insects and everything else. Though drivers like climate change are becoming more and more important, the main cause is still the classic villain in nature conservation – the 'evil quartet' proposed by scientist and historian Jared Diamond in the 1980s

[14]. This evil quartet – loss of natural habitats, overharvesting, invasive species, and extinction chains – were claimed by Diamond to be the main drivers of species loss.

More recent research has found that these remain the primary drivers of extinction risk for many species, despite climate change becoming a more important driver of loss (which is likely to increase substantially over the coming decades) [15]. The journey towards Nature Positive is not solely about investment in conservation: first, and crucially, it is about quantifying the scale of our negative impacts on biodiversity (caused by the evil quartet, and other impacts), and finding a way to reduce those impacts as much as possible, before moving on to proactive investment in conservation [16]. To some extent, this will require (in some cases, though certainly not all) trade-offs to be made between economic activities and nature conservation [17].

This is one of the most important challenges we will have to deal with on the road to Nature Positive: are there some things that we will simply have to do less of? For instance, how much plastic can we make, how much metal can we mine, how much meat can we eat, if we want a nature-positive global economy? The answer is partly an objective scientific one [18] and is partly guided by subjective decision-making; particularly given that it is so closely linked to human well-being and associated trade-offs. But we do know that when it comes to nature there are some fundamental components of biodiversity that physically need to be retained; for example, the insects we need if we want our crops to be pollinated. And this is key: the nature we set aside and retain does not only represent a loss to the economy; it is also a contributor to the economy as an investment in natural capital.

The environmental challenges associated with humanity's current system are essentially associated with lifestyle: the size and type of production and consumption, and the amount of space and other natural resources we consequently demand. Nature Positive is about quantifying these patterns, seeking the greatest possible efficiencies in them, at times making difficult trade-offs, and ultimately moving towards sustainable and restorative use of natural systems. It is something we must achieve at a global scale; it is not necessarily the case that every country, company, or actor on any other scale must be Nature Positive [19] – but that is where we must land in the aggregate.

What can science tell us about the road to Nature Positive? As previously mentioned, nature (and even that focal subset of nature, biodiversity) is a broad concept for which a clear and precise problem statement does not exist as it does for other issues. This is not because we don't understand nature, nor because there are not longstanding truths about it – it is more that Nature Positive encompasses a multitude of social and environmental goals.

For instance, the 'planetary boundaries' framework clearly defines limits that should not be crossed from the perspective of the physical biosphere, including ecological boundaries [20].

PERSPECTIVE 5.1   Nature Positive and Planetary Boundaries

By Johan Rockström, Director of the Potsdam Institute
for Climate Impact Research

Human activities including deforestation, pollution, overexploitation of resources, and, increasingly, climate change, are causing unprecedented damage to nature. The Nature Positive goal was formulated in response to the worldwide accelerating decline of biodiversity and ecosystems and aims to safeguard and enhance the diverse ways in which nature supports human well-being and life on Earth in general. Becoming Nature Positive is a normative goal with the key objective of enhancing nature's ability to provide these critical contributions to people.

The Planetary Boundaries framework (PBs) – the latest update of which you can see in the figure below – does not directly address benefits that ecosystems

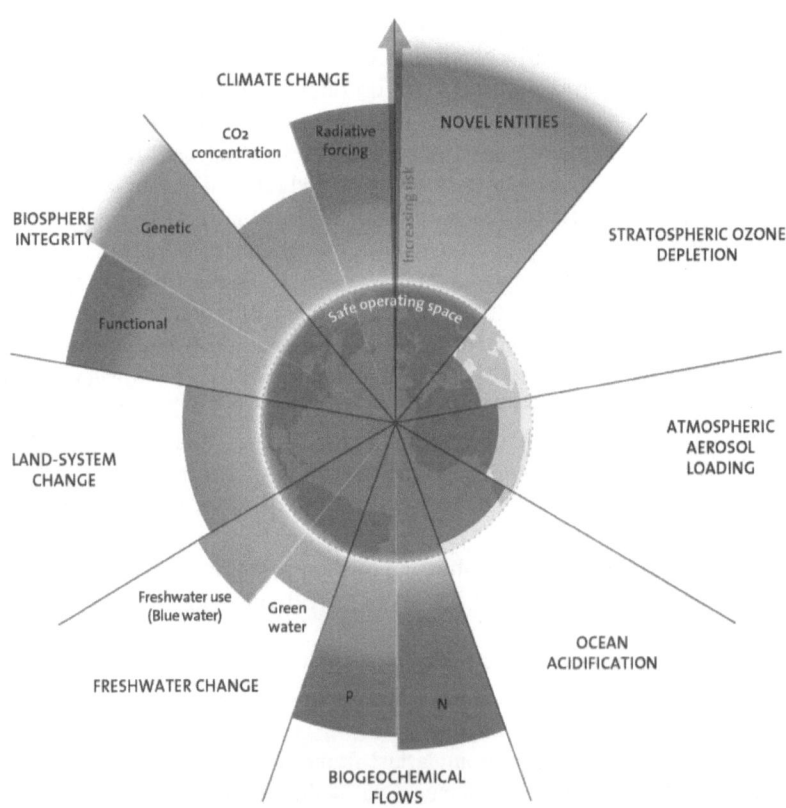

Figure 5.1 The 2023 update to the Planetary Boundaries. Licensed under CC-BY-NC-ND 3.0.

Source: Azote for Stockholm Resilience Centre, based on analysis in Richardson et al 2023 [21]

provide to humans but defines safe limits for nine critical Earth systems that regulate and maintain the stability and resilience of the entire planet. These nine boundaries collectively define a safe operating space for humanity to keep the Earth system in a Holocene-like condition which is the only state of the planet we know for certain can support the modern world as we know it (see below).

Nature Positive and the Planetary Boundaries framework share the common goal of restoring the Earth's ecological balance to ensure resilience of the planet's life-support systems. Earth system resilience refers to the ability of the Earth's interconnected systems, such as atmosphere, biosphere, and oceans, to absorb disturbances and recover from changes while maintaining their essential functions. A resilient planet can withstand climate change, pollution, and habitat loss without losing its capacity to support life and provide vital services such as water and food.

The relationship between the Planetary Boundaries framework and Nature Positive is synergistic. The Planetary Boundaries framework provides the scientific underpinning and global context, while Nature Positive offers a clear, action-oriented target for reversing biodiversity loss. Together, they form a powerful approach for addressing environmental challenges and ensuring a sustainable future for both nature and humanity. By combining these concepts, we can create a more holistic and effective strategy for environmental conservation and restoration, guiding human development within the safe operating space of our planet.

Both the Planetary Boundaries framework and Nature Positive emphasise the urgency of reducing human pressures on natural systems and encourage a paradigm shift in how economies, businesses, and societies value nature. In this way, respecting planetary boundaries supports achieving the Nature Positive goal, aligning ecological health with human development for long-term sustainability.

Meanwhile, proposed climate change mitigation scenarios typically require at least some nature-based solutions (such as carbon sequestration through habitat restoration) [22, 23], which will generally be aligned with efforts towards Nature Positive [24]. Indeed, the IPCC Assessments in recent years have captured how the use, conservation, and restoration of various ecosystems provide the largest options for net global emissions reductions, alongside those linked to energy.

Beyond practical issues, such as ecosystem functioning linked to carbon sequestration, Nature Positive is about safeguarding natural heritage, recognising the long-term role of Indigenous Peoples in the stewardship of nature [25], and needs to be compatible with many other goals (e.g. the UN Sustainable Development Goals). Science can instruct road maps for all these separate issues – the challenge is bringing them all together under one cohesive strategy.

PERSPECTIVE 5.2   The ecosystems we cannot afford to lose

By M. Sanjayan, CEO, Conservation International

Climate change and biodiversity loss are two sides of the same crisis. The science is clear: rising temperatures accelerate nature loss, while nature loss releases additional planet-warming carbon into the atmosphere. These problems must be addressed in tandem – and that starts with conserving the world's vital ecosystems.

All nature has value. But recent science suggests that some ecosystems are truly irreplaceable. Around the world – in places like the old-growth forests of North America, the rainforests of Amazonia, the peatlands of the Congo Basin, and the mangroves of Indonesia – there are incredibly vast reservoirs of sequestered carbon. We call this carbon "irrecoverable," because these high-biodiversity ecosystems are relatively slow-to-regenerate. If they are destroyed, the lost carbon cannot realistically be recaptured in time to avert the worst climate scenarios.

In 2022, a coalition of scientists led by Conservation International developed an atlas of the world's irrecoverable carbon. In total, 139 gigatonnes are stored away in pockets around the world, equivalent to 15 years of global fossil fuel emissions. If this carbon were to be released, it could send global temperatures soaring past the 2°C threshold.

Many of these ecosystems are already endangered by creeping agriculture, mining, and urban development. But fortunately, they are highly concentrated, making them an ideal target for conservation efforts. A mere 14% of the world's land contains 75% of all irrecoverable carbon – and 91% of terrestrial vertebrate species. These same ecosystems support hundreds of millions of livelihoods. Protecting these places at scale is a win-win-win.

In a world constrained by finite resources and capital, the science of irrecoverable carbon brings additional precision to the field of conservation, allowing us to do more with every dollar invested. With targeted efforts, we can make an enormous difference for climate, biodiversity, and human wellbeing – and minimise trade-offs as we all embark on the nature-positive transition.

Returning to those longstanding truths, including the well-established challenge to biodiversity of the evil quartet: mitigating these four impacts (as well as the rapidly growing threat of climate change) will be crucial to any nature-positive trajectory. Connected to this is the fact that, fundamentally, wild nature generally needs to have physical space and to be left alone, so one goal must be to set aside some space for the world's remaining wild places. Similarly, one of the more dependable approaches for making progress towards Nature Positive will always be to reduce impacts in the first place, for example in terms of levels of production and consumption [26–29].

Even though the negative impacts of economic activities on nature will likely never be reduced to zero, preventing them provides one of the more certain ways to achieve net outcome goals (in comparison to compensating for losses). On the other hand, sometimes longstanding 'truths' do get overturned. A classic example would be ecologist Garret Hardin's previously well-established concept of the 'Tragedy of the Commons' [30]. In categorising a public resource (such as a park) as commons, Hardin argues that when individuals access these, they will tend to deplete the resource in their own self-interest [31].

This economic theory was overturned by Elinor Ostrom in her Nobel Prize–winning empirical work, showing that people can, and do in some cases, work cooperatively to safeguard common pool natural resources [32]. Similarly, for a long time 'overpopulation' of humans was seen as the key driver of biodiversity loss; the contemporary view is more that it is overconsumption by some, rather than consumption by too many. The latter points start to inform what Nature Positive might look like from a more socio-economic perspective; although in doing so it will be necessary to draw upon an extensive literature into the plurality of views on nature and conservation [33], how priorities have changed through time [34], and how we want nature to look in the future [35].

Taking this into account, what are the questions that remain to be answered by science?

Foremost among these must be the challenge of how we even measure and define progress towards Nature Positive in general, and particularly in a private sector context; that is, what biodiversity metrics should we use? Again, progress towards nature-positive outcomes is typically tracked by some measurement of biodiversity, as it is a key aspect of nature. In fact, there are many ways to measure biodiversity, and no single universally 'correct' metric to use – rather, there are metrics that have more or less utility in different situations [36, 37].

Recently, a lot of progress has been made in applying the extensive life cycle impact assessment literature to quantitatively track net biodiversity outcomes across value chains in terms of wildlife species threat [38], although in doing so it will be crucial to account for uncertainties [39]. Alternatively, we have habitat based metrics; for example, in England, we measure the type, 'distinctiveness', and physical condition of different habitats as part of the 'Biodiversity Net Gain' policy (mandatory requirements on developers to deliver a net biodiversity gain of 10% in 'biodiversity' during projects, introduced in 2024). It is known that this doesn't capture all aspects of biodiversity; for instance, in terms of insect species diversity and abundance [40] (another good example of invertebrates being left out of metrics), and it does not even begin to consider value chain impacts on nature (where biodiversity impacts are often embedded) [41]. Nonetheless, it provides a means for constraining marginal losses of habitat in England alongside ongoing development activities and, potentially, some investment into local nature recovery networks.

We can measure biological diversity at the level of genes, species, ecosystems, natural processes, and measure each of those in many ways. Also, it is likely that (instead of selecting just one), we use a suite of indicators that capture different aspects of biodiversity to measure progress towards Nature Positive. One example of that might be the so-called 'Essential Biodiversity Variables' [42]. In any case, science can answer the question: which indicators are most effective to use to track a certain aspect of biodiversity that we value? Importantly, the lack of a universal metric (or at least ubiquitous ones, such as those we see analogously in climate policy) is not an excuse for inaction; and again, this is the value of Nature Positive. The target to achieve a net-positive outcome for nature is a 'science-based target' in the absence of a single metric or a clearly defined and constrained problem statement.

Another key question is about *strategic* options for Nature Positive. To what extent can an organisation, sector, or actor make progress toward Nature Positive through adjustments to existing models or extensions of 'business as usual', versus requiring a complete transformation of the business model? The challenge of dealing with impacts from mission critical activities to achieve positive net outcomes for biodiversity is known (for example if you are a university and some of your largest impacts are associated with the materials you procure to enable laboratory research) [43]. However, these will be different for different sectors and organisations. Similarly, operational pathways towards transformative change for business are beginning to be explored [44], but many technical questions for how to get there remain unanswered. Beyond broader issues of policy and intent, questions include those that are eminently concrete, such as the extent to which oil and gas companies should transition to renewable energy, or whether agriculture should shift toward lower meat and dairy production.

Equally, there is the issue of technical trade-offs in moving towards transformative change for businesses, sectors, and whole economies. If we transition towards primarily renewables-powered grid electricity as part of climate change mitigation efforts, what metals will we need to mine to get us there, and how will those mines overlap with high value extant biodiversity worldwide? [45] This is before we even begin to consider the degree to which social scientists need to be consulted, on strategic nature-positive interventions that might be sufficiently well received by the public, and lead to actual change in outcomes [46]. Multiple related questions – the mix of instruments to implement, governance and finance mechanisms, and appropriate framing – are a major component of the nature-positive research agenda going forward [47].

Principles for a Nature-Positive approach in the private sector

Adapted from The "nature-positive" journey for business: A conceptual research agenda to guide contributions to societal biodiversity goals [48].

In 2023, a group of my colleagues attempted to lay out principles governing the decisions businesses need to make to contribute to the nature-positive

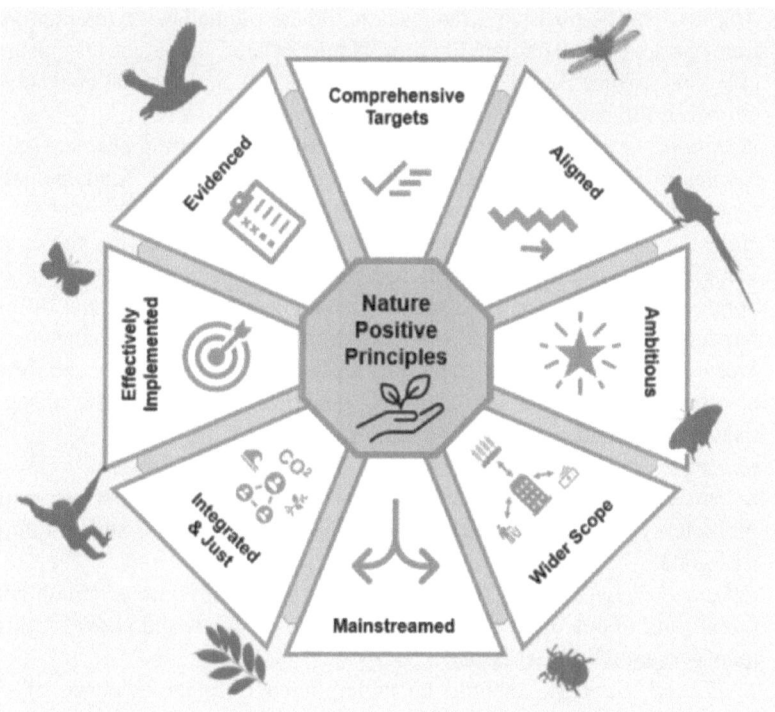

Figure 5.2 Principles for a nature-positive approach to biodiversity management in the private sector.

Source: The Nature Positive Journey for Business, reproduced from Booth et al. [48]

journey [49]. A figure summarising extracted from their findings can be found above.

When crafting a nature-positive strategy, it is suggested companies follow these principles. Their strategy, objectives, and targets should be:

• *Comprehensive*: Targets and goals should be SMART (Specific, Measurable, Achievable, Relevant, and Time-Bound), and developed in line with the 'Mitigation & Conservation Hierarchy' [51] (meaning that an overarching goal is set, with a timeline and a baseline; then scaled down to specific targets for different sectors, locations and actors; and where relevant implementers are used to support the planning of actions to implement targets, monitor outcomes, review and revise actions; and outcomes are integrated across scales, impact types, and actors, and progress towards the goal is assessed).

- *Aligned*: Nature-positive strategies should be aligned with international frameworks (e.g., The Taskforce on Nature-related Financial Disclosures (TNFD), Science Based Targets Network (SBTN), and societal goals (e.g., Global Biodiversity Framework).
- *Ambitious*: Commitments should be aspirational, delivering absolute gains for nature measured against a quantified, static baseline; fundamentally reducing material footprints; and pursuing transformation.
- *Have a wider scope*: Strategies should address impacts across the entire value chain – prioritising direct operations and upstream impacts, incorporate proportional contributions to address historic, indirect and diffuse impacts, and contribute to sector-wide efforts to drive systemic change.
- *Mainstreamed*: Nature should be fully embedded in all processes and forms of organisational decision-making via governance, strategy, risk management and measurement.
- *Integrated and just*: Approaches should be integrated across all relevant dimensions of natural and social systems to ensure socially just outcomes, promote synergies with climate and other societal goals, and minimise trade-offs.
- *Effectively implemented*: Actions should be effectively implemented, with monitoring of outcomes to track and disclose progress, and allow adaptive management where necessary.
- *Evidenced*: Strategies should be underpinned by clear evidence of the impacts and costs of proposed actions, including an analysis of how individual actions will add up to overall net gain for nature.

You can read more about the Nature Positive agenda of companies in Chapter 8 of this book, which is dedicated entirely to this topic.

A third set of questions relate to the effectiveness of nature-positive interventions. Nature conservation is more generally thought to typically be effective [52], so we are clearly not doing enough of it. We might ask, how effective are the tools to prevent and mitigate biodiversity loss, which as discussed is the foundation for ultimately moving towards achieving a nature-positive world? Biodiversity offsets, for instance – the final stage of the aforementioned mitigation hierarchy – though a substantial source of financial flows into biodiversity [53], and which have been shown to be successful in some cases [54], are certainly not always effective in compensating for ecological impacts [55]. Offsets in general (particularly voluntary carbon offsets which are different to biodiversity offsets but share some conceptual similarities) have come under heavy fire as a concept in recent years, after weaknesses in some systems were exposed in high-profile cases. This was not just due to perceived fallibilities in biodiversity offsetting itself, and from an ecological perspective, but is also about whether offsetting sufficiently considers people and social values related to nature in its implementation. Making offsets (and indeed any other mechanisms for biodiversity impact mitigation or conservation) not only effective but also equitable and just is an entire field of study [56]. There are a whole host of other mechanisms beyond impact mitigation and offsetting that might help us towards a nature-positive world, such as biodiversity credit markets [57], corporate action on nature [58, 59], and other emerging mechanisms, if they can deliver [60]. In some respects, the question underpinning all of this is 'what do we want a nature-positive economy to look like?' But once we know that, science can provide us with the technical details of how we get there, and how we measure and monitor progress towards it.

The well-defined concept of Nature Positive is fully aligned with the international principle – agreed by most nations on Earth – to seek to 'halt and reverse' biodiversity loss in absolute terms. It is our 'Paris Agreement' for nature, and one of the few truly science-based targets we can set for nature overall. The need to achieve this goal is clear. There are scientific questions outstanding, and real technical progress is being made towards these, but many of these are technical details. To return to insects: though there are so many species we have yet to describe, we know just how important they are, and how dramatically they have declined. We know about the evil quartet, and how that continues to capture the key drivers of biodiversity loss. We increasingly understand the mechanics of climate change, and the implications it will have for ecosystems. Our traditional conservation interventions, overall, are effective across regions and taxonomic groups, and there is a growing body of research on emerging conservation interventions that could bring in considerably more funding to support conservation. The major challenges now are less about the fundamental science, and more about the issues associated with people, governance, compliance, monitoring, and finance. We know what Nature Positive is, technically, and the general consensus is that we need to get there somehow – the question is, how do we do so socially and economically?

ADDITIONAL PERSPECTIVES

PERSPECTIVE 5.3   The role of Earth observations in supporting Nature-Positive outcomes: the Global Ecosystems Atlas

By Yana Gevorgyan, Director of Secretariat,
Global Earth Observation Network

Earth observation (EO) technologies are at the forefront of science-driven conservation efforts, providing essential data for achieving global nature-positive goals. These technologies enable continuous, large-scale monitoring of the Earth's ecosystems, offering critical insights into ecosystem extent, condition, and changes over time. As global commitments to halt and reverse biodiversity loss grow stronger, the demand for accurate, up-to-date information on ecosystems has never been greater.

Despite the clear need for comprehensive ecosystem data, significant gaps remain in our understanding of global ecosystem distribution and condition.

Currently, over half of the world's ecosystems are inadequately mapped, making it challenging to implement effective conservation strategies and measure progress towards nature-positive outcomes. This is where the Global Ecosystems Atlas, a groundbreaking initiative convened by the Group on Earth Observations (GEO), comes into play.

The Global Ecosystems Atlas ('the Atlas') addresses the *existential* issue of nature loss by leveraging *exponential* capabilities through the integration of advanced EO technologies, artificial intelligence (AI), ecological field data, and local knowledge. By combining these tools with existing high-quality ecosystem data and maps, the Atlas will provide an *essential* resource for identifying and filling critical data gaps. The Atlas will align all data to the IUCN Global Ecosystem Typology,[1] ensuring comparability across scales and regions and offering a unified framework for monitoring, reporting, and verifying conservation and restoration goals globally.

By providing harmonised and consistent data on ecosystems, the Atlas will support net-zero and nature-positive agendas by enabling more accurate measurement, management, and reporting of ecosystem health and services. This robust foundation of data will not only facilitate strategic decision-making for conservation but also enhance the integrity of carbon and biodiversity credit markets. Through this comprehensive approach, the Atlas will help dive transformative change in how we manage and protect the planet's natural resources, aligning investments and policies with sustainable development goals.

Through the use of cutting-edge technologies, integration of existing data, and a collaborative, multi-stakeholder approach, the Global Ecosystems Atlas exemplifies how EO can be harnessed to support the Nature Positive goal, enabling all sectors of society to make informed decisions that promote a resilient and sustainable future for our planet.

Note

1 The IUCN Global Ecosystem Typology has been endorsed by the UN Statistical Commission as an international statistical classification and has been recommended as the basis for ecosystem indicators by the CBD. Similarly, Taskforce on Nature-related Financial Disclosures (TNFD), has integrated the Global Ecosystem Typology into guidance on the identification and assessment of nature-related issues: the LEAP (Locate, Evaluate, Assess and Prepare) approach.

References

1 Wagner, D.L., et al., 2021. 'Insect decline in the Anthropocene: Death by a thousand cuts', PNAS, 118(2), p. e2023989118.
2. Hallmann, C.A., et al., 2017. 'More than 75 percent decline over 27 years in total flying insect biomass in protected areas', PLoS One. Available at: https://doi.org/10.1371/journal.pone.0185809

3 USDA, 2024. Thousands of Commercial Honey Bee Colonies Are Transported Long Distances to Pollinate California Almonds'. Available at: www.ers.usda.gov/data-products/chart-gallery/gallery/chart-detail/?chartId=107088

4 Troudet, J., et al., 2017. 'Taxonomic bias in biodiversity data and societal preferences', Scientific Reports, 7(9132).

5 Convention on Biological Diversity, 2022. Available at: www.cbd.int/gbf

6 Nature Positive Initiative, 2024. Available at: www.naturepositive.org/

7 Milner-Gulland, E.J., et al., 2021. 'Four steps for the Earth: Mainstreaming the post-2020 global biodiversity framework', One Earth, 4(1).

8 Bull, J.W., et al., 2020. 'Net positive outcomes for nature', Nature Ecology & Evolution, 4(1).

9 Convention on Biological Diversity, 2022. Available at: www.cbd.int/gbf

10 IUCN, n.d. The IUCN Red List of Threatened Species. Available at: www.iucnredlist.org/

11 IUCN Red List, n.d. Background & History. Available at: www.iucnredlist.org/about/background-history

12 Milner-Gulland, E.J., et al., 2021. 'Four steps for the Earth: Mainstreaming the post-2020 global biodiversity framework.' One Earth, 4(1).

13 Maron, M., et al., 2024. 'Nature positive' must incorporate, not undermine, the mitigation hierarchy.' Nature Ecology & Evolution, 8(1).

14 Diamond, J.M., 1984. '"Normal" extinction of isolated populations', In M.H. Nitecki (ed.), Extinctions, pp. 191–246. Chicago, USA: Chicago University Press.

15 Maxwell, S.L., et al., 2016. 'Biodiversity: The ravages of guns, nets and bulldozers', Nature, 536.

16 Leclère, D., et al., 2020. 'Bending the curve of terrestrial biodiversity needs an integrated strategy.' Available at: Nature 585, 551–556. https://doi.org/10.1038/s41586-020-2705-y.

17 Bull, J.W., et al., 2022. 'Analysis: The biodiversity footprint of the University of Oxford', Nature, 604(7906).

18 Sonter, L.J., et al., 2020. 'Renewable energy production will exacerbate mining threats to biodiversity', Nature Communications, 11.

19 Maron, M., et al., 2020. 'Global no net loss of natural ecosystems.' Nature Ecology & Evolution, 4(1), pp. 46–49.

20 Rockstrom, J., et al., 2009. 'Planetary boundaries: Exploring the safe operating space for humanity', Ecology and Society, 14(2), 32.

21 Richardson, K., Rockström, J., Bai, X., Donohue, I., Galaz, V., Gudmundsson, L., Leach, M., Liverman, D., Steffen, W. and Winkelmann, R., 2023. 'Earth beyond six of nine planetary boundaries', Science Advances, 9, p. eadh2458. Available at: https://doi.org/10.1126/sciadv.adh2458

22 Locke, H., et al., 2021. A Nature-Positive World: The Global Goal for Nature. Available at: https://www.nature.org/content/dam/tnc/nature/en/documents/NaturePositive_GlobalGoalCEO.pdf

23 Seddon, N., 2022. 'Harnessing the potential of nature-based solutions for mitigating and adapting to climate change', Science, 376(6600).

24 Bull, J.W., et al., 2020. 'Net positive outcomes for nature', Nature Ecology & Evolution, 4(1).

25 Dawson, N. M., et al., 2021. 'The role of Indigenous peoples and local communities in effective and equitable conservation.' Ecology and Society, 26(3), 19. Available at: https://doi.org/10.5751/ES-12625-260319

26 Phalan, B., et al., 2018. 'Avoiding impacts on biodiversity through strengthening the first stage of the mitigation hierarchy', Oryx.

27 Leclère, D., et al., 2020. 'Bending the curve of terrestrial biodiversity needs an integrated strategy.' Available at: Nature 585, 551–556. https://doi.org/10.1038/s41586-020-2705-y.

28 Maron, M., et al., 2024. 'Nature positive' must incorporate, not undermine, the mitigation hierarchy', Nature Ecology & Evolution, 8(1).

29 Leclère, D., et al., 2020. 'Bending the curve of terrestrial biodiversity needs an integrated strategy.' Available at: Nature 585, 551–556. https://doi.org/10.1038/s41586-020-2705-y.

30 Hardin, G., 1968. 'The tragedy of the commons', Science, 162, pp. 1243–1248.

31 Hardin, G., 1968. 'The tragedy of the commons', Science, 162(3859), pp. 1243–1248. Available at: https://doi.org/10.1126/science.162.3859.1243

32 Ostrom, E., 2009. 'A general framework for analyzing sustainability of social-ecological systems', Science, 325(5939).

33 Pascual, U., et al., 2023. 'Diverse values of nature for sustainability', Nature, 620, pp. 813–823.

34 Mace, G.M., 2014. 'Whose conservation?', Science, 345(6204).

35 Pereira, L.M., Davies, K.K., den Belder, E., Ferrier, S., Karlsson-Vinkhuyzen, S., Kim, H., et al., 2020. 'Developing multiscale and integrative nature–people scenarios using the Nature Futures Framework', People and Nature, 2(4), pp. 1172–1195. Available at: https://doi.org/10.1002/pan3.10146

36 Addison, P.F.E., et al., 2020. 'Bringing sustainability to life: A framework to guide biodiversity indicator development for business performance management', Business Strategy and the Environment, 29(8), pp. 3303–3313.

37 Locke, H., et al., 2021. A Nature-Positive World: The Global Goal for Nature. Available at: https://www.nature.org/content/dam/tnc/nature/en/documents/NaturePositive_GlobalGoalCEO.pdf

38 Bull, J.W., et al., 2022. 'Analysis: The biodiversity footprint of the University of Oxford', Nature, 604(7906).

39 Bromwich, T., White, T. B., et al., 2025. Navigating uncertainty in life cycle assessment-based approaches to biodiversity footprinting. Methods in Ecology and Evolution, 00, 1–18.

40 Duffus, N., et al., in prep. Metrics based on Habitat Area and Condition as Proxies for Invertebrate Biodiversity. Available at: https://www.biorxiv.org/content/10.1101/2024.10.02.616290v1

41 Zu Ermgassen, S.O., Howard, M., Bennun, L., Addison, P.F., Bull, J.W., Loveridge, R., et al., 2022. 'Are corporate biodiversity commitments consistent with delivering "nature-positive" outcomes? A review of "nature-positive" definitions, company progress and challenges', Journal of Cleaner Production, 379, p. 134798. Available at: https://doi.org/10.1016/j.jclepro.2022.134798

42 Pereira, H.M., et al., 2013. 'Essential biodiversity variables', Science, 339(6117).

43 Bull, J.W., et al., 2022. 'Analysis: The biodiversity footprint of the University of Oxford', Nature, 604(7906).

44 Booth, H., et al., 2024. 'Operationalizing transformative change for business in the context of Nature Positive', One Earth, 7(7), pp. 1235–1249.

45 Sonter, L.J., et al., 2020. 'Renewable energy production will exacerbate mining threats to biodiversity', Nature Communications, 11.

46 Taylor, I., et al., 2023. 'Nature-positive goals for an organization's food consumption', Nature Food, 4(1), pp. 96–108.

47 White, T.B., et al., 2024. 'The "nature-positive" journey for business: A conceptual research agenda to guide contributions to societal biodiversity goals', One Earth, 7(8). Available at: https://www.cell.com/one-earth/fulltext/S2590-3322(24)00328-2

48 Booth, H., et al., 2023. 'Operationalizing transformative change for business in the context of nature positive', One Earth, 7(7), pp. 1235–1249. Available at: https://doi.org/10.1016/j.oneear.2023.xxxxxx.

49 Booth, H., et al., 2023. 'Operationalizing transformative change for business in the context of nature positive', One Earth, 7(7), pp. 1235–1249. Available at: https://doi.org/10.1016/j.oneear.2023.xxxxxx.

50 Booth, H., et al., 2023. 'Operationalizing transformative change for business in the context of nature positive', One Earth, 7(7), pp. 1235–1249. Available at: https://doi.org/10.1016/j.oneear.2023.xxxxxx.

51 Interdisciplinary Centre for Conservation Science, undated. The Mitigation and Conservation Hierarchy. Available at: https://iccs.org.uk/the-mitigation-conservation-hierarchy/ (Accessed: 6 November 2024).

52 Langhammer, P.F., et al., 2024. 'The positive impact of conservation action', Science, 384(6694), pp. 453–458.

53 Deutz, A., et al., 2020. 'Financing nature: Closing the global biodiversity financing gap', The Paulson Institute, The Nature Conservancy, and the Cornell Atkinson Center for Sustainability.

54 Devenish, K., et al., 2022. 'On track to achieve no net loss of forest at Madagascar's biggest mine', Nature Sustainability, 5, pp. 498–508.

55 zu Ermgassen, S.O.S.E., et al., 2019. 'The ecological outcomes of biodiversity offsets under "no net loss" policies: A global review', Conservation Letters, 12(6), p. e12664.

56 Griffiths, V.F., et al., 2019. 'No net loss for people and biodiversity', Conservation Biology, 33(1), pp. 76–87.

57 Wunder, S., et al., 2024. Biodiversity Credits: Learning Lessons from Other Approaches to Incentivize Conservation. Available at: chrome-extension://efaidnbmnnnibpcajpcglcl efindmkaj/https://osf.io/preprints/osf/qgwfc_v1

58 zu Ermgassen, S.O.S.E., et al., 2022. 'Are corporate biodiversity commitments consistent with delivering "nature-positive" outcomes? A review of "nature-positive" definitions, company progress and challenges', Journal of Cleaner Production, 379, p. 134798.

59 Lamont, T.A.C., et al., 2023. 'Hold big business to task on ecosystem restoration', Science, 381(6662).

60 Lovqvist, S., et al., 2023. 'Incentives and barriers to private finance for forest and landscape restoration', Nature Ecology & Evolution, 7, pp. 707–715.

6 Building a Nature-Positive society

Harvey Locke and Leroy Little Bear with Éliane Ubalijoro, Brigitte Baptiste and Fuwen Wei

Introduction

A Nature Positive society is one in which humanity lives in harmony with nature. It is a worthy dream for the troubled twenty-first century. There are glimmers of hope that this dream could come true.

We begin with a story of halting nature loss in what might seem an unlikely place. The Province of Alberta, Canada, is home to the oil sands (also known as the tar sands), which are so big they have made Canada the world's fourth largest oil producer. The province, which is home to almost five million people, is one of the world's wealthiest jurisdictions, with per capita incomes in league with those of the richest OECD countries [1].

Alberta is also home to the Canadian Rockies, one of the world's most iconic mountain landscapes, including world-famous Banff and Jasper National Parks and adjacent provincial parks. But most of the mountains are under provincial jurisdiction without any formal legal protection. Known as the Eastern Slopes, these lands are also very beautiful and much loved by the public. The provincial lands and the national parks are the headwaters of the rivers that flow across the Canadian prairie to Hudson Bay, watering vast ranchlands, one of the world's most important cereal cropping regions, and several cities.

The Eastern Slopes also contain large accessible deposits of metallurgical coal that can be mined for making steel. Nearby, in British Columbia, some of the world's largest open pit coal mines are generating a great deal of revenue for that province from the same geology. They also have a very large land use impact and have polluted the river downstream from them. However, in Alberta, for over 40 years there was a policy in place against open pit coal mining in the Alberta Rockies due to their environmental values [2]. Some within the government and the mining sector were waiting for an opportunity to change that. Leases were acquired. Arrangements were made.

Then the COVID-19 pandemic hit. People were very worried about the economy. In April 2020 the price of oil in some places was less than zero [3]. But metallurgical coal remained valuable on the global market [4]. Here was the moment the coal industry and the government had been waiting for, so in May 2020 the Alberta government rescinded the policy restrictions on coal mining in the Eastern Slopes [5].

DOI: 10.4324/9781003474043-8

To their great surprise the public reacted angrily, with 70% of Albertans opposed across all parts of the political spectrum [6]. Albertans valued the Rockies for their natural beauty, recreation, wildlife and as a source of freshwater. Recession or no recession, these values were more important to Alberta society than money from damaging resource extraction that put them at risk. After intense and enduring expressions of public outrage, the government had to reverse the decision and started to apologise for their mistake [7].

The Eastern Slopes story shows that awareness of environmental benefits and costs can shift a society towards Nature Positive decision-making. Albertans have a Nature Positive view of their beloved mountains. But four years later, the Alberta government tried again to open the mountains to coal mining, arguing coal development will create jobs and bring new investment [8]. Predictably, this reinstatement of coal mining provoked another angry response [9]. While the resolution of this issue is unknown at the time of writing, the saga of coal mining in the headwaters of Alberta's Eastern Slopes demonstrates how hard it is for us to make enduring Nature Positive decisions with our current mindset.

This is largely the case everywhere. There are some special natural places set aside, but mostly our economic activities come first; nature is a distant secondary concern, a 'nice to have' rather than a 'must preserve'. The result is a few happy stories among an alarming general trend that is causing a massive and dangerous decline in nature.

6.1   How our world became nature negative

How we see the world is the fundamental driver of the condition of nature. In the story that opened this chapter, we talked about how Albertans view the Eastern Slopes of the Rockies as special, associating the mountains with who they are, their place in the world and the way they want to live their lives. Yet even when they reject destructive coal mining and embrace conservation, their government tries again, because like most human institutions in the twenty-first century it sees nature as a commodity for us to exploit indefinitely. We have therefore collectively overexploited nature in ways that have dangerously undermined the stability of the Earth system we all depend on.

Not only have our attitudes become exploitative, but massive population growth, driven by advances like antibiotics and fertilisers, has caused the human population to increase sevenfold in just 125 years. This population growth, combined with rampant consumerism, is leading the world into uncharted territory, with dangerous tipping points. A striking reminder of this is the fact that global biomass is now dominated by humans and the animals we have domesticated. The proportion of domestic animal mass to wild animal mass in 1900 was three to one; by 2000 it was 25 to one. During the same period, human biomass quadrupled [10]. And as of 2020, all human-built structures and products outweighed all the living things in the world – forests and other plants included [11].

Western science makes it clear that we will continue to threaten our own existence unless we deliberately shift course to halt and reverse nature loss to restore the stability of the Earth system [12]. We need to change our perception of the economy and human well-being as competing interests. Instead, we should adopt a more holistic view, recognising the environment as the foundation for all life, with human society as part of it, and the economy existing solely to serve human society [12]. We need to think of ourselves as being in a relationship with the world's natural processes, species and ecosystems, and not view ourselves as their master with an unbridled right to disregard or exploit them. Western science tells us this is an urgent imperative [12]. We need a Nature Positive shift now. The good news is that some positive changes are underway that will help build the momentum we need to get us there.

6.2   We are positioned for a global transition

There are several encouraging signs that a Nature Positive shift is possible. In December 2022, 196 country parties to the Convention on Biological Diversity (CBD) agreed to halt and reverse biodiversity loss by 2030 and to live in harmony with nature by 2050 through the Kunming-Montreal Global Biodiversity Framework (the GBF) [13]. Then, at the UNFCCC Climate COP, 198 countries agreed in the Dubai Stocktake Agreement that we have to protect, conserve and wisely use nature in line with the GBF if we are to meet the Paris Agreement goals of limiting climate change-induced temperature rises [14].

For many years, Western medicine has been focused on the individual. Now it tells us that we must think of our health in the context of nature's health. The World Health Organization (WHO) has developed the concept of 'One Health', which recognises that the health of humans is inextricably bound to the health of other species and the planet. In 2023, a large group of medical journals called on the world to recognise the concept of One Health and for a shift whereby human health, climate change and biodiversity conservation are seen as aspects of one common problem that require joint action [15].

Many countries recognise that humans have a right to a healthy environment. Faith communities have also raised the moral imperative behind nature protection and climate action. For example, in 2015 Pope Francis issued the encyclical letter *Laudato Si*: *On Care for Our Common Home* [16], in which he called for an ecological conversion, noting that we have wrongly "come to see ourselves as the Earth's lords and masters, entitled to plunder her at will." He called this a violence present in our wounded hearts, reflected in the symptoms of sickness evident in the soil, water and air and in all forms of life.

Patriarch Bartholomew of the Greek Orthodox Church has similarly spoken out, stating:

> For human beings . . . to destroy the biological diversity of God's creation;
> for human beings to degrade the integrity of the Earth by causing changes
> in its climate, by stripping the Earth of its natural forests or destroying its

wetlands; for human beings to contaminate the Earth's waters, its land, its air, and its life – these are sins [17].

To commit a crime against the natural world is a sin against ourselves and a sin against God, he added.

The United Kingdom's King Charles III, who is also the head of the Church of England, called for the world to be Nature Positive and to achieve net-zero carbon goals when he spoke at the opening of COP 28 in Dubai, UAE, in December 2023. And before he became King, he created the Sustainable Markets Initiative (SMI) to engage business and encourage companies to do the same. SMI is informed by a Terra Carta which seeks to put "Nature, People and Planet at the heart of global value creation – one that will harness the precious, irreplaceable power of Nature combined with the transformative innovation and resources of the private sector" [18].

Many enlightened private sector participants have engaged with SMI or other similar initiatives. Bolstered by Target 15 of the GBF, the Taskforce on Nature-related Financial Disclosures (TNFD) calls on companies to report their impacts and dependencies on nature [19]. Over 100 central banks, meanwhile, have grouped together to create the Network for Greening the Financial System [20].

While these threads of hope in the Western world have not yet been woven into a Nature Positive tapestry, the potential is there. But the world is more than that. Cultures vary widely. One is not superior to another. A search for shared beliefs and commonalities across cultures reveals sufficient alignment with the Nature Positive concept, offering a common vision that can inspire all of humanity.

6.3 The societal shift we know we need to make

The Nature Positive shift will only be a valid Global Goal if it aligns with other ways of seeing the natural world. In providing advice during the GBF process, the Intergovernmental Science-Policy Platform on Biodiversity and Ecosystem Services (IPBES) noted that perceptions of "human-nature relationships vary across cultures and knowledge systems due to differing worldviews and cosmologies" [21]. It highlighted the importance of a shared conceptual framework to align various knowledge systems, acknowledging that nature encompasses diverse categories and holistic concepts across different cultures worldwide. Nature Positive could serve as that unifying framework.

Traditional value systems see the world differently to the consumption-driven West. They recognise the responsibility they have towards the world around them, and that their health is tightly bound to the health of the environment. A stand-out example of a culture that brought its traditions forward into the twenty-first century is the Himalayan Kingdom of Bhutan, whose constitutional goal is 'Gross National Happiness' (GNH). Instead of measuring GDP, Bhutan measures GNH [22], using nine categories: psychological well-being; health; education; time use; cultural diversity and resilience; good governance; community vitality; ecological diversity and resilience; and living standards. Its constitution includes a requirement to always maintain 60% forest cover, and the country has designated over half its land to a system of protected areas that are interconnected with ecological corridors. Bhutan is a net carbon sink rather than a source of CO_2 and has therefore already surpassed the Global Goal of achieving carbon neutrality by 2050.

Up to now, secular, Enlightenment-based Western science has been the primary foundation of global environmental policy, rather than the traditional values of other cultures. However, Western Earth systems science now calls for a shift in perception, recognising that humanity is embedded within nature, not separate from it, and that we are entirely dependent on the Earth's systems to maintain conditions suitable for human life – aligning it more closely with traditional worldviews [12].

This alignment raises the possibility of a unifying vision, supported by both modern science and the revival of culturally diverse Nature Positive values. We can consider this possibility by examining traditional East Asian (Chinese), South Asian (Indian), African (Bantu) and Indigenous thought. The following simplified discussion of complex worldviews and traditions shows that Nature Positive, with its hierarchy of humans embedded in nature and its focus on the flow of abiotic and biotic processes across species, ecosystems and natural systems, aligns with other thought traditions, even if reached through different methods. (The authors acknowledge that there are other Asian traditions including Jainism, Buddhism, Sikhism and Shinto which provide rich insights into the relationship between humans and nature, but we do not discuss these here. There are, of course, other important religions and traditions in other parts of the world.)

The following discussion of four complex worldviews and traditions is intended to highlight what has been called a "multi-perspectival cross section of the symbolic richness regarding attitudes towards nature within the world's religious traditions," [23] and to demonstrate that there are commonalities with Nature Positive.

Traditional Chinese cosmology is based on creative energy. In Taoism, the Tao begets the reason for being (the One), the One begets the Two (yin and yang), the Two beget the Three (Heaven, Earth and Humans/yin, yang and qi) and the Three beget the Ten Thousand (everything). Heaven and Earth were born with me, and the 10,000 things and I are one. Humans are part of this creative process which moves ceaselessly through the flux of cyclical time, and back and forth between opposites (*yin* and *yang*) and through the five phases of Metal, Wood, Water, Fire and Earth. These phases apply to the human body as well as the seasons of the year and the cycles of the moon [24].

At the landscape scale this energy is reflected in the concept of Shan Shui (mountain and water), which refers to the flow of energy across the landscape from the mountains to the ocean. At the household level, this orientation towards energetic flows is expressed as Feng Shui (wind and water) and the flow of energy through the home. In terms of human health, Chinese medicine is based on energy flows, with treatments touching one part of the body to affect another. This traditional view of the unity of nature and humanity [25] underpins the constitutionally adopted idea of 'eco-civilization', which involves living within and respecting the laws of nature as expressed by President Xi Jinping [26]. This idea was also reflected in China's presidency of CBD COP15, which chose the theme 'Eco-civilization – creating a shared future for all life on Earth'.

Hinduism is a major world religion, centred in South Asia, which also gives a prominent role to the flow of energy. The three faces of the divine (the Trimurthhi) are each in charge of one aspect of the cycle of creation, with Brahma as creator, Vishnu as preserver and Shiva as destroyer [27]. Hindus believe that the Universe is in an eternal cycle of creation and destruction [28]. Vishnu is the preserver, protector, sustainer and guardian of the Earth (patterns). Shiva embodies the energy that both destroys and regenerates, symbolising the Universe's energetic potential to undergo continuous cycles of creation, life, destruction and regeneration (process). This divine energy exists within celestial bodies, human beings, all other living beings and 'inanimate' objects (which are considered animate) [28]. This has much in common with the theory of evolution and quantum physics [29, 30]. Humans and nature are so intertwined that the most prominent household god, Ganesha, son of Shiva, is half elephant and half human, and there are thousands of other animal gods like this. Vishnu's first four incarnations on earth involved fusion of humans and animals, and the reciprocal relationship between tigers and the health of forests is discussed in the sacred text of the Mahabharata [31].

In Bantu African philosophy, the concept of Ubuntu, meaning 'I am because you are', expresses the mutuality and interdependence of all humans and all living beings. The dignity of a refined person is defined by a triple relationship: the first with God, spirits and ancestors who never die; the second with the self and all other people; and the third with the natural world. All beings are forces and are in continuous motion ordered by Hantu, the concept of space and time, that situates spatial and temporal phenomena and every event and motion [32]. This had much in common with space-time thinking in Western physics [33]. Traditional African communities often identify with a particular animal or plant that symbolises their clan's identity, reinforcing their deep reverence for nature, of which they see themselves as an integral part. Similar concepts to Ubuntu are found in Yoruba philosophy with 'Iwa', which represents character, ethical living and alignment with the natural order, and in Akan culture with 'Suban,' which refers to a person's moral character and conduct, emphasising virtues that maintain harmony within society and with the environment [29].

While Indigenous Peoples are widespread and speak many languages, one thing they all have in common is their worldview regarding humanity's relationship with the natural world [34, 35]. The distinction between Indigenous and non-Indigenous can be stark in some regions and more subtle in others, and even in societies with a clear colonial history, it is not always straightforward. However, self-identified Indigenous Peoples are widespread across the globe and share significant commonalities in their worldview.

Indigenous or Aboriginal philosophy believes that there is a constant flux of energy waves, sometimes called spirit, that gives rise to creation in its present but inherently unstable form. Everything is animate, including rocks, the Earth, the Sun, the Moon, animals and people. Since all of them are animated by the same spirit, there is no distinction between abiotic and biotic elements [36, 37]. The land is foundational to Indigenous Peoples' self-identity, but it too is seen as dynamic. It is the most stable manifestation of the flux, changing over longer periods of time than other aspects of the dynamic world. This worldview has much in common with quantum physics [38, 39] and geological theories such as plate tectonics and the rock cycle.

Meanwhile, the land is regarded as the Mother, with people seen as active participants within nature, rather than separate from it. This gives rise to the concept of 'all my relations', because humans are related to everything [36]. For example, the 50 sovereign First Nations and tribes of the Great Plains of North America that are signatories to the Buffalo Treaty consider the presence of Plains bison (*Bison bison*) critical to a holistic way of life. Plains bison have a name in each of their languages, but they all refer to them as Buffalo in English. To them, Buffalo are a family member and a teacher, a source of food and shelter, central to religious practice and the health of both people and the land. Their presence and survival are vital to both culture and nature, which are indivisible [40].

For Indigenous Peoples, when humans participate as embedded ecological actors on land or ocean, the aim is to maintain balance and harmony. The relationship is one of responsibility to all my relations, not a right to exploit. 'All my relations' have an interest in the land because it does not belong exclusively to people [36].

Elsewhere, the Māori people and the New Zealand government have entered into a deed of settlement that creates a new legal framework for the Whanganui River. The agreement is centred on the legal recognition of the river, from the mountains to the sea, and incorporates its tributaries and all its physical and metaphysical elements as an indivisible and living whole: Te Awa Tupua [41]. This also confers personhood on the river [42, 43].

In keeping with these cultural views on the human relationship with nature, the Nature Positive goal involves a conceptual shift away from seeing human development as something that competes with nature. Instead, it emphasises paying attention to the flow of natural processes over time and across different scales. Thus, Nature Positive is not only conceptually aligned with other traditions and knowledge systems but moves a Western, science-based goal into line with the goals of other traditions and their understanding of the proper, normative relationship of belonging within nature's processes. Like those traditions, Nature Positive involves shifting mindsets from rights to responsibilities, driven by the need for personal survival, intergenerational concern and respect for all forms of life.

6.4   Unresolved questions

To be a Nature Positive society we need to fix three big things: the market failure that makes nature worth more dead than alive; the need for equitable sharing and a personal understanding of 'how much is enough', which currently fuels consumerism; and diminishing confidence in domestic governments and global agreements as effective vehicles for collective action.

Under our current global economic system, waste is an externality and nature itself is viewed as a static natural resource that has been put there for humans to exploit, rather than as life forms and dynamic processes worthy of our care, attention and affection. Because of this, nature is worth more dead than alive: for the most part an intact wild forest is worth nothing until it has been reduced to lumber, a fish in the ocean is worth nothing until it has been caught, water running in a wild river is worth nothing until it has been dammed and its energy taken away. Even ecosystem service methodologies which seek to attach a value to nature's many benefits tend to discount nature that does not serve people – nature for its own sake is not an ecosystem service. Yet it is the functions of nature that drive the health of the Earth system on which humanity fundamentally depends. This is a gross market failure. We need to find a way to economically incentivise protection of intact nature and the restoration of degraded areas. Although the first is by far most important, we have only made progress on the second.

All the UN Sustainable Development Goals (SDGs) depend on a healthy bio-sphere [12]. Each of the SDGs also require its own focus within the context of the Nature Positive world. Notably achieving SDGs 5 (gender equality), 8 (decent work and economic growth), 10 (reducing inequality) and 16 (peace, justice and strong institutions) are essential to the integrated dream of an equitable, Nature Positive and carbon-neutral world. This requires not only improvements in gender equity and resource sharing within countries but also on an international scale. Thus, it is essential that the Global North keeps to the promises it made to the Global South in the form of UN agreements to provide financial support for the energy transition, to compensate for loss and damage caused by climate change and to preserve and restore biodiversity [44, 45]. Furthermore, the Global North will need to recognise that solutions may come from other cultures.

Declining faith in global agreements and institutions is a serious concern [46.]. While ecosystems occur on local scales, biomes cross boundaries, and global-scale natural processes like climate and rainfall do not respect borders. It is not sufficient to only act at the local and national level. We must all work together to make the world Nature Positive. International agreements require our collective support and commitment. A shared vision of an equitable, Nature Positive and carbon-neutral future can act as a rallying cry that brings us all together to create a liveable twenty-first century.

Conclusion

The mindset of western society can be shifted towards a Nature Positive perspective in place of the prevailing view of our environment as a commodity to exploit. At the international level, collective efforts, based on the recognition that humans live in relationship with the natural world and have responsibilities to care for it, could influence and ultimately unify worldviews, cultural values and institutions towards a common goal. Aligning policies and societies with Nature Positive principles would make this shift achievable.

Nature Positive is a collective societal project, anchored in present-day science and diverse cultural traditions that must also be supported by a sense of personal responsibility. While there are hopeful examples on our current path, the general view of the future is one of nightmarish scenarios of climate change, natural disasters, extinctions and refugee crises. It need not be so. Aligned actions of individuals, businesses and governments, civil society and Indigenous Peoples across diverse societies, united by a common goal of living in harmony with nature, could change things for the better. Such collective action, combined with a shared determination to halt and reverse nature loss by 2030, can deliver a Nature Positive future. It is a hopeful dream for the twenty-first century that is within our grasp.

SOCIETY: ADDITIONAL PERSPECTIVES

PERSPECTIVE 6.1   The right to a healthy environment

David Boyd, Associate Professor, University of British Columbia,
and Former UN Special Rapporteur on human rights and environment.

The right to a clean, healthy and sustainable environment is perhaps the most important human right of the 21st century, for without a livable planet, what good are other rights? In 2022, the United Nations General Assembly finally recognised the right to a healthy environment as a universal human right. More than 160 nations recognise the right to a healthy environment in law, through constitutions, legislation and regional human rights treaties. This right includes clean air, safe and sufficient water, healthy and sustainably produced food, non-toxic environments, a safe climate, and, critically, healthy ecosystems and biodiversity. The right also guarantees access to information, inclusive public participation in decision-making, and access to justice with effective remedies.

Of vital importance is that the right to a healthy environment can be used by activists, civil society, communities, and Indigenous Peoples to hold governments accountable for fulfilling their human rights obligations, which are informed by their international environmental commitments. For example, people in Costa Rica used their right to a healthy environment to stop fishing for endangered hammerhead sharks and to protect critical habitat for sea turtles and scarlet macaws. In Portugal, the right to a healthy environment was harnessed to protect Iberian wolves. In Mexico, to protect coral reefs and mangrove ecosystems. In the Philippines and Brazil, to protect tropical rainforests. The right to a healthy environment can also be deployed by governments to defend themselves against industry lawsuits challenging measures intended to protect the environment. Argentina relied on the right to defeat a business attacking a law that protected glaciers, while other nations have used the right to overcome industry lawsuits against regulations restricting plastic use.

In the face of a planetary environmental emergency, the right to a healthy environment is one of the most powerful tools at our disposal. It has immense potential for contributing to an equitable, carbon-neutral and Nature Positive future for all.

PERSPECTIVE 6.2. A 'Nature Positive' world is possible

Gavin Edwards, Executive Director,
Nature Positive Initiative Secretariat

If asked the question 'what do you imagine a 'Nature Positive' world would look like?', we can all access our old memoires and lived experiences of nature, or perhaps call on memories of wildlife shows we have been enthralled by on TV. We might extrapolate those memories out and imagine a world where forests are teeming with wildlife. We may envisage farmers who are producing lower yields to help restore some nature yet still feeding the world thanks to vast reductions in food waste and more healthy diets. We can picture coastal mangroves that are providing livelihoods for local fishers while buffering nearby cities from storms and flooding. And we might be inspired by healthy whale populations that are once again navigating the world's oceans. We may also imagine our place in that world – all 8 billion of us – living in much more sustainable cities and towns with nature on our doorstep.

But what are the major levers of changes that would create tipping points such that living in harmony with nature becomes inevitable? Governments can and must play a leading role in driving this. For example, measuring and regularly reporting nationally on whether biodiversity is recovering or declining and doing so in a comparable way will provide important recognition for countries leading on nature's recovery, as well as heaping pressure on those countries that are continuing to let nature unravel. This could be backed up by net-gain policies which ensure that any new developments avoid impacting intact and highly valuable nature, and that any nature that is impacted is more than compensated through restoration efforts to recover previously degraded nature.

Imagine financially rewarding communities and Indigenous People for stewarding the last of the world's intact nature and rewarding farmers for restoring wildlife habitats on some of their lands. All of this would be funded because halting and reversing nature loss would no longer be seen as a luxury or a cost of development and progress, but instead as something in the interests of national and global security. A necessity to help tackle climate change, and necessary to ensure that pollinators thrive and keep crops yielding year after year. And all for a fraction of the financial cost of wars and military spending.

Imagine a world where every company and financial institute reports its impacts on nature and contributes to nature's restoration through setting science-based targets. Imagine private sector innovations in plant-based diets such that most of the world embraces them for much of their weekly diet, in turn significantly lowering the need for cattle rearing which in turn takes much pressure off the natural world.

In fact, all these examples of policies, innovations and changes in mindset do not need to be imagined – they exist somewhere in the world today. Costa Rica pays many of its communities for stewarding nature and the key ecosystem services nature provides. A UK government policy is designed to ensure developers put 10% more nature back than they lose, while avoiding important natural habitats. Hong Kong builds its city vertically as much as possible instead of horizontally, ensuring that more than 40% of its territory is protected for nature and is available for restoration. Innovations in plant-based diets are exploding, and healthier diets are more available and are beginning to be embraced. Hundreds of companies are beginning to report their impacts and dependencies on nature, and that will soon become thousands.

All these best-practice examples do not yet add up to nature in recovery. The reality today is one where wildlife populations continue to shrink, the collapse of ecosystems continue at an accelerating pace, and the benefits that nature has provided to human development will begin to diminish. We need to ensure that these Nature Positive examples are not too-little, too-late. Time is against us.

Meanwhile the world is now beginning to embrace an energy revolution by decarbonising its power generation and electrifying transportation, with an eye on ensuring this happens at the scale needed. While there is still much to do to bend the curve downwards on greenhouse gas emissions, this energy revolution is finally beginning to reach important tipping points such that ridding the world of dangerous levels of greenhouse gas emissions may one day be within reach.

Similarly, we need a Nature Positive revolution, with an equal focus on massively scaling breakthrough solutions. A Nature Positive world is necessary, and the positive examples I touch on here underline that nature's recovery is also possible.

References

1 The Conference Board of Canada, 2016. Income Per Capita. Available at: www.conferenceboard.ca/hcp/income-per-capita-aspx-2/

2 Fletcher, R. and Omstead, J., 2020. 'Alberta rescinds decades-old policy that banned open-pit coal mines in Rockies and Foothills', CBC News, 22 May. Available at: www.cbc.ca/news/canada/calgary/alberta-coal-policy-rescinded-mine-development-environmental-concern-1.5578902

3 www.statista.com/statistics/466293/lowest-crude-oil-prices-due-to-covid-19/

4 Iea, 2020. Coal 2020 Prices and Costs. Available at: www.iea.org/reports/coal-2020/prices-and-costs

5 Fletcher, R. and Omstead, J., 2020. 'Alberta rescinds decades-old policy that banned open-pit coal mines in Rockies and Foothills', CBC News, 22 May. Available at: www.cbc.ca/news/canada/calgary/alberta-coal-policy-rescinded-mine-development-environmental-concern-1.5578902

6 Villani, M., 2021. 'Majority of Albertans opposed to expanded coal mining operations: Poll', CTV News, 8 February. Available at: https://calgary.ctvnews.ca/majority-of-albertans-opposed-to-expanded-coal-mining-operations-poll-1.5299906.

7 Knopff, R., 2021. Oil and Gas+ But Not Coal: Thoughts on Alberta's Mining Imbroglio, School of Public Policy, University of Calgary. Available at: www.policyschool.ca/oil-and-gas-but-not-coal-thoughts-on-albertas-mining-imbroglio/

8 Alberta Energy Regulator, 2025. 'Coal Industry Modernization Initiative'. Available at: https://www.alberta.ca/coal-industry-modernization-initiative

9 Farrell, J. and Johnson, L., 2025. 'Alberta Government Lifts Coal Mining Moratorium, Critics Concerned It's "open season"'. Available at: https://globalnews.ca/news/10969621/alberta-government-lifts-coal-mining-moratorium-aer/

10 Smil, V., 2011. 'Harvesting the biosphere: The human impact', Population and Development Review, 37(4), pp. 613–636.

11 Elhacham, E., Ben-Uri, L., Grozovski, J., Bar-On, Y.M. and Milo, R., 2020. 'Global human-made mass exceeds all living biomass', Nature. Available at: www.nature.com/articles/s41586-020-3010-5

12 Locke, H., Rockström, J., Bakker, P., Bapna, M., Gough, M., Hilty, J., Lambertini, M., et al., 2021. 'A Nature Positive World: The Global Goal for Nature'. Available at: https://library.wcs.org/doi/ctl/view/mid/33065/pubid/DMX3974900000.aspx

13 Conservation International, 2024. UN Biodiversity Negotiations. Available at: www.conservation.org/events/biodiversity-negotiations#:~:text=In%20December%202022%2C%20196%20countries,a%20healthy%20planet%20and%20human

14 United Nations Climate Change, 2023. COP28 Opens in Dubai with Calls for Accelerated Action, Higher Ambition against the Escalating Climate Crisis. Available at: https://unfccc.int/news/cop28-opens-in-dubai-with-calls-for-accelerated-action-higher-ambition-against-the-escalating

15 Abbasi, K., Ali, P., Barbour, V., Benfield, T., Bibbins-Domingo, K., Hancocks, S., Horton, R., Laybourn-Langton, L., Mash, R., Sahni, P., Sharief, W.M., Yonga, P. and Zielinski, C., 2023. 'Time to treat the climate and nature crisis as one indivisible global health emergency', BMF, 383, p. 2355. Available at: https://doi.org/10.1136/bmj.p2355

16 Bartholomew, P., 1997. Address at the Environmental Symposium, Santa Barbara, California. Cf. Chryssavgis, J., 2012. On Earth as in Heaven: Ecological Vision and Initiatives of Ecumenical Patriarch Bartholomew, Bronx, New York. Available at: https://apostolicpilgrimage.org/the-environment/-/asset_publisher/9b108Swk2KIh/content/the-fragile-beauty-of-the-world-opening-address/3200849a7.html

17 His Majesty King Charles III, the Then Prince of Wales, 2020. Sustainable Markets Initiative 2020. Available at: www.sustainable-markets.org/terra-carta/

18 Taskforce on Nature-Related Financial Disclosures, n.d. Home. Available at: https://tnfd.global/

19 Network for Greening the Financial System, n.d. Origin and Purpose. Available at: www.ngfs.net/en/about-us/governance/origin-and-purpose

20 Diaz, S., 2015. 'The IPBES conceptual framework-connecting nature and people', Environmental Sustainability, 14, pp. 1–16.

21 Gross National Happiness Commission, 2008. Gross National Happiness Index. Available at: www.gnhcentrebhutan.org/gnh-happiness-index/

22 Trithart, A. and Case, O., 2023. 'Do people trust the UN? A look at the data', IPI Global Observatory. Available at: https://theglobalobservatory.org/2023/02/do-people-trust-the-un-a-look-at-the-data/

23 Tucker, M.E. and Grim, J., 1997. 'The challenge of the environmental crisis', In Series Forward in Religions of the World and Ecology Book Series. Harvard Divinity School, distributed by Harvard University Press.
24 Schipper, K., 1993. The Taoist Body. Berkeley: The University of California Press.
25 Ma, T., 2021 'Unity of nature and man', National Science Review, 8(7), p. nwaa265.
26 Xi, J., 2021. 'Keynote speech', 15th Meeting of the Conference of the Parties to the United Nations Convention on Biological Diversity (COP15). Available at: www.youtube.com/watch?v=i1r8d6JRvNw
27 Trimurti, 2014. Encyclopaedia Britannica. Available at: www.britannica.com/topic/trimurti-Hinduism
28 Hari, D.K. and Hema Hari, D.K., 2011. Understanding Shiva. Barath Gyan and Sri Sri Publications Trust.
29 Chidambaram, R., 2011. Foreword to 'Understanding Shiva'. Baratha Gyan and Sri Sri Publications Trust.
30 Capra, F., 1983. The Tao of Physics: An Elaboration of the Parallels between Modern Physics and Eastern Mysticism. Shambhala Publications.
31 Locke, H., 2014. 'Leave me alone: India's opportunity to lead the world in recognizing Nature Needs Half', Sanctuary Asia, 2, pp. 22–27.
32 Ramphele, M., 2021. 'Ubuntu: The dream of new planetary community', In P. Clayton, et al., The New Possible: Visions of Our World Beyond Crisis. Cascade Books, pp. 27–34.
33 Lokanga, E., 2021. 'The concept of space and time: An African perspective', International Journal of Advancement in Physics, 10, pp. 1–3.
34 Fisher, M.P. and Luyster, R., 1991. Living Religions. Englewood Cliffs, NJ: Prentice Hall.
35 Little Bear, L., 2000. 'Jagged worldviews colliding', In M. Battiste (ed.), Reclaiming Indigenous Voice and Vision. University of British Colombia Press, pp. 77–85.
36 Little Bear, L., 2004. 'Aboriginal paradigms: Implications for relationship to land and treaty making' In K. Wilkins (ed.), Advancing Aboriginal Claims: Visions/Strategies/Directions. Purlich Publishing, pp. 26–38.
37 Van der Post, L., 1961. Heart of the Hunter: Customs and Myths of the African Bushman (Khoisan). William Morrow & Company.
38 Peat, F.D., 1994. Blackfoot Physics: A Journey into the Native American Worldview. Secaucus, NJ: Carol Pub. Group.
39 Little Bear, L., 2004. 'Preface in Bohm, D.', In L. Nichol (ed.), On Creativity. London: Routledge.
40 Buffalo Treaty, 2014. The Buffalo: A Treaty of Cooperation, Renewal and Restoration. Available at: www.buffalotreaty.com/treaty
41 Te Arawhiti The Office for Māori Crown Relations, 2014. Whanganui Iwi (River Settlement). Available at: www.govt.nz/browse/history-culture-and-heritage/treaty-settlements/find-a-treaty-settlement/whanganui-iwi/whanganui-iwi-whanganui-river-deed-of-settlement-summary
42 Hutchinson, A., et al., 2014. 'The Whanganui River as a legal person', Alternative Law Journal, 39(3). Available at: https://journals.sagepub.com/doi/10.1177/1037969X1403900309
43 New Zealand Parliament, 2017. Innovative Bill Protects Whanganui River with Legal Personhood. Available at: www.parliament.nz/en/get-involved/features/innovative-bill-protects-whanganui-river-with-legal-personhood/
44 United Nations, 2022. Developed Countries Must Deliver on Climate Change, Finance Commitments, Delegates Stress. Available at: https://press.un.org/en/2022/gaef3566.doc.htm
45 Campaign for Nature, 2024. Funding Nature: The Essential Role of Governments and the Illusion of Biodiversity Credits, 9 January. Available at: www.campaignfornature.org/funding-nature-essential-public-finance

7 The role of business in a Nature-Positive economy

Eva Zabey

Introduction

> "I have a question for you: do you want the Global Biodiversity Framework to actually change anything? Because if you do, this is your chance."

The Kunming-Montreal Global Biodiversity Framework (GBF) was adopted by 196 governments in December 2022 [1]. It was a pivotal moment to reset our relationship with nature and highlighted just how critical business action was to achieving the mission of halting and reversing nature loss by 2030, in other words achieving our shared Nature Positive goal. It was unprecedented in the history of the UN Convention on Biological Diversity (CBD) that over 1,000 business and finance representatives showed up and spoke up at the UN Convention on Biological Diversity (COP15) at which the Global Biodiversity Framework was negotiated and agreed. These private sector participants helped push for more policy ambition and drive a historic agreement. At previous COPs there were only a few dozen business participants, and business-related sessions were definitely not centre stage.

In Montreal, I was there, representing many of the more nature-conscious and forward-looking businesses around the world in my capacity as CEO of Business for Nature, a global coalition created for exactly this purpose. And in one moment, I felt that our whole reason for being was vindicated. It occurred during the negotiations of so-called Target 15 in the Global Biodiversity Framework [2], where governments asked businesses and finance to take action for nature, and assess and disclose their nature-related impacts, dependencies and risks. Governments present at the meeting were negotiating exactly how strong the business-focused target should be, fearing perhaps that if they went too far, businesses would object. But the companies I represented had precisely the opposite concern.

"Do you want the Global Biodiversity Framework to actually change anything?" I asked the negotiators when I was invited to speak from the floor in my capacity as observer. "Because if you do, this is your chance." I knew I was speaking on behalf of many businesses – and many dedicated individuals working for those businesses – when I asked that question. Through an incredibly bold campaign we ran ahead of COP15 to influence the ambitious Target 15, more than 400 companies and financial institutions urged governments to "make it mandatory" for them

DOI: 10.4324/9781003474043-9

to assess and disclose their risks, impacts and dependencies on nature [3]. Now, in the tense negotiation room, we had a chance to drive the message home. It worked. The fact that so many businesses were present, and asked for more, not less ambition, proved to be a tipping point. Governments recognised the role of business in addressing nature loss, and conversely, businesses started to realise their essential role and that what they brought to the table – innovation, experience, knowledge and expertise of addressing issues like climate change and, increasingly, nature loss – was valuable.

But without decades of key organisations and individuals – from outside and inside business – tirelessly pushing nature onto the business agenda and holding companies accountable, businesses would not have been ready at COP15 to advocate for nature so effectively. They would not have been ready to embrace the Nature Positive Global Goal today. And without scaling and speeding up business action now, we won't see the transformation that is necessary to remain in the safe operating space for humanity.

'Business' is not a homogenous group. There are differences around sectors and sizes, geographies and regulatory requirements – and I am aware my own insights are influenced by having worked with many European ones. There are companies that are publicly traded, family-owned or state-owned, those with different leadership, culture and history, different shareholder, stakeholder and employee demands. But in my view, the one thing that really differentiates them regardless of what I listed earlier is: are they part of the system transformation that is preparing to succeed in a new norm; or are they digging in their heels and delaying any change to profit from 'Business as Usual' for as long as possible? The challenge with the global nature loss crisis is that we need all companies – from *all* sectors and geographies and of all sizes – to contribute to achieving a nature-positive economy for all by 2030.

A Nature Positive goal places business and finance within a collective partnership that is necessary for nature recovery, moving businesses and financial institutions towards an appreciation of the interconnectedness of nature-positive outcomes. Every business has a role to play, in line with their abilities, level of impact and influence, and their responsibilities. Businesses don't operate in a silo. They operate in multiple systems and value chains that they can – and must – influence, despite the fact that their indirect impacts cannot be solely and fully attributed to them. Therefore, individual companies and financial institutions must adopt strategies across all their spheres of influence to contribute to the shared goal of a nature-positive planet by 2030. Achieving this ambitious goal requires systemic change. We need a fundamental shift in the economic system, driven by regulations, investors and consumer demand. While progress has been made, accelerating action is crucial. Businesses must transition from simply playing the game differently to playing a different game. The foundation for business action exists; the next step is exponential action and collaboration to achieve a nature-positive future.

This chapter tells the story of how nature got onto the business agenda in the first place, to being as high as it's ever been today and with the biggest challenges yet to come. That could only happen with decades of effort from all parts of society. One key demonstration of the growing business momentum and interest was at COP15, which was delayed due to COVID-19. Was there heightened awareness in employees who were in lockdown around the world about people's (re) connection to nature and a wake-up call for humanity during the pandemic? Did CEOs suddenly have more time available – without all the travel and face-to-face meetings – to reflect deeply and meaningfully on how they wanted to lead their companies to have a positive impact on society and the planet? Were we all a bit emboldened to step outside our comfort zones and agree something much bigger together? I think it was all of this rolled into one. "Never let a good crisis go to waste", as the adage goes [4, 5].[1]

With the Global Biodiversity Framework, the business world has the overarching political roadmap that gives it certainty to innovate, invest and shift business models. Businesses now also has frameworks and support to drive action. But whether nature remains high on the agenda isn't only up to businesses, it's up to all of us. For those of us lucky enough to decide for ourselves: How do we spend our money, as consumers and investors – including in our pension plans? How do we decide to spend our time? Who and what do we vote for? Businesses operate according to the rules of current economic and financial systems that recognise and reward them – or at least mostly – based on their financial performance, but without taking into account their environmental and social impact. If we are to reach our collective Nature Positive goal, that will need to change, and there are positive signs it is starting to do so.

Before we look forward to what still needs to be done, let's have a look at how we got here and recognise how far we have come. This chapter aims to take you on a journey from when business entered the sustainability movement, to when pioneering businesses started to assess the value of nature, to when a collective momentum was built calling for ambitious policy and how companies now need to scale and speed up action.

7.1 The early days: business starts to engage in the sustainability agenda

The origin story of the Western environmental movement traditionally begins when the American writer Rachel Carson published *Silent Spring* in 1962 [6], showing the harmful effects of pesticides on the environment. The book, which was featured in *The New Yorker* before it went to press, completely altered the way many people viewed humanity's impact on nature – and it blamed large chemical companies and insufficient government action.

Prior to Carson's exposé, not many people had given too much thought to how pesticides affected plants and animals beyond the insects they were meant to kill. But when it became clear just how devastating the impact of DDT and other chemicals were, the companies responsible for it, such as DuPont and Monsanto, tried to undermine the findings, rather than investigate, acknowledge and repair the damage. *Silent Spring* birthed the environmental movement and led to the creation of government agencies such as the Environmental Protection Agency (EPA) in the US [7].[2] In turn, some companies realised that they no longer would have 'carte blanche' to ignore their negative environmental impacts and started lobbying against environmental standards. But, as I wrote in the introduction to this chapter, business is not a homogenous group, and some new trailblazing companies emerged around the same time in the 1970s that explicitly placed environmental and social aims at the core of their businesses, including Patagonia, The Body Shop and Ben & Jerry's.

The first signs that large corporations could also be on the right side of environmental protection came a few decades later, in 1987, when the Montreal Protocol international treaty was signed. The UN treaty, signed by the world's governments, was meant to protect the planet's ozone layer by phasing out the production of "ozone depleting substances" (ODCs) [8]. The treaty became necessary because it was clear that the protective ozone layer was at risk of disappearing due to the excessive use of ODCs. If the hole in the ozone layer grew, it would unleash the Sun's harmful ultraviolet radiation, scientists found, causing severe damage to human health and the environment [9]. Governments negotiated and ratified the treaty, but businesses would prove central to its successful implementation.

Companies used their advances in science and innovation to rapidly develop and adopt non-ozone depleting alternatives to the ODCs often used in refrigerators and other products, and the ozone layer started to recover. By 2021, the UN Environment Programme (UNEP) said the results were "dramatic" [10] and that "around 99 per cent of ozone depleting substances have been phased out and the protective layer above Earth is being replenished" [11]. For the first time, a global science-based policy drove well-defined action and reached the desired outcome. For the first time, business managed to go from being a part of the problem to being a central part of the solution.

A second moment of progress came with the United Nations Conference on Environment and Development, nicknamed the 'Earth Summit', which took place in Rio de Janeiro, Brazil, in June 1992 [12]. The Earth Summit happened amid euphoria about the end of the Cold War, and brought together 179 governments in a "massive effort to focus on the impact of human socio-economic activities on the environment", as the UN put it [13]. It led to the creation of three major so-called 'Rio Conventions': one on climate (the UN Framework Convention on Climate Change, the UNFCCC), one on biodiversity (the UN Convention on Biodiversity, the CBD) and one on desertification (the UN Convention to Combat Desertification, the UNCCD).

At the Earth Summit, business also entered the scene. Maurice Strong, the Canadian diplomat who led the summit [14], said business needed to be part of the discussion, a notion hitherto uncommon in the UN. Strong reached out to the Swiss businessman Stephan Schmidheiny and challenged him to spread the concept of sustainable development among the world's business leaders and companies ahead of the summit [15]. Schmidheiny delivered: he brought together a group of businesses from around the world and published a report called *Changing Course* [16]. It showed how business could contribute to the overarching goal of sustainable development, and coined the term 'eco-efficiency' as "creating more goods and services with fewer resources and less environmental impact." At the time, 'eco-efficiency' was a disruptive idea. Today, it reads like a no-brainer. And the argument of being more eco-efficient would be even more compelling if the true value of natural resources (like freshwater) and negative environmental impacts (like pollution) were properly accounted for.

I think it is fair to say these events in 1992 were the start of the corporate sustainability movement, as the business gatherings ahead of and after the Earth Summit ultimately led to the creation of the World Business Council for Sustainable Development (WBCSD) in 1995, which has been a leader in corporate sustainability ever since [17].[3]

Yet, when the corporate sustainability movement started around the Earth Summit, it did still have a rather 'nice to have' feel to it in those early days, with well-meaning companies seeing it more of a philanthropic or feel-good part of their work, rather than as central to their business model. The term 'Corporate Social Responsibility' (CSR) became more mainstream, with more companies understanding that they had a 'social license to operate' which could be at risk unless they embarked on a sustainability journey (for example, to reduce the risks of boycotts or reputational issues and increase their chances to secure their next permit for a mine, factory or development project).

It is insightful to look where – and how high up – in the company any sustainability work was happening and which budgets were being used to finance it. Sustainability was often a part of the communications or marketing function, or embedded in the company's foundation. How things have changed! More progressive companies' sustainability activities today can fit in core strategy, innovation, procurement and increasingly under the Chief Financial Officer's responsibilities. But before we got to that point, businesses first had to understand that nature has a value not only in its own right but also to society and businesses themselves – I'll dive into that in the next section.

PERSPECTIVE 7.1   The Nature-Positive journey of the WBCSD and its member companies

By Peter Bakker, CEO, World Business Council for Sustainable Development (WBCSD)

WBCSD and its member companies support the global goal to halt and reverse nature loss by 2030 – the goal is explicitly called out within our membership criteria. Companies increasingly understand how their success depends on nature, whether because they require access to raw materials, ecosystem services, or flood protection to name just a few. And they understand the impacts that they have on nature. More and more of our member companies are developing robust nature approaches, identifying their material negative impacts and taking action to address these, protect nature and promote nature-positive strategies.

Companies now need support in unpacking nature impacts and risks, setting robust nature commitments and embedding these in their accountability systems. Nature is a much more complex issue than climate, and it's three to five years behind it in terms of adoption. But much of what has been put in place for climate action and accountability can be repurposed for nature – for instance, the Taskforce on Nature-related Financial Disclosures (TNFD) carries over the Task Force on Climate-related Financial Disclosures' (TCFD) four pillars of governance, strategy, risk and impact management, and metrics and targets.

My own journey exemplifies that of the broader business community. I led a large international company, TNT, before joining WBCSD. When I attended a conference on biodiversity for the first time, in the 2010s, I quickly realised I couldn't make much sense of the terms that were used in these discussions, such as ecosystem services. Surely that was my fault, I told my team. But as a business person, we simply did not look at the world in that way.

My first 'aha moment' on nature arose when my team started framing nature as a capital. Natural capital, like financial capital, they told me, is a type of capital that any company needs to do what it does. Talking about it in that way helped me relate to 'nature' much better. Ultimately, it helped lead WBCSD and partners to kickstart the 'Natural Capital Protocol' on behalf of the Capitals Coalition, launched back in 2016, as a means to connect business with nature.

A second big step forward in the consciousness of the global business community regarding nature came with COVID–19. The pandemic made the business community realise just how closely related the fates of nature and people are. The risk of zoonotic epidemics and pandemics continues to increase as a result of deforestation, climate change, urbanisation and global connectedness. Nature loss is both an enormous societal risk and an incredible opportunity for a more resilient world.

Even though we don't have all the answers, any company, regardless of their maturity or sector, can get started today on their journey. WBCSD co-developed the foundations for any corporate nature journey with the high-level business actions to Assess, Commit, Transform and Disclose (ACT-D) framework (see text box on ACT-D later in this chapter), so that companies can identify where action is needed and work towards credible nature goals, and has developed Roadmaps to Nature Positive [18]. The next important advances are focused on prioritising the metrics that companies use to measure impacts and set targets, aligned with mandatory and voluntary frameworks, and reflecting companies' positions within value chains. This is the focus that is required to get from hundreds to thousands of companies working together to restore nature and build resilience back into planetary, social and economic systems. This will unlock the agenda, and scale up collective action, ensuring all companies (not just the leading ones) can advance on the nature-positive agenda.

7.2   Pioneering corporate efforts to recognise nature's value

After the Earth Summit, avant-garde businesses started to recognise the importance of corporate sustainability. But to understand a concept is one thing; being able to measure and act on it, is another. And so, in the 15 years or so after Rio, corporate sustainability mostly developed in two interconnected 'spaces': 1) at the UN and other international organisations that defined and informed what corporate sustainability means and how it related to the broader environment, nature and biodiversity agenda; and 2) at forward-thinking companies voluntarily exploring why and experimenting how to manage what the sustainability agenda meant to their business.

In the early years of the new millennium, one major step forward came when the concept of 'ecosystem services' entered the vocabulary of international organisations and was introduced to global businesses active in sustainable development. It originated in the year 2000, when then UN Secretary-General Kofi Annan called for a major global study of human impact on the environment. This report, the *Millenium Ecosystem Assessment* (MA), was completed in 2005 [19]. Despite a limited awareness among the general public and companies, which remains to this day, the *Millenium Ecosystem Assessment* was at the time the biggest worldwide assessment of the health and status of ecosystems worldwide. No fewer than 1,300 expert contributors from 95 countries came together to complete it [20].

The *Millenium Ecosystem Assessment* conceptualised and quantified 'ecosystem services', which were defined as "the benefits societies obtain from ecosystems" such as fresh water, fibre, food as well as erosion control, climate regulation, water filtration, pollination and more [21]. Importantly, the report showed that a majority of ecosystem services were degrading and, as a result, imperiling the benefits humans and businesses received from these ecosystems. The report's findings drew some business attention. At the time, most businesses limited their focus on compliance-driven Environmental Impact Assessments (EIA), Life-Cycle Assessments and International Organization for Standardization (ISO)-guided environmental management plans, all of which were restricted to the key impacts these businesses had on the environment. However, with the MA, it started to dawn on some businesses (the ones that were paying attention) that they had much broader ecosystem impacts and – perhaps even more importantly, dependencies on these ecosystem services. What would happen to their bottom line if this full suite of services provided by nature – for free – including those that are essential inputs to their business, could one day dry up?

Building directly off the *Millennium Ecosystem Assessment*, in 2008, the World Resources Institute (WRI), a non-profit research institute, collaborated with WBCSD and the Meridian Institute to deliver the very first 'Corporate Ecosystem Services Review' (ESR) [22]. The ESR helped companies understand that they too had both impacts (negative and positive) and dependencies on ecosystem services, and added the logical step of identifying what related business risks and opportunities should be addressed. It was the first simple checklist for companies to know which ecosystem services they had an impact on, and which they depended on.[4]

The logic behind the ESR was a mindset shift. Prior to this, companies were thinking only of – essentially negative – impacts on some specific things like water, land and air pollution. They were often required to carry out structured impact assessments and adapt their management systems. But they were focused purely on the negative impacts, and few would even consider that they could have any positive impacts, like restoring or reforesting degraded ecosystems and replenishing groundwater. The real shift, however, related to the dependencies. Fundamentally, if you are, say, an agribusiness, you rely on ecosystem services like pollination, pest control, freshwater and soil health. If you lose them, your business will be affected negatively. But equally, if you invest in them, you will have a more resilient business in the future. Back in 2008, this concept of ecosystem services, and how you can harm or foster them, was still kind of a novel idea in the corporate world. Businesses – and the economic system of growth – were based on the idea of infinite resources, that there would be an abundance of natural resources. Now, little by little, that was starting to change.

Around the same time (approximately 2004–2008), organisations like the International Union for Conservation of Nature (IUCN) and the Business and Biodiversity Offsets Program (BBOP) started to work with mining companies such as Rio Tinto and oil companies such as BP to develop methodologies around "no net loss and net positive impact approaches to biodiversity" [23]. They demonstrated how sectors with the biggest impact (such as mining, oil and gas, cement, agri-chemicals) should and could be the first to understand their impacts and take early action. It was a start, even if these efforts alone weren't going to drastically change business models and reflect the interconnected nature, climate and societal challenge.[5]

In 2010, the corporate world took the next step in putting sustainability higher on its agenda, and this time it was because of a negative event. On 20 April, the BP-owned oil rig Deepwater Horizon exploded in the Gulf of Mexico. The subsequent oil spill was the largest in marine history, and it happened just a few dozen kilometers off the Louisiana coast [24]. This caused great environmental damage, affecting almost all fish and bird species in the area, and also the inhabitants of the Southern states close to the spill. Nature and the affected families and inhabitants paid the heftiest price, with BP in a distant second place. The company was ordered to pay $20 billion in damages to affected communities (the largest ever such fine); it was held responsible for the clean-up of the disaster [25]; and Tony Hayward, then BP CEO, was forced to resign in the aftermath of the crisis. If there was any silver lining, it was that BP acknowledged responsibility for its negative nature impact in a very open way, including by plastering a photo of the disaster on the front cover of its sustainability report [26], which was the first time I saw such a massive company with no choice but to take full responsibility for such a terrible event.

The way the company managed the disaster was a signal of how sustainability was also starting to become a material issue at the board and management level, specifically for those working in enterprise risk, finance and legal functions. BP, however, did finally bounce back, even if some of the damage it caused in the

Gulf of Mexico was permanent. It was a harsh reminder that in our current global economic system, there often aren't any existential consequences to businesses for unacceptable environmental or social performance in the longer term. Bad environmental and social incidents mustn't be brushed off as anomalies that might at best drive a wake-up call. They need to be fully accounted for, if we want to achieve the transformational change to our economic system that is so importantly needed.

In the years following Deepwater Horizon, the corporate sustainability agenda also continued to take structural steps forward. In 2011, for example, WBCSD released its *Guide to Corporate Ecosystem Valuation* (CEV) [27]. It took the next step in informing business decision-makers about nature, by not only *measuring* but also *valuing* the impacts and dependencies on ecosystem services, and better understanding risks and opportunities. The guide helped companies put either a qualitative, quantitative or monetary value on impacts and dependencies to help drive decision-making. A dozen pioneering companies piloted the guide, and their CEOs, including from Holcim, Rio Tinto, Energias de Portugal, Hitachi and Dow, signed a message in which they voiced their support for the approach. "We see the value of ecosystem valuation," they said [28].

Unfortunately, there was no sustained uptake of the CEV by companies beyond the pilots. But it *was* the precursor that set in motion the 2016 Natural Capital Protocol [29], which I led the development of on behalf of the organisation now known as the Capitals Coalition, which is seen as a recognised authority on measuring and valuing natural, social and human capital.[6]

PERSPECTIVE 7.2   The Natural Capital Protocol and the all-important 'Plumbing' of capitals accounting

By Mark Gough, CEO, Capitals Coalition

To deliver the ambitious goal of a nature-positive world, we must understand and explain our relationship with nature. For many years, people have been using the concept of natural capital to do this, but it is only in recent times that this has really started to gain serious traction. One of the reasons for this is that the global community has been actively working together.

In 2013 there were nearly 40 natural capital methods to choose from. I was involved in the development of one of these and was therefore responsible for adding to the confusion. Instead of arguing over whose approach was best, this unique community did something new. United by their common purpose to bring clarity to the field, they joined forces. Through their leadership and this collaborative spirit, they produced a game-changer: The Natural Capital Protocol, an internationally accepted framework.

It provided a consistent way to produce robust and actionable information on people's relationship with nature and was evidence of the power of collaboration, and a story of how diverse voices coming together can create something truly groundbreaking.

Figure 7.1 Viewing biodiversity through a natural capital lens – guidance for business.
Source: Capitals Coalition, 2016

The Protocol set a new baseline and gave birth to innovative ideas and experimentation. From nature-based solutions to the Taskforce for Nature-related Financial Disclosures, the concept of natural capital has been an essential underpinning force. Words such as 'impact' and 'dependency', which were rarely considered or used before the Protocol, are now ubiquitous and included in the Global Biodiversity Framework, and increasingly in finance and accounting approaches.

So, what next? Part of the reason we have seen such a loss of nature and biodiversity is because we made them invisible in our decision-making processes. Whilst natural capital is crucial in helping rectify this oversight, we must also include the other capitals, social, human and produced, when making decisions. I know this sounds complicated, but we cannot ignore complexity. We must instead embrace it as we have done in so many other parts of our lives.

The Capitals Coalition is breaking ground in this area and producing new integrated guidance and requirements across all the four capitals. As Nature Positive takes hold, many recognise these connections and aim for a climate-neutral, nature-positive, and equitable world. Capitals have the potential to provide the infrastructure (the plumbing) needed to turn this ambition into action.

You can read more about the Natural Capital Protocol, the Framework for Integrated Decision-making and more on capitalscoalition.org

Around the same time, experimental ecosystem valuation efforts also got underway at other businesses, including the following:

• Sportswear company Puma performed a world-first Environmental Profit & Loss (EP&L) [30] in 2011. It calculated the total negative environmental impact of the company's direct and supply chain operations, which were previously unaccounted for, at 145 million euro. Puma continues to use an updated version of the EP&L today, as does Kering [31], its former holding company.

- Around 2010–2011, Brazilian cosmetics firm Natura (now part of Natura &Co) started paying local communities in the Amazon to keep trees in the forest standing rather than cutting them down to be sold as timber. For Natura, the indigenous ucuuba tree, whose seeds have moisturising properties, is significant, as it uses its seeds as input for its beauty products. But the programme benefits the environment as well, as it supports the conservation of 635,000 acres of forest and 25 native species, and provides thousands of families with additional income [32].
- An older case, that now serves as a cautionary tale, is that of Vittel water. In the 1990s, Nestlé, the owner of the Vittel brand, used so-called Payments for Ecosystem Services (PES) to secure its mineral water bottling business. At the time, agricultural intensification in the aquifer Vittel used posed a real risk of nitrate contamination. To prevent contamination, Nestlé started to pay farmers upstream in the catchment to change their farming practices so Vittel could secure its business and keep the label 'natural mineral water'. Indeed, if it had to treat the contaminated water, it would have to ditch the 'natural mineral water' label, and therefore the water would lose financial value. In 2006, a research paper on Payments for Ecosystem Services found that the Vittel case was a "success" that should be replicated elsewhere [33].

  In January 2024, however, the success story had a twist. Press reports revealed that Nestlé – along with many other water bottling companies – had been treating and purifying its French 'spring' waters [34], including that of Vittel, meaning the label 'natural' water was *de facto* misleading. The French government allegedly knew about the illegal practice for some years and had loosened the regulatory requirements, absolving the company of any legal wrongdoing. This example highlights the need for strong accountability and transparency, as well as clear, consistent and enforced regulation, while also demonstrating the real value to the business of natural water purification as an ecosystem service. Despite the incident previously that Nestlé admitted to, the company remains a widely recognized sustainability leader[7] that is shifting its business model towards regenerative agriculture.

These early experimental initiatives were useful as pilots, and allowed nascent sustainability teams at companies to learn new skills, identify where gaps existed and raise awareness inside and outside their organisation about their company's complex dependencies and impacts on nature. But as the ownership of these projects often resided with sustainability teams, and the projects tended to be very technical, they did not often top the broader executive management agenda. Nevertheless, in my view, two legacies stand out from this era:

- Firstly, awareness arose of the skills gap that existed within companies to understand the value of nature. The discipline of environmental valuation had been around for decades, but in the early 2010s, it was still new to most businesses, and still is today. There was a real need for companies to develop capacity building programmes, a gap that has since been filled by projects such as the Business

Ecosystems Training from WBCSD, the We Value Nature initiative supported by the European Commission, A-Track (see a-track.info), the 'Valuing Nature and People' Coursera course from the Capitals Coalition, the Cambridge Institute for Sustainability Leadership and UNEP-WCMC, among others.

• Secondly, the term 'natural capital' entered the glossary of sustainability managers at companies. It is defined as the "stock of renewable and non-renewable natural resources (e.g. plants, animals, air, water, soils, minerals) that combine to yield a flow of benefits to people which make human life and a thriving economy possible" [35]. In 2012, the UNEP Finance Initiative also brought together the CEOs of more than 40 financial institutions from around the world to sign a declaration, which, among other things, said that "ecosystem goods and services from natural capital underpin productivity and the global economy" [36].

In other words, some leading and pioneering companies were starting to consider nature, but what was needed next was a complete shift to get nature up the mainstream business agenda, and to use the global negotiations of the Global Biodiversity Framework as the hook to build the biggest business momentum on nature to date. We'll look at that evolution next.

7.3   Collective business momentum drives policy ambition for nature

By around the mid-2010s, two trends occurred which allowed nature – and sustainability more broadly – to enter the mainstream conversation in business. On the one hand, the international organisations and NGOs working on sustainable development, including the UN, began to realise that their efforts would be insufficient if the private sector wasn't more actively involved. On the other hand, the businesses that had been engaging with sustainability initiatives managed to rally together and bring on board a much broader set of businesses to create momentum for the nature and climate policy agenda.

The realisation in the UN came about in large part because of the flailing Millennium Development Goals (MDGs). Agreed upon in 2000 by the UN member states, the MDGs were a set of eight ambitious targets set by the UN in 2000 to improve the lives of millions by 2015. But by the time of the final assessment in 2015, then UN Secretary-General Ban Ki-moon acknowledged that "inequalities persist and progress has been uneven" [37].

As the international community reflected on the successes and failures of the MDGs, one conclusion was that next time, business should get a seat at the table. At the time of the MDGs, business was not seen as a key stakeholder, and subsequently, they didn't have an active role in achieving the MDGs. In the future, that would have to change: the private sector is crucial to achieving progress on sustainable development. Several organisations like the UN Global Compact and the International Chamber of Commerce helped create platforms between business and UN stakeholders. And so the next two new UN-led programmes actively invited business to the table. First, the Sustainable Development Goals (SDGs), which replaced the MDGs. And second, the Paris Agreement on Climate Change. Both were developed with active involvement from business, and their adoption in 2015 was seen as important for business too.

Getting businesses to engage in these processes didn't happen by magic – it was a concerted effort that took significant time, resources and collaboration. Ahead of COP21 in Paris in 2015, several business leaders stepped up to indicate what they hoped to see in a climate agreement, and what contributions they were willing to make. That was the case, for example, of Tim Cook, CEO of Apple, who helped launch the We Mean Business Coalition (WMBC) in 2014 [38]. The IKEA Foundation funded the Coalition from the get-go, and other businesses also stepped up. H&M, Nestlé and BT were among the first to commit to science-based targets on their carbon emissions that year. On climate change, businesses felt the responsibility to advocate for ambitious policy. Ahead of COP21 in Paris, WMBC released eight policy 'asks' and worked with its partners – BSR, Ceres, CDP, Corporate Leaders Group, Climate Group, the B Team and WBCSD – to mobilise business support for a climate agreement. This meant leading businesses were ready to contribute their part when an agreement was reached in December 2015.

The Paris Agreement and the years that followed served as a catalyst for those working on the business-nature nexus. Paris showed how much more was possible when business played an active and positive role in the forging and delivery of international agreements. It wasn't just that businesses got involved ahead of the Paris Agreement. From 2015 onwards, many businesses also started to work on introducing science-based climate targets to their operations, while identifying how their activities could contribute to the achievement of the SDGs.

A few years later, in 2019, a new wave of energy started bubbling up around an upcoming 'Paris moment' for nature. Indeed, another 'Conference of the Parties' (this time on biodiversity under the UN Convention on Biological Diversity) was supposed to take place in October of 2020 in Kunming, China. As for climate, the UN member states' governments were expected to adopt a major new agreement on nature, called the Global Biodiversity Framework. The omens, however, weren't good. The last time a similar 10-year framework on biodiversity was agreed, in Aichi, Japan, in 2010, the targets it set failed miserably. One reason for that, many people like myself believed, was that the 2010 Aichi agreement was a traditional intergovernmental framework without sufficient multi-stakeholder involvement, in particular from business and finance. As a result, how the targets would be achieved, and who was responsible for doing so, was not clear. The Aichi Biodiversity Targets also didn't acknowledge the need to transform the systems – like the global economy – in which we operate. So, there was a real concern the next 10-year framework would be a repeat – unless its leadership got *all* stakeholders onboard, including business. The early signs weren't encouraging. The 'zero draft' of the framework [39], shared in January 2020, didn't include anything on the role of business.

We needed to quickly mobilise the business community to influence an ambitious Global Biodiversity Framework, just like the We Mean Business Coalition had done. But most companies weren't as familiar or comfortable with the broader nature agenda as they were with climate, where they had clear accounting and reporting through the GHG Protocol for over a decade. This meant it was a challenge to get business leaders to speak up and call for policy ambition: how could we do this if nature was a new issue for them to consider, and if they didn't yet have their own house in order?

As we were faced with an urgent planetary crisis, we needed to push for both in parallel: drive credible business action at the same time as calling for policy ambition. When we dug into existing business efforts, we estimated that over 1,000 companies around the world were already taking action and making commitments in different ways to help tackle nature loss, for example through addressing freshwater consumption or pollution, soil health and deforestation or by investing in circular and more sustainable products and services. And of course, many efforts went hand in hand with a company's climate plans, including restoring ecosystems to sequester carbon, protecting mangroves from climate impacts like storms and hurricanes. But clearly, these voluntary and fragmented efforts were not enough. It was crucial to use the influential, progressive business voice to push for policy ambition and level the playing field. This would then ensure fair competition and incentivise the right action and therefore drive even more positive impact on the ground.

We needed to get organised. Key visionary leaders such as Paul Polman (then CEO of Unilever), André Hoffmann (Vice Chair of Roche), Nigel Topping (then CEO of We Mean Business Coalition), Mark Gough (CEO of the Capitals Coalition, see above), Akanksha Khatri (Head of Nature and Biodiversity at the World Economic Forum) and Marco Lambertini (then Director General of WWF International) used their convening power ahead of and at the World Economic Forum's Davos meeting in January 2019 to get this COP15 opportunity onto the business agenda. WWF was instrumental in providing *pro bono* staff time to act as an interim secretariat for what then became the Business for Nature coalition. Meanwhile, the Gordon and Betty Moore Foundation committed some seed funding via the World Economic Forum, and WBCSD stepped up to offer to host the coalition through its legal entity. All founding partners, including Capitals Coalition and We Mean Business Coalition, chipped in to help pull together funding proposals, pay for early communications support and host meetings. These organisations launched Business for Nature mid-2019, and then looked to recruit someone to lead and run the coalition.

PERSPECTIVE 7.3   Pushing Nature up the mainstream economic agenda

By Akanksha Khatri, Head, Nature and Biodiversity,
World Economic Forum

Nature moved up the mainstream business and economic agenda in a relatively short amount of time at the World Economic Forum, which is to say, less than five years. The fact that sustainability more broadly had previously moved up the CEO agenda probably certainly played a role. Concerns on climate change, extreme weather, and natural disasters topped the list of business risks by the time nature entered the conversation.

In September 2018, the World Economic Forum, along with WWF International (with Marco Lambertini as its then Director General), Unilever (with Paul Polman, then CEO) and Forum Trustee André Hoffmann, hosted the first dialogue on 'How to catalyse a public-private movement for Nature in the run up to COP15 and beyond'. In attendance were UN CBD Executive Secretary, Ministers and leading scientists. All participants recognised that nature needed to be taken beyond the environment movement and anchored right at the centre of all economic decision-making. This was followed by the first ever 'Nature dinner at Davos' in 2019 and the formation of a group of Champions for Nature, committed to raising ambition and catalysing action for a future that lives in harmony with nature. The history of what happened next at Davos and beyond, is well described by Marco in a previous chapter of this book.

In 2025, the World Economic Forum continues to deeply believe that economy is embedded in society which in turn is embedded in the ecosystem. Therefore the health of species and spaces we inhabit is critical to our wellbeing and prosperity. 2030 is the goalpost for multiple global commitments such as the Sustainable Development Goals, the Global Biodiversity Framework, the Paris Agreement, and more. Decisions we take today bind us to the future we must accept. Hence the urgency to avoid and mitigate risks of nature loss; but equally critical is making investments into nature-positive opportunities. The World Economic Forum and its partners are committed to ensuring that the way we produce, consume, finance, build cities and provide energy access to all is both nature and people positive. Since that first meeting in September 2018, the portfolio of work at the Forum has encompassed the food systems, urban transformation, industry transition and investment portfolios. That work continues going forward, until a nature-positive world is achieved.

Or, to quote one of my favorite authors: "Another world is not only possible, she is on her way. On a quiet day, I can hear her breathing." – Arundhati Roy

---

At the time, I was leading WBCSD's work on measurement, valuation and reporting of natural, social and human capital as part of the Redefining Value programme. But with a passion for the environment from a young age, a background in ecology and having led pioneering efforts with business along the way, I knew this was my time to step up and lead. So I put my name in the hat and was honoured and thrilled to be selected. In November 2019 I started what I thought (and what I had promised my husband) was going to be an intense one-year position leading up to COP15 in October 2020. Then COVID-19 happened.

As a small, new and remote team, we were incredibly agile. While many larger organisations were figuring out how to manage the very real challenges caused by the COVID-19 pandemic, including the suffering of staff members, logistical problems and office closures, we were lucky to have been able to navigate these challenges relatively well because we could adapt as we were still in 'set up' mode.

In the few months leading up to lockdown, we managed to squeeze in a public consultation of key policy asks which were launched at a high-level session in Davos in January 2020. I then went to Brussels to have a private meeting with Commissioner for the Environment, Oceans and Fisheries, Virginijus Sinkevičius [40]. His role as lead negotiator, together with our shared ambition for the agreement at COP15, made him an excellent partner for the conversation. I told him that CEOs and companies were stepping up on nature and this should give policymakers the confidence required to adopt ambitious policies, in particular a strong Global Biodiversity Framework. He was encouraged by this. However, he advised us that we also needed to show that many more companies, well beyond only the leaders, were part of the agenda too. Our advocacy strategy then became a two-pronged approach of both business leadership and momentum to provide the courage and comfort to policymakers that they would be supported. We had a good set of leaders; we now needed the evidence that we had the momentum.

To further expand our reach, we used strategic campaigns and communications to influence negotiators so the agreement would be relevant to business. We therefore jumped on the opportunity to start securing the business momentum we needed by using CEO leadership just as people were reconnecting with nature and re-evaluating what mattered most to them. However, with COVID-19 and the subsequent lockdowns, COP15 was postponed repeatedly. Some people began to question whether or not the meeting would ever take place. However, the silver lining was that, because of COVID-19, we gained two years to work towards a landmark agreement.

During this period, resilience overtook efficiency as a prime focus for business. We organised our first ever global virtual event on 15 June 2020, which was called 'Building business resilience: How collective leadership will reverse nature loss' [41], which was attended by over 2,200 people. I am pretty sure that at the time it was the largest online business event on nature ever, and it was the start of a suite of strategic events and campaigns spearheaded by Business for Nature's Communications Director, Lucy Coast. Another pleasant surprise was having to worry about the sheer number of CEOs who wanted to speak and fitting them into the short amount of time allocated.

During the event, we heard live from a variety of CEOs in position at the time, including CEO of Danone France, Emmanuel Faber, and AXA CEO, Thomas Burbel. Asian conglomerates were also represented by Fosun International Chairman Guo Guangchang and Sintesa Group CEO Shinta Widjaja Kamdani. Unilever CEO Alan Jope and Natura &Co CEO Roberto Marques, meanwhile, provided their expertise on sustainability in consumer goods companies, as well as in the cosmetics industry.

We launched the 'Nature is Everyone's Business' Call to Action [42] with a video that included support from CEOs, CFOs and Chairs from, for instance, fragrance companies L'Occitane Group and Firmenich. Other businesses represented included Brazilian companies BRK Ambiental, Electrobras, as well as fertiliser and sustainable foods suppliers Yara and Ofi. Finally, a range of perspectives regarding nature in the insurance industry and electric utilities were provided by MS&AD Insurance Group Holdings, EDF, Nature Home and many more. Today, about 1,500 companies have signed up to 'Nature is Everyone's Business' from a wide range of sectors, sizes and geographies – collectively calling for policy ambition.

Collectively, we were getting nature – and COP15 – onto the business agenda, but we also needed to contribute in a helpful and constructive way to the negotiations themselves, and this meant having clear technical recommendations that were developed collaboratively with our partners and leading companies. The mastermind leading this effort was Maelle Pelisson, Business for Nature's Advocacy Director.

As the leading business voice, we had three top priorities for the Global Biodiversity Framework going into COP15:

- The mission had to be clear – to halt and reverse nature loss by 2030, that is, Nature Positive. The fact businesses supported this gave additional weight and showed that a long list of targets without the overall global goal would not help businesses keep the overarching goal in mind.
- There needed to be clear and ambitious requirements for businesses and financial institutions to assess and disclose their risks, impacts and dependencies on biodiversity by 2030, which is now in Target 15. Initially, the zero draft of the framework had no business target. However, in what was agreed by 196 governments at COP15, we can find some of the most explicit text ever that governments have agreed in a multilateral agreement on what governments expect from business and finance. We managed to influence the Target 15 text via a major campaign, working especially with CDP – the central organisation to which thousands of companies have been voluntarily disclosing their climate, forest and, increasingly, water-related information – alongside Capitals Coalition and other partners, where we rallied more than 400 business and finance institutions from 52 countries to support our 'Make it Mandatory' campaign [43].
- Governments needed to address the elephant in the room – namely, environmentally harmful subsidies [44]. If companies depend on these subsidies, they represent hidden risks for companies but also opportunities for those pushing for a transition away from nature-negative towards nature-positive outcomes. We therefore co-led a campaign with The B Team calling for reform of these subsidies, estimated to be worth $1.8 trillion per year (this figure has since been updated in 2024, and is estimated at $2.6 trillion). The campaign, which garnered support from key global figures such as Christiana Figueres, former Executive Secretary of the UNFCCC, and Richard Branson, Founder of the Virgin Group, helped to significantly strengthen what is today an ambitious Target 18 in the Global Biodiversity Framework and, more broadly, contributed to the growing awareness and need to include subsidy reform in any discussions around shifting financial flows.

By uniting businesses around a shared narrative and level of ambition, together we helped give courage and comfort to policymakers and secured a successful outcome. Finally, in December 2022, countries from around the world gathered in Montreal where the Kunming-Montreal Global Diversity Framework (GBF) was adopted by 196 governments. On our side, by working with our partners, we got about 1,000 business and finance representatives to show up at COP15, which was unprecedented and sent a strong signal that the stakes were higher than ever for governments to adopt a strong agreement. The landmark framework has become the Paris climate agreement-equivalent for nature. Overall, the Framework gave businesses and governments the key elements they needed to begin working towards a nature-positive future.

Interestingly, the courage of business leadership to call for policy ambition continued after COP15, despite geopolitics, elections, conflict, the cost of living crisis and other challenges. This wasn't a given as many companies, even the most advanced, are sometimes overwhelmed by complexity and regulatory pressures. In many cases, sustainability teams find themselves under-resourced and in the middle of heated, misinformed and increasingly politicised debates around sustainability, ESG and anti-wokism. However, businesses have continued to raise their voice and signed open letters and statements, for example in support of the EU Nature Restoration Law [45] that was in jeopardy, and the Corporate Sustainability Reporting Directive (CSRD) [46]. And, in October 2024, ahead of the Biodiversity COP16 in Cali, Colombia, over 230 businesses and financial institutions signed an open letter to heads of state, calling for renewed policy ambition to implement the Framework and halt and reverse nature loss this decade [47]. While some progress was made at COP16, the pressure to deliver was called out, not only by conservation groups, youth, Indigenous Peoples, scientists and others but also business. Thanks to our open letter, we had the backing that those signatories expected more.

The EU has introduced strong policies that directly affect business, such as the Directive on Corporate Sustainability Due Diligence, which entered into force in July 2024 [48]. The CSRD, meanwhile, which was adopted in 2022, marked a world first in terms of requiring large companies [49] and listed small and medium-sized enterprises to disclose sustainability information on both the environment's impact on the company as well as the company's impact on the environment (known as double materiality). The directive includes guidance on biodiversity and ecosystems [50], which effectively kick-started the implementation of Target 15 of the Global Biodiversity Framework (GBF). Even with attempts to improve the practical implementation of these policies, they continue to provide strong direction that will drive business action.

Also worth noting are the UK's biodiversity net gain policy and Australia's Nature Repair Market. In the UK, the policy requires developers to leave habitats in a better state than they were before development. And the Australian Nature Repair Market is expected to be the world's first-ever framework for a voluntary nature and biodiversity market. It aims to incentivise actions to restore and protect the environment. The Chinese government has pledged that by 2030, all key companies would assess and disclose their nature-related dependencies, impacts and risks. As a result, in February 2024, China's three largest stock exchanges put forward mandatory nature-related disclosure requirements for large, listed companies, starting as early as 2026. Finally, environmental damage may face stricter legal consequences, including criminal penalties. The European Union was the first international body to criminalise wide-scale environmental damage. Other countries such as Mexico are looking to go down the same route. On the other hand, the US is encouraging market-driven innovation with incentives through the Inflation Reduction Act.

The next few years will determine how effective these policies are being implemented and enforced. If they are, there is the potential to scale and speed up business action. But how can companies help to achieve the Nature Positive goal? Let's look at this next.

7.4   Scaling and speeding up business action for nature

Calling for policy ambition is clearly not enough. Companies must take immediate action – they should have started in the 1960s with *Silent Spring*, or at least when the GBF was agreed, but certainly by now, in the mid-2020s, even in an imperfect regulatory environment, they can and must transition from being part of the problem to part of the solution. For companies to take action, we must, collectively, drive convergence, consistency, clarity and collaboration in an increasingly complex environment, and this is a key piece of work that has been led by Michael Ofosuhene-Wise on my team.

Each organisation has its own value-add and way of explaining what companies should do for the nature agenda. But all these individual actions should in aggregate be complementary and aligned. It is why Business for Nature in 2022 teamed up with a suite of organisations to agree on the high-level business actions – also called the 'ACT-D' framework (see below). By following the ACT-D framework – **A**ssess, **C**ommit, **T**ransform and **D**isclose – companies can ensure their actions for nature are congruent with what is needed to achieve the Nature Positive goal at the aggregate level [51].

In 2022, Business for Nature worked with the Science Based Targets Network (SBTN) and the Taskforce on Nature-related Financial Disclosures (TNFD), as well as WBCSD, the World Economic Forum, WWF, Capitals Coalition and others, on the overarching high-level actions companies should take with regard to nature: The ACT-D framework that came out of this collaboration provides companies with the key actions they can take to signal that they are making meaningful contributions to help reverse nature loss. They are the following:

*   *Assess*: Measure, value and prioritise your impacts and dependencies on nature to ensure you are acting on the most material ones.

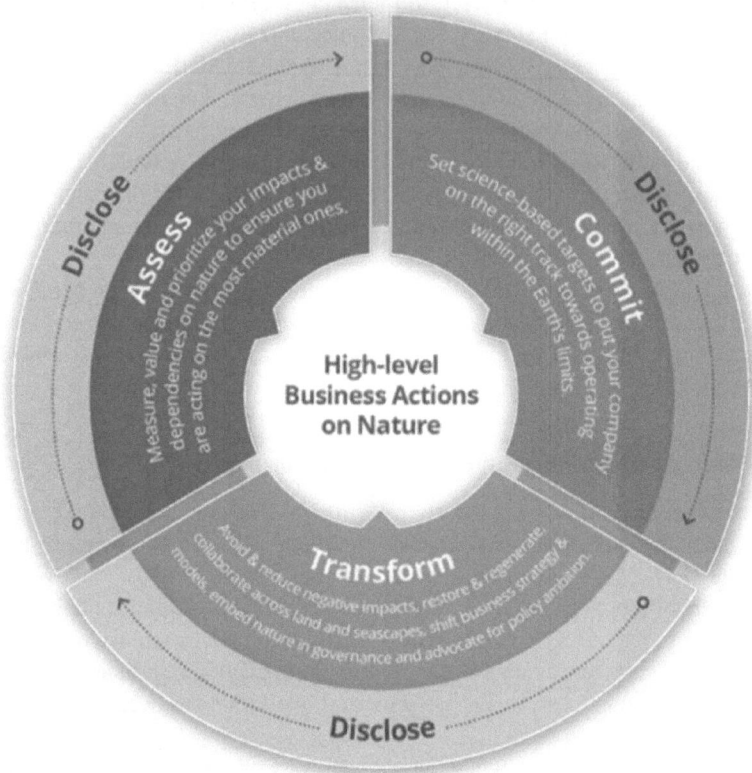

Figure 7.2 The ACT-D framework.

Source: Business for Nature

- *Commit*: Set transparent, time-bound, specific, science-based targets to put your company on the right track towards operating within the Earth's limits.
- *Transform*: Contribute to systems transformation by avoiding and reducing negative impacts, restoring and regenerating, collaboration across land, seascapes and river basins, shifting business strategy and models, advocating for policy ambition and embedding your strategy within your corporate governance.
- *Disclose*: Publicly report material nature-related information throughout your journey.

PERSPECTIVE 7.4   The corporate goalposts for nature action

By Erin Billman, Executive Director,
Science Based Targets Network (SBTN)

What does it concretely mean for a company to do its part to operate within earth's limits? How much is enough? The Science Based Targets Network (SBTN) [52] exists to provide companies and their stakeholders with answers to these questions, and it is why I was drawn to lead this initiative. So that companies can know the goalposts to aim for. So that if they set and meet their science-based targets, they and their stakeholders will know that they are taking enough of the right action, in the right places, at the right time, in order to do their part to reverse and halt nature loss by 2030, in support of the Global Biodiversity Framework.

The corporate guidance for science-based target setting is measurable, time-bound, and grounded in the latest scientific research with backing by leading NGOs working together to define those goalposts. This ensures the targets are robust, credible, and capable of delivering meaningful outcomes. I recognise that to drive the urgent action needed, the targets must also be pragmatic for companies. SBTN's approach involves ongoing engagement with companies to understand their challenges and needs, including 160 who are preparing to set science-based targets for nature. It also worked with the first pioneering companies to have set science-based targets in October 2024 – luxury group Kering, biopharma company GSK, and building materials company Holcim [53]. This engagement allows SBTN to tailor the guidance to be both scientifically rigorous and practically applicable.

I have focused my career on system change, and bring those key tenets to SBTN's ways of working. The Network listens, learns and adapts, in collaboration with a diverse range of stakeholders to ensure relevance, impact and alignment with the evolving scientific and regulatory landscape. In turn, SBTN's guidance includes how companies should engage with – and recognise the knowledge and potential contributions of – local stakeholders including Indigenous Peoples and local communities affected by companies' activities and value chains and therefore the target-setting process.

As companies increasingly recognise their impacts and dependencies on nature, there is a growing call for transparency and accountability. One way we are meeting this demand is by partnering with the Global Commons Alliance Accountability Accelerator, which will conduct independent validations of science-based targets, thus safeguarding the integrity of the targets and providing a clear, trusted framework for companies to follow. This not only enhances corporate credibility but also drives industry-wide progress by setting a benchmark for credible environmental action. It means the goalposts won't move, unless the science says they need to. And if they do need to be moved, they will be moved for all.

Having ACT-D was a critical achievement for a number of reasons:

- It provided a shared – unbranded – structure for all of us to use when we spoke with companies about what actions they should take, while giving flexibility to every organisation to expand on their own specific approach within ACT-D.
- It focused the conversation on taking the action, rather than companies getting stuck on trying to define or claim to be 'nature positive' themselves.
- It gave us a common foundation to build on and provided consistency as we developed widely accepted priority actions for 15 sectors [54]. This sector-specific work was co-led by the World Economic Forum, WBCSD and Business for Nature, and we engaged with many others including The Fashion Pact, One Planet Business for Biodiversity (OP2B), Planet Tracker, Animondial, the Textile Exchange, the Global Cement and Concrete Association and others. ACT-D is also the backbone of the Nature Strategy Handbook, available in multiple languages, that supports the 'It's Now for Nature' campaign that I will introduce later [55].
- It was essential to ensure our policy asks were robust by showing how, for example, Target 15 could be implemented, and it gave governments a widely-recognised structure to use as well in their own private sector engagement.

We had made good progress on agreeing what companies should do; however, the main question remained: how can we scale and speed up business action and demonstrate collective progress towards a nature-positive world by 2030? After multiple workshops and input from our partners and close companies, we developed and then launched a joint global business action campaign called 'It's Now for Nature' [56] that centered around companies having a strategy for nature, that is, a forward-looking plan of how they will contribute to a nature-positive world by 2030. We asked them: "Where is nature in your business strategy?" Because without a strategy, evidence tells us, meaningful action is less likely. We launched the campaign in November 2023, and it will run until 2030. Companies need to develop, publish and then submit their strategies to the platform, and they will be reviewed based on their strategy meeting minimum criteria. The first strategies were announced in May 2024, including GSK and Kering, and more and more have since joined [57]. The idea is that just as governments need to have their National Biodiversity Strategies and Action Plans (NBSAPs) as part of the CBD, so do business and financial institutions if they are going to meaningfully contribute. This doesn't mean that all these strategies are perfect – because a perfect strategy doesn't exist – but they still show that business momentum is growing and action is happening.

In addition, businesses will likely be expected to need nature transition plans in place, just as they have for climate. In 2025, initial guidance for how to do this is in development, which includes embedding nature in climate transition plans. Having a corporate strategy for contributing to a nature-positive world will be a critical piece of any transition plan in the future and help companies proactively navigate through the evolving reporting and regulatory requirements.

As we strive for evermore ambitious and impactful action to reach our collective nature-positive goal, our next challenge will be to ensure that company strategies, including their targets and actions, are in practice leading to the desired result. But what do we mean by that? A discussion paper that I co-authored, titled *How business and finance can contribute to a nature-positive future* [58], explains:

> Individual companies and financial institutions must adopt strategies across all their spheres of influence to contribute to the shared goal of a nature-positive planet by 2030. Theoretically, if a business or financial institution contributes more to restoring, regenerating and enhancing nature, than to harming it, across their value chains and portfolios, they could be nature positive. After all, the only way we can reach the global goal to be nature positive by 2030 is if all the components add up to that target. However, achieving this in practice is not as straight forward as one could imagine. It is highly unlikely and needs to be measured against strict spatially explicit criteria.
>
> Businesses should, therefore, aim to be as comprehensive as possible in the actions they take to assess, commit, transform and disclose their impacts and dependencies on nature, and in doing so highlight their contributions towards a nature-positive future rather than claiming to be nature positive themselves. The urgency of today's nature emergency demands that companies act immediately. They can learn and adjust their approach along the way as supporting regulations and methodologies evolve to provide more standardised guidance globally. What companies cannot do is hold back until the route ahead is clearly marked. By then, it will be too late.

As businesses are encouraged – and will be mandated – to assess and disclose their nature-related information, and as they publish strategies and targets, they risk accusations of greenwashing, which is when companies overstate their sustainability efforts. Greenwashing now also has a friendly cousin called 'greenwishing', which applies when the greenwashing is unintentionally unrealistic.

Greenwashing is a real issue because it hampers meaningful action and erodes trust. Regulation like the EU's Greenwashing Directive [59] is starting to crack down on it, and this will be key to making the rules fair for all companies. However an unintended consequence is that many companies – even industry leaders – have started to respond to the risk of greenwashing by resorting to 'greenhushing' – when companies limit what information they share for fear of being criticised. But in order for governments, regulators, investors and consumers to assess – and then reward or penalise – a company's performance, we need companies to be transparent and to disclose the right information. Companies that disclose (even if their performance is not great) should still not be disadvantaged over those that prefer to stay quiet. As these dynamics evolve, I recommend following this guidance: "Don't make nature-positive claims or statements that can't be evidenced. Nothing undermines the credibility of a business's pro-nature credentials more than making claims that are exaggerated, misleading, or false."

7.5   Looking forward

The journey to achieving the Nature Positive goal means transforming the global economic system and to embed the value of nature into business decision-making and disclosure. In June 2023, Frank Elderson, a member of the Executive Board of the European Central Bank, said that nature loss is an "existential [threat] for the economy and the financial system" [60]. It was a signal of change. I completely agree and would even go a step further by saying that *financial performance is irrelevant on a dead planet.*

System-wide change is essential for meaningful progress. Business and financial institutions operate within interconnected networks and value chains that they can and must influence but whose impacts cannot be fully attributed to them and they cannot be fully accountable for. By taking shared responsibility, being accountable and working together, we can drive positive change across value chains, landscapes and seascapes towards the collective Nature Positive goal.

Without transforming the way we do business and the way performance is recognised and rewarded by governments, regulators, financial institutions and consumers, businesses will not be able to contribute at scale and speed to deliver a nature-positive economy for all.

What can businesses anticipate and plan for?

a.   *Enhanced accountability, governance and transparency*

Businesses will increasingly be evaluated and held accountable for their nature-related performance, including dependencies, impacts, risks and opportunities. Rather than get stuck in the greenwashing, greenwishing or greenhushing trap, companies should use standardised guidelines to assess and disclose their material information. We can expect that it will be increasingly clear which companies are leaders, followers or laggards. As a result, we will see which ones will benefit from a stronger accountability system that recognises – and rewards or penalises – performance.

Companies must adapt their own governance. A way to do so is by assigning responsibility within the board, perhaps even considering giving a seat to 'nature', and adapting the company's articles of association to include mission-aligned legal language, which is a requirement for B Corp companies. A prime example of these changes is in France, which has a legal option for companies to become an '*societé à mission*', or a purpose-driven company. Danone became the first publicly listed company to adopt this status, which was embedded in its Articles of Association in 2020 [61]. Companies can also use their Annual General Meetings to get shareholders to agree to their climate transition plans, mission and strategy, like Unilever and others have done.

b. Evolving and unevenly-distributed regulatory landscape

COP15 created policy momentum for nature, and over the last two years, there have been multiple encouraging nature policy developments all over the world, sending a strong signal to businesses to start putting nature at the top of their agenda, explore how to assess and disclose their nature impacts and to develop strategies to reduce their negative impacts and increase positive impacts on nature.

National and regional nature policies will directly impact how companies operate. Businesses need to stay updated on these regulations and understand the implications for their own operations and supply chains. Multinational companies will need to navigate through political complexity and differences across the world, as described in Section 7.3 (Collective business momentum drives policy ambition for nature).

c. Increased investor pressure and standardised methodologies

Investors are increasingly considering environmental factors in their investment decisions. Businesses that prioritise nature conservation, restoration and sustainable use will likely attract more capital. Increasingly, investors and other financial institutions, pension plans and central banks are recognising the importance of including nature-related risks in their portfolios. Initiatives like Nature Action 100 are mobilising investors to engage with companies and drive greater action on nature. Alongside this increasing demand is the need for more standardised guidance and methodologies that will provide confidence and enable companies to move further, faster. This rising standardisation will not only make it easier for businesses to progress on their nature-positive journey; it will also increase transparency within and across sectors, so that companies can compete on performance using trusted – hopefully financial-grade – data.

d. Innovation boost and shifting business mindsets

Technological innovations will play a crucial role in addressing nature-related challenges. 'Nature Tech' means the broad set of technologies that support the implementation, acceleration and growth of nature-based solutions including using AI, satellite and spatial data for biodiversity tracking in real time and at low cost, and technologies to reduce environmental impact across the supply chain [62].

There also needs to be a mindset shift: individual companies and financial institutions must adopt strategies across all their spheres of influence to contribute to the shared goal of a nature-positive planet by 2030 [63]. This requires a strategic shift in focus from doing less damage to creating regenerative and restorative business models aligned with Nature Positive. All these transformations intrinsically imply a decoupling of business activity from natural resource use, including through the circular economy and requiring an absolute reduction in natural material consumption and production.

e. *Collaboration and partnerships*

Over the coming years, we need to see radical sector collaboration while leveraging the influence of trade associations. The guidance for the 15 sectors represents a strong start, but we also need greater pre-competitive collaboration to address shared challenges, particularly given the complexity and interconnectedness within specific sectors [64]. Companies working alone are unlikely to meet their sustainability goals, whereas those working together will ultimately deliver long-term value within and across their sectors.

Halting and reversing nature loss by 2030 requires an 'all hands on deck' approach. The business and finance community will have a major role to play and need to take accountability for their actions and influence. Nevertheless, they are not on this journey alone. They are supported by leading organisations working on methodologies, tools, frameworks and guidance, while governments are gearing up to deliver the policy frameworks that incentivise and enable businesses to be even more ambitious.

*****

Conclusion

As company leaders consider the journey we have been on so far and what to expect as we get closer to 2030, it should be clear that with greater accountability, action and ambition comes the need to invest in skills development and attracting and retaining talent. Companies will need to invest in developing their employees' skills in sustainability, biodiversity and nature-based solutions. Young professionals are increasingly prioritising sustainability in their career choices. Businesses that demonstrate a commitment to nature protection, restoration and conservation will be more attractive to top talent.

A key message to businesses on their journey to contribute to a nature-positive economy is: don't delay and don't let the perfect be the enemy of the good. There is no silver bullet for the best way companies can contribute to halting and reversing nature loss by 2030. But good, credible tools and approaches are out there, have been tested and are ready to be deployed. Today's nature emergency demands that companies act immediately. Companies shouldn't spend two years doing a perfect materiality assessment – they must already start taking priority actions now.

As I conclude this chapter and reflect on the beginning in the 1960s, I cannot end without being clear that progress in business has been far too slow. We now need an exponential increase in action. With nature risk rising up the business agenda, it becomes more urgent for companies to act, to plan, to be methodical and strategic and to collaborate with others to ensure success. The costs of inaction will be much higher. Even so, uncertainty and complexity mean mistakes will almost certainly lie ahead. But companies can learn and adjust their approach along the way. What companies cannot do is hold back until the route ahead is clearly marked. By then, it will be too late. It's *now for nature.*

ADDITIONAL PERSPECTIVES

PERSPECTIVE 7.5   Standards alignment is enabling accountability for biodiversity impacts

By Eelco Van Der Enden, former CEO,
Global Reporting Initiative (GRI)

Biodiversity on a global scale is threatened by unprecedented pressures – and the repercussions extend well beyond the environment. The reality is that planetary and financial wellbeing are inter-linked; a 2020 World Economic Forum analysis found more than half of global GDP is at least moderately or highly dependent on healthy ecosystems [65].

Yet a lack of consistent and comparable reporting has meant organisations often do not fully understand how their activities impact biodiversity, and where they need to make changes. It is no surprise that multiple stakeholders – including governments, investors, and consumers – expect companies to do more. Not only to disclose their impacts but also how they are managed, mitigated, and reduced.

Against this backdrop, we published a major update to the GRI Biodiversity Standard in January 2024 [66]. Setting a new benchmark for corporate accountability on biodiversity, 'GRI 101' enables any organisation to assess its impacts throughout the value chain, with location-specific reporting that gets to the heart of the drivers of biodiversity loss.

A fragmented reporting landscape is in no one's interest. While it can seem confusing – the so-called alphabet soup of standards – the reality is quite different. When it comes to how a company discloses its' impacts, on biodiversity and other sustainability topics, GRI provides reassurance through our engagement with other standard setters and key organisations. While we may come from different perspectives, it's important to ensure approaches are aligned as much as possible.

As such, we did not develop GRI 101 in isolation – the drafting saw cooperation with the European Financial Reporting Advisory Group (EFRAG), on their regulatory EU biodiversity standard, and it was shaped by input from the Taskforce on Nature-related Financial Disclosures (TNFD) and the Science Based Targets Network (SBTN). Most recently, GRI and the International Financial Reporting Standards (IFRS) Foundation announced deeper collaboration to optimise how the GRI and ISSB standards can be used together [67]. to equally address impacts as well as risks and opportunities. A first outcome is a pilot under which the IFRS Biodiversity and Ecosystem Services project will build on GRI 101.

Well-aligned standards on biodiversity are good news for companies and data users, ensuring more globally relevant reporting data. The increased transparency this leads to is crucial if we are to see improved corporate decision-making that fully accounts for impacts on the natural world.

PERSPECTIVE 7.6   The IUCN contribution to Nature Positive

By Grethel Aguilar, Director General,
International Union for Conservation of Nature

The International Union for Conservation of Nature (IUCN) has been an active and engaged member of the Nature Positive Initiative. We share the conviction that contributions to the Kunming-Montreal Global Biodiversity Framework (GBF) by companies, in particular, are crucial in enabling society to improve the status of biodiversity. Our focus has been on defining pathways for companies to identify and deliver verified, quantified contributions to the GBF, through the use of metrics and toolkits based around IUCN data sets. These contributions need to be delivered with the utmost urgency if the GBF goals and targets are to be met, so the approach is designed to be as efficient as possible, consistent with the delivery of unequivocally positive impacts.

IUCN's Measuring Nature Positive approach enables companies to screen for opportunities to reduce species extinction risk and ecosystem collapse risk across their landholdings, value chains and portfolios. At its simplest, the approach takes a company on a journey through screening for biodiversity impact risk and opportunity, using the Species Threat Abatement and Restoration Metric (STAR), through a focus on the sites or value chains with maximum potential for positive change.

In these places or value chains, the Calibrated STAR validation process allows companies to set baselines of threat intensity and targets that focus action on mitigating the most important threats in particular places. These threats are identified through the IUCN Red List of Threatened Species and the Red List of Ecosystems, and quantified reductions in these threats contribute to the GBF goals and targets, particularly Goal A.

IUCN has conducted a Union-wide consultation on the approach and is developing a web-based application of the approach that will allow companies to move along their transition pathway in the most efficient and productive manner. This application is being piloted with a range of companies and civil society actors across the world and is being improved with each step. We plan to have an operational version available to companies, together with guidance for participation in the process by government and civil society, by early 2025.

PERSPECTIVE 7.7   Building the Nature-Positive future:
a unified approach to climate, biodiversity, and economy

By Lucy Almond, Chair, Nature4Climate

Droughts, floods, fires and food. Like air, water, fire and earth, climate issues seem to cover every element of our modern life with added complexity. The interconnected crises of biodiversity loss, climate change, and desertification demand interconnected solutions. Climate change has become the third leading driver of biodiversity loss worldwide, and for communities on the frontlines, the divide between climate and nature targets is non-existent. These interconnected crises demand a unified, urgent response. How to tackle such a grand crisis? Where to find the real solutions?

The solution lies in the very systems that have supported life for millions of years: nature itself. Forests, wetlands, drylands and the ocean have, for millions of years, managed to weave the thread that brings balance to the planet and all those that live in it. Natural ecosystems, when left intact or restored, can absorb carbon from the atmosphere. This helps stabilize global temperatures, while allowing biodiversity to thrive and promoting sustainable economic systems that provide for people's health and livelihoods.

Many people think that intact nature and economic growth cannot co-exist – and to date, we haven't done a great job at proving the opposite. However, healthy ecosystems are the essential foundation of resilient economies. A nature-positive economy not only fosters environmental recovery but also ensures sustainable livelihoods – creating jobs, securing food sources, promoting health, and driving long-term economic development. Nature can provide up to a third of the climate mitigation needed by 2030 and significantly enhance resilience to climate change. This saves money from devastating impacts, protects businesses' sustainability and increases our chances to provide for a growing population.

Business-as-usual does not serve us anymore. Currently, harmful financial flows towards nature destruction outsize investments in nature-based solutions by 35 times. This needs to change if we are to meet our climate and biodiversity goals. Financial institutions, governments, and investors face operational, reputational, and regulatory risks if they fail to address ecosystem destruction.

However, investing in nature-based solutions provides a strategic and cost-effective way to address the interconnected challenges of the Rio Conventions: climate change, biodiversity loss, and land degradation, while at the same time making tangible headway towards the Sustainable Development Goals.

By shifting away from activities that cause deforestation and land degradation, businesses can benefit from new markets, financial approaches, and more resilient investment portfolios. Technologies like mapping, transparency tools, and supply chain innovations are de-risking nature-based solutions projects, attracting investors, and democratising data access, enhancing nature-based investments' credibility.

In 2017, the Nature4Climate coalition (N4C) was formed to raise awareness about the potential of nature-based solutions in providing up to one-third of the carbon mitigation required to achieve the Paris Agreement goals by 2030. From the 'Forgotten Solution' to 'Nature Positive', past N4C campaigns and advocacy efforts have highlighted the role of nature-based solutions in carbon mitigation and their fundamental importance alongside net-zero strategies. Change is underway. But it needs fast-paced ambitious action.

Our whole economy relies on natural resources, yet the destruction of nature seems to follow business-as-usual. We must transition from an extractive economic model to one that revalues natural ecosystems, ensuring long-term wellbeing rather than short-term gains.

There can be no lasting climate or economic solution without nature. Governments must embed nature into their policies, businesses must halt and reverse nature loss, and investors must seize the economic opportunities that come from protecting and restoring our ecosystems. Success depends on all of us working together – with nature.

Notes

1 In the modern era, it was Rahm Emmanuel, then Chief of Staff to President-Elect Barack Obama, who said in 2008 that "you never want a serious crisis to go to waste. And what I mean by that is an opportunity to do things that you think you could not do before." (See endnote 1). Centuries before that, Machiavelli, the Italian author, made a similar point in his treatise *The Prince*; namely that "princes become great when they overcome the difficulties and obstacles by which they are confronted" (See endnote 2).

2 The EPA acknowledged this in its "The Birth of EPA" article in 1985. It read: "Silent Spring played in the history of environmentalism roughly the same role that Uncle Tom's Cabin played in the abolitionist movement. In fact, EPA today may be said without exaggeration to be the extended shadow of Rachel Carson."

3 As a side note, Schmidheiny in the years prior to the Earth Summit had experienced firsthand the importance of sustainability, as during the 1980s he led the family-owned company Eternit, which used asbestos in many of its products. By his own account, he recognised the adverse health effect of the product early, and by 1976, "he began to push for an end to asbestos processing, leading the way in the world in this regard", ahead of Swiss and European bans on the product. But the fallout from the prior use of asbestos by his company, Eternit, haunted him until 2024, including in the Italian courts, where he was held liable for the deaths of workers and citizens who were affected by the asbestos produced by the company that was once owned by his family and was led by him.

4 Along with Craig Hanson, Janet Ranganathan, James Griffiths, Charles Iceland, John Finisdore, Suzanne Ozment and others, I was lucky to be part of the team working on the ESR. And we must have got that very framing of nature-related "impacts, dependencies, risks and opportunities" right, because it has now become mainstream through the influential work of the Taskforce on Natural-Related Financial Disclosures in 2023.

5 This era was also the start of my personal journey in corporate sustainability. In 2004, I supported the WWF to develop a biodiversity study for Lafarge, now Holcim, a world leader in construction materials.

6 To push the concept of ecosystem and natural capital valuation into business required vision, technical expertise and perseverance. The efforts would not have been possible without the significant role of many (too many to name all of them) including Mikkel Kallesoe, Will Evison, Emily McKenzie, James Spurgeon, Mark Gough, Pavan Sukhdev, Josh Bishop, Rosimeiry Portela, Holly Dublin, Hannah Brooke, Gerard Bos, Steve Lang, Richard Spencer and Richard Mattison.

7 Nestlé's efforts have been recognized globally, for example, through the World Bench-marking Alliance's Nature Benchmark where the company ranks second out of 816 businesses assessed. https://www.worldbenchmarkingalliance.org/publication/nature/companies/nestle-6/

References

1 Convention on Global Biodiversity, 2022. Kunming-Montreal Global Biodiversity Framework. Available at: www.cbd.int/gbf
2 Convention on Global Biodiversity, 2024. GBF Home, Target 15: Businesses Assess, Disclose and Reduce Biodiversity-Related Risks and Negative Impacts. Available at: www.cbd.int/gbf/targets/15
3 Business for Nature, 2022. More Than 330 Businesses Call on Heads of State to Make Nature Assessment and Disclosure Mandatory at COP15, 25 October. Available at: www.businessfornature.org/news/business-call-for-mandatory-nature-assessment-and-disclosure-at-cop15
4 The Wall Street Journal, 2008. Rahm Emanuel on the Opportunities of Crisis, 19 November. Available at: www.youtube.com/watch?v=_mzcbXi1Tkk
5 Machiavelli, N., 1532. The Prince, Antonio Blado d'Asola. Available at: www.gutenberg.org/files/1232/1232-h/1232-h.htm
6 Carson, R., 1962. Silent Spring, Houghton Mifflin Company. Available at: www.rachel-carson.org/silent-spring
7 US Environmental Protection Agency, 1985. 'The Birth of EPA' – by Jack Lewis. Available at: www.epa.gov/archive/epa/aboutepa/birth-epa.html
8 UN Environment Programme, n.d. About Montreal Protocol. Available at: www.unep.org/ozonaction/who-we-are/about-montreal-protocol
9 European Environment Agency, 2024. What Is the Current State of the Ozone Layer? Available at: www.eea.europa.eu/en/topics/in-depth/climate-change-mitigation-reducing-emissions/current-state-of-the-ozone-layer
10 UN Environment Programme, 2021. Rebuilding the Ozone Layer: How the World Came Together for the Ultimate Repair Job. Available at: www.unep.org/news-and-stories/story/rebuilding-ozone-layer-how-world-came-together-ultimate-repair-job
11 UN Environment Programme, 2021. Rebuilding the Ozone Layer: How the World Came Together for the Ultimate Repair Job. Available at: www.unep.org/news-and-stories/story/rebuilding-ozone-layer-how-world-came-together-ultimate-repair-job
12 Britannica, 2024. United Nations Conference on Environment and Development. Available at: www.britannica.com/event/United-Nations-Conference-on-Environment-and-Development
13 United Nations, n.d. United Nations Conference on Environment and Development, Rio de Janeiro, Brazil, 3–14 June 1992. Available at: www.un.org/en/conferences/environment/rio1992
14 UNEP, 2015. The World Mourns One of Its Greats: Maurice Strong Dies, His Legacy Lives On. Available at: www.unep.org/news-and-stories/story/world-mourns-one-its-greats-maurice-strong-dies-his-legacy-lives
15 Our History (wbcsd.org) 'The birth of WBCSD'.
16 Stephan Schmidheiny with Business Council for Sustainable Development, 1992. Changing Course: A Global Business Perspective on Development and the Environment. Available at: www.wbcsd.org/resources/changing-course/
17 SRF, 2024. Stephan Schmidheiny zu zwölf Jahren Gefängnis verurteilt, 8 June. Available at: www.srf.ch/news/schweiz/wegen-fahrlaessiger-toetung-stephan-schmidheiny-zu-zwoelf-jahren-gefaengnis-verurteilt
18 WBCSD, n.d. Roadmaps to Nature Positive. Available at: www.wbcsd.org/actions/roadmaps-to-nature-positive

19 Millennium Ecosystem Assessment, 2005. Overview of the Millennium Ecosystem Assessment. Available at: www.millenniumassessment.org/en/About.html

20 World Resources Institute, 2005. Millennium Ecosystem Assessment: Ecosystems and Human Well-Being, 1 March. Available at: www.wri.org/research/millennium-ecosystem-assessment-ecosystems-and-human-well-being

21 Reid, W., Carpenter, S., Mooney, H., Chopra, K., a.o., 2005. Ecosystems and Human Well-Being, Millennium Ecosystem Assessment. Available at: www.researchgate.net/publication/40119375_Millenium_Ecosystem_Assessment_Synthesis_Report

22 The Corporate Ecosystem Services Review | World Resources Institute (wri.org), 2008. WRI, WBCSD and Meridian Institute.

23 IUCN, 2015. No Net Loss and Net Positive Impact Approaches to Biodiversity. Available at: www.iucn.org/resources/publication/no-net-loss-and-net-positive-impact-approaches-biodiversity

24 Britannica, 2024. Deepwater Horizon Oil Spill. Available at: www.britannica.com/event/Deepwater-Horizon-oil-spill

25 Britannica, 2024. Deepwater Horizon Oil Spill. Available at: www.britannica.com/event/Deepwater-Horizon-oil-spill

26 BP, 2010. Sustainability Review 2010. Available at: www.bp.com/content/dam/bp/business-sites/en/global/corporate/pdfs/sustainability/archive/archived-reports-and-translations/2010/bp_sustainability_review_2010.pdf

27 WBCSD, 2011. Guide to Corporate Ecosystem Valuation. Available at: www.researchgate.net/publication/356427693_Guide_to_Corporate_Ecosystem_Valuation_About_the_World_Business_Council_for_Sustainable_Development_WBCSD

28 WBCSD, 2011. Guide to Corporate Ecosystem Valuation. Available at: www.researchgate.net/publication/356427693_Guide_to_Corporate_Ecosystem_Valuation_About_the_World_Business_Council_for_Sustainable_Development_WBCSD

29 Capitals Coalition, 2016. Natural Capital Protocol. Available at: https://capitalscoalition.org/capitals-approach/natural-capital-protocol/?fwp_filter_tabs=guide_supplement

30 Kering, 2024. Kering Digital EP&L Platform. Available at: www.kering.com/en/sustainability/measuring-our-impact/our-ep-l/

31 Kering, 2024. Kering Digital EP&L Platform. Available at: https://kering-group.opendatasoft.com/pages/home/

32 Natura Brasil, n.d. The Standing Forest. Available at: www.naturabrasil.com/blogs/blog-do-brasil/the-standing-forest

33 Perrot-Maître, D., 2006. The Vittel Payments for Ecosystem Services: A "perfect" PES Case?, International Institute for Environment and Development, London, UK. Available at: www.cbd.int/financial/pes/france-pesvittel.pdf

34 Le Monde, 2024. Revealed: France's Bottled Water Plants Widely Used Fraudulent Purifying Techniques, 30 January. Available at: www.lemonde.fr/en/environment/article/2024/01/30/revealed-france-s-bottled-water-plants-widely-used-fraudulent-purifying-techniques_6477927_114.html

35 UNEP, 2012. Natural Capital Declaration. Available at: www.unepfi.org/natural-capital-declaration/

36 UNEP, 2012. Natural Capital Declaration. Available at: www.unepfi.org/natural-capital-declaration/

37 United Nations, 2016. The Millennium Development Goals Report 2015. Available at: www.un-ilibrary.org/content/books/9789210574662/read

38 We Mean Business Coalition, 2024. Annual Report 2023. Available at: www.wemeanbusinesscoalition.org/wp-content/uploads/2024/07/We-Mean-Business-Coalition-Annual-Report-2023.pdf

39 Convention on Biological Diversity, 2020. Zero Draft of the Post-2020 Global Biodiversity Framework, January. Available at: www.cbd.int/doc/c/efb0/1f84/a892b98d2982a829962b6371/wg2020-02-03-en.pdf

40 European Commission, n.d. Virginijus Sinkevičius. Available at: https://commissioners. ec.europa.eu/virginijus-sinkevicius_en
41 Zabey, E., 2020. Nature Is Everyone's Business: A Call for Collective Action to Reverse Nature Loss, Business for Nature. Available at: www.businessfornature.org/news/ nature-is-everyones-business-a-call-for-collective-action-to-reverse-nature-loss
42 Business for Nature, n.d. Business for Nature's Call to Action. Available at: www.busi- nessfornature.org/call-to-action
43 Business for Nature, n.d. Make It Mandatory Campaign. Available at: www.businessfor- nature.org/make-it-mandatory-campaign
44 Business for Nature, n.d. Reform $1.8 Trillion Yearly Environmentally Harmful Sub- sidies to Deliver a Nature-Positive Economy. Available at: www.businessfornature.org/ news/subsidy-reform
45 Business for Nature, 2023. CEOs and Executives from More Than 80 Companies and Financial Institutions Urge the EU to Adopt Environmental Legislation to Address the Nature and Climate Crises Together. Available at: www.businessfornature.org/news/ euenviroregulation
46 Business for Nature, 2023. Businesses and Organizations Support Ambitious EU Reporting Standards to Scale and Speed up Nature Action. Available at: www.business- fornature.org/news/csrd
47 Business for Nature, n.d. Business Statement. Available at: www.businessfornature.org/ business-statement
48 European Commission, n.d. Corporate Sustainability Due Diligence. Available at: Cor- porate sustainability due diligence – European Commission (europa.eu)
49 Holger, D., 2023. 'At least 10,000 foreign companies to be hit by EU sustainability rules', Wall Street Journal, 5 April. Available at: www.wsj.com/articles/at-least-10-000- foreign-companies-to-be-hit-by-eu-sustainability-rules-307a1406
50 European Commission, 2023. Annex to the Commission Delegated Regulation (EU) Supplementing Directive 2013/34/EU of the European Parliament and of the Coun- cil as Regards Sustainability Reporting Standards. Available at: www.efrag.org/sites/ default/files/sites/webpublishing/SiteAssets/ESRS%201%20Delegated-act-2023- 5303-annex-1_en.pdf
51 Business for Nature, n.d. High-Level Business Actions on Nature. Available at: www. businessfornature.org/high-level-business-actions-on-nature
52 Science Based Targets Network, n.d. Home. Available at: https://sciencebasedtar- getsnetwork.org/
53 Science Based Targets Network, 2024. SBTN Announces First Companies Publicly Adopting Science-Based Targets for Nature, 30 October. Available at: https://science- basedtargetsnetwork.org/wp-content/uploads/2024/10/SBTN-October-announcement. docx-1.pdf
54 Business for Nature, n.d. Sector Actions towards a Nature Positive Future. Available at: www.businessfornature.org/sector-actions
55 Now for Nature, 2023. The Nature Strategy Handbook, November. Available at: https:// nowfornature.org/read-the-handbook/
56 Now for Nature, n.d. First Business Nature Strategies Published. Available at: https:// nowfornature.org
57 Now for Nature, 2024. First Nature Strategies Featured as Part of 'It's Now for Nature'. Available at: https://nowfornature.org/news/first-nature-strategies/
58 Business for Nature, 2022. How Business and Finance Can Contribute to a Nature Positive Future Now. Available at: www.businessfornature.org/news/nature-positive- discussion-paper
59 EU Directive Empowering Consumers for the Green Transition through Better Protec- tion against Unfair Practices and Better Information. Official Journal of the European Union, 2024. Directive (EU) 2024/825 of the European Parliament and of the Council

of 28 February 2024. Available at: https://eur-lex.europa.eu/legal-content/EN/TXT/PDF/?uri=OJ:L_202400825

60 Elderson, F., 2023. 'The economy and banks need nature to survive', The ECB Blog, 8 June. Available at: www.ecb.europa.eu/press/blog/date/2023/html/ecb.blog230608~5cffb7c349.en.html

61 Danone, n.d. Danone Became the First Publicly Listed Company to Adopt the "Société à Mission" Status. Available at: www.danone.com/about-danone/sustainable-value-creation/danone-societe-a-mission.html

62 Nature4Climate, 2024. Nature Tech Hub. Available at: https://nature4climate.org/nature-tech/

63 Business for Nature, 2022. How Business and Finance Can Contribute to a Nature Positive Future Now. Available at: www.businessfornature.org/news/nature-positive-discussion-paper

64 Business for Nature, n.d. Sector Actions towards a Nature-Positive Future. Available at: www.businessfornature.org/sector-actions

65 World Economic Forum, 2020. Nature Risk Rising: Why the Crisis Engulfing Nature Matters for Business and the Economy. Available at: www3.weforum.org/docs/WEF_New_Nature_Economy_Report_2020.pdf

66 GRI, 2024. Transparency Standard to Inform Global Response to Biodiversity Crisis, 25 January. Available at: www.globalreporting.org/news/news-center/transparency-standard-to-inform-global-response-to-biodiversity-crisis/

67 IFRS, 2024. GRI and IFRS Foundation Collaboration to Deliver Full Interoperability That Enables Seamless Sustainability Reporting, 24 May. Available at: www.ifrs.org/news-and-events/news/2024/05/gri-and-ifrs-foundation-collaboration-to-deliver-full-interoperability/

8 Nature-Positive finance

Dorothy Maseke[1]

Figure 8.1 The sun sets on Rusinga Island, Kenya

Photo credit: Norman Abade

Introduction: why I work in nature finance

I will not pretend that I have been in this field for many years. I lay no claim whatsoever to any expertise on nature, nor will I claim to have made any significant contributions to it. I am new to this field – my focus on nature beginning with what I can only call discontent. Discontent at what I saw happening around me, and which led me to pursue this passion project. My 'discontentment' journey has been peppered with many experiences, big and small, that have altered the lens through which I view the world around me.

At the same time, I also realise how privileged I am to pursue this as a mainstream career, following a mid-career shift, and how privileged I am to be able to write about it. When I reflect on the work of zoologist, primatologist, and anthropologist Jane Goodall, having followed her keenly for years, I now see the profound impact her work has had on me. I think of Sir David Attenborough – whose voice I have listened to on the BBC since I was a child – and I suddenly understand what spurred this deep passion for what I do. These voices and their

DOI: 10.4324/9781003474043-10

writings are the voices that fostered this deep passion in me ever since I was a little girl.

So here is the reason for my discontent. You see that tree over there? That tree has a story. I call it my 'discontentment visual'. I have it framed in my study. A special picture that serves as a reminder of why I do what I do. You see, as a child, my siblings and I used to climb it. We would run around it, skipping and singing, collecting shells at the lake shore during school holidays when we would be shipped, quite literally, to the countryside of Rusinga Island in Kenya. Sadly, that tree moves further into the lake every year. As far back as I can remember, that sunset has remained the only constant on that shore. In a few short years, that tree will rot and die. On the island, a building also stands at what's known as the 'Centre'. In many villages across Kenya, there is often such a place referred to as 'Centre' – usually the marketplace, where shops and businesses are concentrated. But on Rusinga Island, the Centre's main building, located on the shore of the island, has been shattered by the encroaching waters. Slowly, the Centre is losing its beach, as rising waters continue to displace businesses and establishments.

That is my village.

When I first published those pictures and the story of Rusinga Island, readers came out and told me they too had stories to tell. Stories of the rivers they used to swim in which have since dried up, or the forests and meadows they would hide in while playing as children which have since disappeared. Therefore, every time I wake up in the morning, though I am hundreds of miles away, I am reminded of that old tree, that building in the Centre, and now this beach that is slowly disappearing. I know my village has seen nowhere near the extreme levels of devastation that many places in the world have experienced due to degradation of nature and biodiversity, but it has been the cause of my discontent. I say this with 100% certainty, that we must respect the world we live in. We must protect what remains of it. It is foolhardy to view our present world as having significant time left. One must simply look at the devastating effects we are seeing on nature and our communities.

Many times have I said to myself, "too much to be done, too little time". These words describe exactly how I feel when no words can do these feelings justice. But through this chapter, I will find the words. Words to transform this arduous task, driven by this heavy burden in my heart – this discontent – into a challenging but stimulating experience. Words that will save my country, my continent, and the planet for future generations. Words that will push for greater finance for nature. Let me now introduce you to the history, present, and future of this topic, which is so close to my heart.

8.1   The history of nature finance

While various elements of nature finance have entered the mainstream lexicon of the finance world only recently, some of its foundations go back hundreds of years. Consider, in this regard, the concepts of conservation finance and socially responsible investing.

• *Conservation finance*

Let's start with *conservation finance*. The earliest recorded instance of conservation finance dates to 1634, in the United States. In that year, the residents of Boston voted to impose a six-shilling tax on each household to purchase William Blackstone's farm, which was turned into a communal area used for public, military, agricultural, and recreational purposes. As a form of 'utilitarian' conservation, regulations were put in place to limit the number of cattle each family could graze on the communal area to prevent overgrazing. This example demonstrates how a self-governing community can invest in open spaces for both public and private benefit [1].

Further West in the US, Yellowstone National Park, established on 1 March 1872, is often recognised as the world's first national park. This landmark was protected from private development through legislation known as the Yellowstone National Park Protection Act. The act declared: "The headwaters of the Yellowstone River ... is hereby reserved and withdrawn from settlement, occupancy, or sale ... and dedicated and set apart as a public park or pleasuring-ground for the benefit and enjoyment of the people" [3]. During a period of territorial expansion and post-expedition discoveries, the federal government acknowledged the immense value of this land and moved to preserve it from private exploitation. Similarly, the Hot Springs in Arkansas, protected since 1832, is considered the oldest federal reservation in the United States, originally set aside to conserve and manage its hot water resources for public use.

• *Forerunners of impact investing*

The roots of *socially responsible investing* – a forerunner to modern impact investing – too can be traced far back in time. The concept of Tzedek (which

Figure 8.2 View of the Water Celebration on Boston Common
Source: U.S. National Park Service [2] (Public domain image)

means justice and equality), outlined in Jewish law and written between 1500–1300 BC, established early economic standards. These standards encompassed the rights and responsibilities of individuals and emphasised the importance of preventing both immediate and potential harm to not just people but also the environment [4]. Similarly, the Qur'an, written between 609–632 AD, also set forth principles grounded in Islamic teachings, with a primary focus on preventing exploitation through the Riba standard. This standard not only prohibits the charging of interest but also excludes investments in alcohol, pork, gambling, armaments, and gold and silver [5].

In the United States, the roots of responsible investing can be traced back to the eighteenth century with the Methodists, a Protestant denomination that opposed the slave trade, smuggling, and conspicuous consumption. They also avoided investing in companies involved in the manufacture of liquor or tobacco products or those promoting gambling [6]. This approach evolved in the 1960s when protests against the Vietnam War called for university endowment funds to divest from defence contractors [7]. Furthermore, protests against the Apartheid regime in South Africa led to significant divestments from South African companies and prompted institutional and legislative changes worldwide [8]. Over time, the principles of socially responsible investing became aligned with a consistent investment philosophy reflecting investors' concerns, from opposing the slave trade and conflict to supporting fair trade and addressing modern issues related to environmental, social, and corporate governance (ESG).

- *Debt-for-nature swaps and green bonds*

Building on these roots, the past few decades saw the introduction of true nature finance, notably through debt-for-nature swaps, and green bonds. Let's look at debt-for-nature first.

In 1984, the World Wildlife Fund (now the World Wide Fund for Nature, or WWF) introduced the *debt-for-nature swap* to bolster conservation efforts in developing countries [9]. This concept emerged from the recognition that countries with significant biological diversity often face severe financial pressure from foreign debt. The first such agreement was signed in 1987 between the Government of Bolivia and Conservation International (CI). CI acquired $650,000 of Bolivian external debt for a discounted price of $100,000. In return, Bolivia committed to providing the Beni Biosphere Reserve with maximum legal protection, creating three adjacent protected areas, and allocating $250,000 in local currency for the reserve's management activities. Despite initial controversies and delays, this swap highlighted the potential of debt-for-nature swaps as a conservation financing tool.

Africa's first debt-for-nature swap, meanwhile, took place in August 2023 through a $500 million deal arranged by Bank of America. This agreement reduced the interest rate on Gabon's debt and extended the repayment period, with Gabon pledging to invest at least $125 million to expand a marine reserve and strengthen fishing regulations, aiding the protection of endangered humpback dolphins [10]. Growing in scale, debt-for-nature swaps are increasingly utilised as a financial strategy to alleviate national debt in exchange for environmental commitments [11].

Previously, the European Investment Bank, the EU's lending arm, issued the first *green bond* in 2007, followed by the World Bank a year later. Since then, numerous governments and corporations have entered the market to finance environmentally friendly projects. Initially, the green bond market grew slowly, but it has gained significant momentum due to global initiatives like the Paris Agreement and the UN Principles for Responsible Investing [12].

8.2   Where we are today

• *Why nature finance is needed*

First, let us remind ourselves of the gravity of the issue.

Human activity has drastically transformed 75% of the Earth's land surface and impacted over 66% of ocean areas, according to the Intergovernmental Science-Policy Platform on Biodiversity and Ecosystem Services (IPBES). This has led to the loss of more than 50% of coral reefs and 85% of wetlands, with 25% of the world's animal and plant species now at risk of extinction, and over one million species facing the threat of disappearing entirely [13].

We are at the point of irreversible tipping points for nature and climate with over half of global GDP, $44 trillion, potentially threatened by nature loss yet at the same time moderately or highly dependent on nature. The destruction of natural systems, aside from exacerbating climate change, also introduces significant risks to the global economy [14].

Experts have pointed out that there is an increase in economic and financial stability risks caused by the continued degradation of nature [15], including in *The Economics of Biodiversity: The Dasgupta Review* [16].

The World Economic Forum stated in 2020 that a fundamental transformation is needed across three socioeconomic systems, food, land, and ocean use; infrastructure and the built environment; and energy and extractives [17]. Incidentally, these systems are also responsible for most of the significant business-related pressures on biodiversity, with the IUCN Red List stating that 46,300 species are threatened with extinction, which is still only 28% of what is assessed [18]. These are also the systems with the largest opportunity to lead the co-creation of nature-positive pathways. Business opportunities worth $10 trillion that could create 395 million jobs by 2030 have been identified, paving the way for development that benefits both people and nature, while also building resilience to future shocks. To scale the transition, $2.7 trillion per year through to 2030 is needed [17].

Restoration costs alone, due to their high cost, require the highest levels of investment – $125 billion per year by 2025 and over $177 billion per year by 2030. This is estimated at over half of Nature-based Solutions (NbS) finance each year until 2030 [14].

• *The financing challenge*

Aiming to reduce the financing gap, Target 19 of the Kunming-Montreal Global Biodiversity Framework (GBF) aims to mobilise an additional $200 billion per year for biodiversity from all sources, including $30 billion through international finance [19]. This is a far cry from what is required. Current financial flows to NbS stand at approximately $200 billion – only one-third of the amount required to meet climate, biodiversity, and land degradation targets by 2030. These financial flows need to increase by $236 billion by 2025, with another $342 billion from current figures by 2030 to reach set targets [14].

Despite funding for NbS increasing by only 11% over the past year, there is some growing momentum with 66% of governments committed to restoring and protecting ecosystems as part of their Nationally Determined Contributions (NDCs) to the Paris Agreement, and over 100 governments including natural ecosystems as part of their adaptation plans. Governments continue to be the primary source of NbS funding, contributing around 82% of the total [14].

PERSPECTIVE 8.1   The (almost) $1 trillion nature finance gap

By Jennifer Morris, CEO, The Nature Conservancy

I remember the moment I began to understand, on a deep level, how nature is connected to health and economic wellbeing and how it touches down in our everyday lives. I was living in a rural village in Namibia and saw first-hand the enormous consequences climate change and biodiversity loss pose. Surrounded by drought-stricken fields, women had to spend entire days traveling far distances to collect water and firewood, leaving them exposed to violence and diminishing their chances of finding paying jobs. Three decades on and we are seeing this loss happen at great pace and scale.

The climate emergency and rapid biodiversity loss are the greatest threats of our lifetime and yet, nature is too often not fully valued – and, until recently, was ignored on the balance sheet. At The Nature Conservancy, we identified a financing gap of $722–967 billion per year between what the world spends on economic activity benefiting nature and activity damaging it [20]. To reverse the decline in biodiversity by 2030 and reach our 30x30 commitments, we need to be spending as much as $824 billion per year.

Since identifying this nature finance gap in 2019, there has been tremendous progress. The establishment of the Kunming-Montreal Global Biodiversity Framework was the watershed moment we desperately needed. Subsequent efforts like the 10-Point Plan [21] have laid out a clear vision for how we can close the gap. But while we have made incredible strides, our systems are not currently built for scale.

Governments, philanthropy, and corporations are stepping up but for every $3 dollars of public funding for nature, we only see $1 dollar of private finance. To meet the urgency and the scope of the challenge, we must change and embrace bold ideas that will rapidly and ambitiously increase financial flows to those crucial places and people who need it the most. The path to a nature-positive world supporting long-term stable societies and healthy economies is not simply about reducing harm, it is a fundamental shift in how we value and price nature – and will rely on the ingenuity and innovation needed to rethink how to catalyze private funding and investment and get resources into the hands of communities that need it most.

- *The paradox of negative finance flows*

While there is a significant underfunding of NbS, nature-negative economic and financial systems continue to rapidly degrade nature. Environmentally harmful subsidies, estimated to make up about 2% of global GDP ($1.8 trillion), accelerate the depletion of natural resources via methods of production and consumption, undermining the broader ecosystems which support planetary health [22].

These global financial flows, from both public and private sources, negatively impacting nature are estimated at $7 trillion per year [14]. This figure is based solely on direct impacts, known as Scope 1 emissions, which refer to greenhouse gases released directly from sources owned or controlled by an organisation, such as emissions from fuel combustion or industrial processes.

Despite the minimal private investment in NbS, approximately $5 trillion in private finance has a direct negative effect on nature – 140 times greater than the amount invested in NbS. On the public side, while government commitments abound, environmentally harmful subsidies have risen by 55%. The combined effect of these nature-negative finance flows from both sectors is immensely destructive, significantly undermining efforts to increase financing for NbS [14]. As highlighted earlier, the current levels of private, public, or Official Development Assistance (ODA) financing is insufficient to address biodiversity loss, therefore requiring a transformative shift from harmful to nature-positive incentives.

- *Developments in multilateral institutions and the private sector*

Target 19 of the GBF includes goals and targets to specifically address finance for biodiversity. It emphasises the need to align financial flows with biodiversity goals and to mobilise financial resources from all sources, including private finance, to address the nature crisis. Several actors in the global financial and economic system have taken initial steps to address the nature crisis and answer the GBF's call, as the examples below demonstrate. Still, much work remains to ensure that these efforts translate into significant and lasting impacts on biodiversity conservation and ecosystem health.

Integration of nature in global financial discussions

Nature is increasingly being integrated into global financial and economic discussions. Initiatives like the Taskforce on Nature-related Financial Disclosures

(TNFD), which has developed a framework for organisations to report on nature-related risks, similar to the Taskforce on Climate-related Financial Disclosures (TCFD), has helped raise awareness about the financial risks associated with biodiversity loss and ecosystem degradation.

The OECD policy guide for Finance, Economic, and Environment Ministers calls for the embedding of biodiversity in the financial sector, including recommendations on the alignment of finance flows with biodiversity goals by policymakers, regulators, standard setters, investors, and finance providers to pay greater attention to the impact of finance on biodiversity. It further highlights aspects of the need to align budget and fiscal policy with biodiversity objectives [23].

Over 130 businesses and financial institutions, with combined revenues of $1.1 trillion, have called on heads of state and governments to adopt and enforce nature policies to implement the Global Biodiversity Framework and accelerate decisive corporate action for nature. The business statement, coordinated by Business for Nature, is an urgent appeal from businesses to governments for immediate leadership to strengthen – not weaken – the policies, incentives, and legislation that will drive the necessary business action to halt and reverse nature loss this decade.

It delivers recommendations for nature policies that ensure business and financial actors protect nature and restore degraded ecosystems; sustainable resource use and management to reduce negative environmental impacts; that nature is valued and embedded in decision-making and disclosure; that financial flows are aligned to a nature-positive, net-zero, and equitable economy; and that there is adoption of ambitious global agreements to address key nature loss challenges. If these bold initiatives are implemented, they could unlock new opportunities and create a robust environment conducive to business action. Among these initiatives, the establishment of clear regulatory frameworks and incentive mechanisms are key to enhancing corporate accountability and to ensure a level playing field for businesses and financial institutions [24].

A group of scientists and economists from the Stockholm Resilience Centre, Potsdam Institute for Climate Impact Research, Norwegian Business School, and the Club of Rome have published a book called *Earth for All – A Survival Guide for Humanity* [25], which explores economic policies likely to deliver the most extraordinary turnarounds to achieve prosperity for all within planetary boundaries [26] within a single generation.

Commitments from financial institutions

A growing number of financial institutions are committing to aligning their investments with nature-positive outcomes. For example, the Finance for Biodiversity Pledge has seen over 194 financial institutions representing 29 countries and over 23 trillion euros in combined assets commit to protecting and restoring biodiversity through their finance activities and investments [27]. Similar initiatives are taking place in Africa through the African Natural Capital Alliance which brings together financial institutions and most recently other market players from over 45 African countries.

Development of nature-based taxonomies

Several countries and regions are developing or have developed taxonomies that include nature-related criteria. The EU sustainable finance agenda [28] aims to support companies and the financial sector by encouraging private funding of transition projects and technologies and facilitating financial flows to sustainable investments. The EU Taxonomy for Sustainable Activities [29] includes criteria for activities that significantly contribute to biodiversity and ecosystems. This helps guide investments towards nature-positive projects.

Increased private sector involvement

The private sector is increasingly recognising the financial risks posed by biodiversity loss. Companies and investors are beginning to consider natural capital in their decision-making processes. This is reflected in the rise of sustainable finance products, such as green bonds and biodiversity-focused funds. Details of this will be covered later.

Reform of harmful subsidies

Efforts to reform environmentally harmful subsidies that contribute to biodiversity loss are gaining traction. Various international agreements and national policies are targeting the redirection of subsidies towards activities that support biodiversity and nature conservation. Both G7 [30] and G20 [31] nations have made pledges to reduce subsidies that contribute to environmental degradation, such as fossil fuel subsidies, and redirect funds toward sustainable practices, including Nature-based Solutions.

Global alliances and convenings

Global alliances to scale investment in nature continue to be formed. For example, the Natural Capital Investment Alliance task force, launched in 2021, aims to scale and create synergies between mainstream asset owners and asset managers through a commitment to mobilise at least $10 billion in private capital into natural capital investments. It seeks to promote and demonstrate the opportunity associated with natural capital investment as a new and scalable asset class [32].

- **Government initiatives: a panoramic overview**

Several countries are increasingly recognising the role of finance and the economy in addressing the nature crisis as well and are putting in place efforts ranging from policy reforms and financial guidelines to innovative programmes that align economic activities with nature conservation.

The *United Kingdom*, for example, launched the *Dasgupta Review on the Economics of Biodiversity* in 2021, which emphasised the importance of natural capital in economic planning [33]. Following this, the UK has committed to mandatory biodiversity net gain for developments and the integration of nature-related financial risks into its Green Finance Strategy [34]. The UK has also established the Taskforce on Nature Markets to guide the development of nature-based markets.

Further afield, the *European Union* has adopted the EU Taxonomy for Sustainable Activities [32], which includes specific criteria for activities that contribute to biodiversity and ecosystem health. This taxonomy guides investment towards sustainable projects and is part of the EU's broader European Green Deal, which aims to make Europe the first climate-neutral continent by 2050. The EU is also actively working to reform environmentally harmful subsidies, with plans to redirect these funds towards nature-positive outcomes.

Costa Rica's attempt to recognise the economic value of nature, meanwhile, led the country to implement a 'Payments for Environmental Services' (PES) programme that compensates landowners for conserving forests, which provide vital ecosystem services [35]. This programme is funded in part by a fuel tax and has contributed to a significant increase in forest cover. Costa Rica has also integrated natural capital into its national accounting system, allowing for more informed policy decisions.

China has integrated the concept of 'ecological civilisation' into its national development strategy, emphasising the need to harmonise economic growth with environmental protection [36]. The country has developed Green Finance Guidelines to encourage financial institutions to support environmentally sustainable projects [37]. Additionally, it is piloting natural capital accounting (NCA) in several provinces to better measure and manage its natural resources [38].

New Zealand's approach involves incorporating natural capital into its 'Wellbeing Budget' framework [39]. Its Living Standards Framework includes environmental sustainability as a key pillar, ensuring that nature is considered in all major economic decisions. New Zealand has also been a pioneer in recognising the rights of nature, such as granting legal personhood to the Whanganui River and the Te Urewera Forest [40].

Canada's approach to the infrastructure and built environment socio-economic system includes the Nature Smart Climate Solutions Fund, which invests in Nature-based Solutions that reduce greenhouse gas emissions. It is also making progress in advancing NCA to better value its vast natural resources and incorporate these values into national and corporate decision-making [41].

- *Barriers and opportunities for investment in nature*

Despite the actions taken so far, finance for nature clearly falls short of what is needed to achieve the Nature Positive goals. Consider, in this regard, a finding of the Dasgupta Review we mentioned earlier:

"Collectively, however, we have failed to manage our global portfolio of assets sustainably. Estimates show that between 1992 and 2014, produced capital per person doubled, and human capital per person increased by about 13% globally; but the stock of natural capital per person declined by nearly 40%. Accumulating produced and human capital at the expense of natural capital is what economic growth and development has come to mean for many people."

Why has there been such little progress? The fundamental question that remains unanswered is: How can we not value what makes up half of what the entire world's GDP is moderately or highly dependent on?

The answer is simple, despite global commitments to halt biodiversity loss, our economic and financial systems fail to properly value nature and do not address the market failures that drive its over-exploitation. Instead, these systems often incentivise environmentally harmful economic activities. Ecosystem services have been freely given to us by nature for millennia, and it therefore does not come naturally to value and invest in what has been available for years. Estimating the timing, progression, and extent of their impacts is challenging due to their nature as global public goods. As a result, countries often prioritise more immediate and visible needs [42].

While in some instances nature is priced into the economy via policies and markets, the pricing may not necessarily be correct [43], for example where environmental taxes are insufficient to halt the destruction of nature, as Dasgupta argues [33]. Prices also do not capture all costs, especially those of negative externalities which mostly prove difficult to price [44].

Initiatives such as the G20 Sustainable Finance Roadmap and the Bridgetown Initiative illustrate the much-needed reforms to the global financial architecture that would enable countries to achieve the objectives of the GBF and reverse biodiversity loss. Notably, while efforts are being made by several countries, addressing biodiversity loss is challenging because the economic and financial rules and incentives are not aligned with the goals of the GBF.

Developing countries face an even bigger challenge due to limited market access, while they also lack the fiscal space to mobilise financing at the scale required to avoid the severe negative impacts of biodiversity loss, nature degradation, and reduced ecosystem services. These delayed investments lead them into a vicious circle exposing them to the risk of ecosystem collapse [42].

Additionally, the existing frameworks are voluntary. Although the International Sustainability Standards Board (ISSB) has adopted the IFRS S1 General Requirements for Disclosure of Sustainability-related Financial Information and the IFRS S2 Climate-related Disclosures Standards, nature-related disclosure standards are planned for the next phase of development. Such standards aim to ensure that investors receive high-quality, comparable information about sustainability-related risks and opportunities, and to enable companies to provide such information to their investors efficiently.

Let's have a look now at some more challenges faced by nature finance.

a. Natural capital accounting challenges

Natural Capital Accounting (NCA) is a systematic approach to measuring and valuing the stocks (which track the health of natural assets relative to a baseline) and flows (which track the value of benefits nature provides in a given period) of natural resources and ecosystem services in economic terms. This gives us a way to integrate the value of nature into national and corporate accounting systems, ensuring that the benefits derived from natural assets are recognised and accounted for in decision-making processes.

The primary goal of NCA is to improve the management and sustainable use of natural resources by making their economic value visible and quantifiable. Several challenges hinder its effectiveness in nature finance. One significant issue is the lack of standardised methodologies for measuring and valuing natural assets. This inconsistency leads to difficulties in comparing data across regions and sectors, making it hard to create a unified approach for investors.

Standardisation challenges for businesses include aspects of impact assessment and valuation methods, valuation factor data sets, and reporting frameworks [45], among others. Additionally, NCA often struggles with the complexity of ecosystems, where the interdependencies between various natural elements make it challenging to isolate and quantify individual components. The lack of high-quality, granular data further exacerbates these issues, as current data is often fragmented and inconsistent, limiting the accuracy of NCA outputs. As of 2023, only 90 countries utilise the UN System of Environmental Economic Accounting (SEEA) and maintain at least one natural capital account. Utility and devolution of NCA data also remains a challenge with varying levels of maturity between countries [46]. The most common accounts are for water flows, forests, land, agriculture, and energy – although many other accounts, including for biodiversity and ecosystems, are on the rise.

b. Blind spots

In our rapidly evolving world, there is a growing tendency to overlook the intrinsic role that nature plays in addressing immediate concerns such as climate change, conflict, and pandemics. While climate change is often framed as a critical issue, it is essential to recognise that the health of our natural ecosystems is closely intertwined with the stability of our climate [47]. Similarly, conflicts and pandemics are not isolated from environmental factors – degraded ecosystems can exacerbate both [48], contributing to resource scarcity and the spread of diseases. Ignoring the foundational role of nature in these pressing issues undermines our ability to develop holistic and sustainable solutions.

c. Data constraints

The 'data challenge' is a term that is used frequently in day-to-day discourse on nature finance. These constraints are said to pose significant challenges in nature finance and investment, impeding the ability to make informed decisions that balance financial returns with ecological sustainability. Specific concerns relate to the lack of standardised, high-quality data on the economic value of natural capital and the risks associated with biodiversity loss. The complexities in measuring and valuing ecosystem services further exacerbate this challenge.

The limited availability of robust and reliable data also restricts the development of financial products and mechanisms that could drive investment in nature. For example, the TNFD has identified data constraints as a major barrier to integrating nature-related risks into financial decision-making. It published a scoping study for a global data facility, stating that "high quality, nature-related data is a global public good increasingly demanded by a wide array of public, private and civil society stakeholders everywhere", and that "wherever possible, baseline nature-related data should be accessible to a broad range of stakeholders and not kept behind paywalls or in proprietary systems" [49].

The TNFD proposes a public data facility with the next step being the blueprinting of a preferred governance, funding, and operating model that can work with related climate data initiatives. Without comprehensive data, investors and policymakers struggle to assess the true value of natural assets and the potential returns on investments in conservation and restoration projects, leading to a persistent financing gap in nature-positive initiatives [50]. This situation underscores the urgent need for enhanced data collection, standardisation, and sharing practices to support the scaling of nature finance and to ensure that investments align with broader sustainability goals. There are also arguments on the challenge not being a lack of data but how biodiversity measurements are interpreted as well as elements such as accessibility, quality, format, relevance, scale, and transparency [51].

d. Lack of capital markets development

The development of capital markets dedicated to financing nature and biodiversity remains underdeveloped globally [52].

This significant lack of financial products, instruments, and markets specifically designed to address biodiversity conservation and sustainable natural resource management creates a major gap in efforts to mobilise necessary capital. Part of the reason for this is that nature and biodiversity products do not always provide a financial return or generate positive cash flow, and biodiversity considerations are not fully integrated into the financial models investors use to allocate capital [45]. Other demand and supply side challenges that act as barriers to investment in nature include lack of investible deals, deal structure, small deal size, long-term investment terms, and high associated risks [53]. Due to market failures, financial markets may not adequately reflect the true value of natural assets due to market imperfections, lack of information, or short-term profit motives. This can result in underinvestment in conservation and restoration efforts or, worse, lead to the degradation of natural capital as market forces fail to account for long-term ecological sustainability.

This is why, despite growing interest from the financial sector, the flow of private capital into nature remains painstakingly slow, particularly among institutional investors. Furthermore, while green bonds and sustainability-linked loans have gained traction in recent years, their focus has largely been on climate-related initiatives, with only a fraction directed towards biodiversity and Nature-based Solutions. Climate-related funds are estimated at around $534 billion of assets under management (AUM) [54], a contrast to the current 15 biodiversity funds with $1 billion AUM [55]. This disparity highlights the urgent need for innovative financial mechanisms and regulatory frameworks that can channel private and public capital into projects that support biodiversity conservation and ecosystem restoration on a global scale.

The integration of nature into capital markets should be a necessary evolution in finance. As financial institutions continue to drive for change, and customer preferences shift, the potential for capital markets to support a sustainable and resilient future becomes increasingly attainable.

e. A lack of nature-positive outcomes

Even as we discuss capital market development, a key principle is that any market development mechanism should lead to nature-positive outcomes, which means investments must result in a measurable net gain in natural capital. However, experts have pointed out that not all nature markets are equitable or truly Nature Positive [43]. For instance, certain investments can inadvertently lead to environmental degradation rather than improvement. A well-known example is the replacement of biodiverse native forests with monoculture plantations, which can undermine biodiversity and ecosystem health despite being labelled as a Nature-based Solution. Significant progress has been made in terms of development and the continued adoption of frameworks such as the TNFD (see later in this chapter) which seek to address these issues by providing structured guidance on how businesses and financial institutions can integrate nature-related risks and opportunities into their decision-making. However, the lack of capacity and awareness of both the risks and potential opportunities in nature remains a challenge. Many investors and policymakers still struggle to differentiate between truly nature-positive investments and those that may have unintended negative consequences. To ensure that capital market development genuinely supports nature, there is a need for stronger regulatory frameworks, better data and disclosure mechanisms, and enhanced collaboration between financial institutions, conservation experts, and local communities. Without these measures, the risk remains that nature finance could reinforce harmful practices rather than drive meaningful environmental restoration and resilience.

f. The conflation of valuation and monetisation of natural capital

Scholars argue that the monetisation of natural capital should not be conflated with the valuation of natural capital. Thomas Cuckston contends that while monetisation involves assigning a financial value to elements of nature for the purpose of economic transactions, this process does not necessarily reflect the true ecological value of biodiversity [56]. He emphasises that ecological value is complex and multifaceted, often encompassing aspects that cannot be easily quantified or captured in financial terms. Ian Thomson, meanwhile, argues in a similar manner that monetisation, while useful in some contexts, is not synonymous with valuation [57]. He calls for a more nuanced approach that goes beyond simple economic metrics and one that has an integrated framework that combines multiple forms of knowledge, including ecological science, local community insights, and ethical considerations into environmental accounting and decision-making.

Financial markets, on the other hand, thrive off monetisation, since a dollar figure can be assigned to a transaction, and this monetisation is often seen as corollary to valuation of nature. Regardless of your stance, whether supporting or opposing valuation or monetisation, both present economic opportunities while also raising significant dilemmas. While assigning a monetary value to nature could help address the biodiversity financing gap, this approach raises serious concerns regarding its ecological, practical, and moral implications. Firstly, the monetisation of nature risks reducing complex ecosystems to oversimplified economic metrics [56]. Valuing a forest solely for its carbon sequestration capabilities, as is often the case in carbon markets, neglects other critical functions such as biodiversity, water regulation, and cultural significance. This narrow focus can lead to the prioritisation of certain ecosystem services over others, potentially resulting in harmful practices like the promotion of monoculture plantations at the expense of more sustainable and biodiverse land uses.

Moreover, the equity implications are profound. Wealthy investors and corporations may disproportionately benefit from nature markets, while Indigenous Peoples and Local Communities (IPLCs), who rely on these resources for their livelihoods, could be marginalised. This could not only lead to social conflicts but also undermine trust in nature markets, stalling their progress. Finally, the ethical considerations cannot be overlooked. The commodification of living organisms and ecosystems raises questions about whether it is appropriate to assign economic value to elements of nature that have intrinsic worth beyond monetary terms. Monetisation in this context is therefore seen to denigrate nature's intrinsic value, replacing people's innate love for nature and the moral obligation to protect it with market-based incentives to act in their own financial self-interest [59].

In their article "The Fabrication of Environmental Intangibles as a Questionable Response to Environmental Problems", Eve Chiapello and Anita Engels [60] critically examine the commodification of environmental intangibles, such as carbon credits and ecosystem services. The main argument presented by Chiapello is that the process of turning environmental elements into marketable commodities – referred to as "fabrication" – is a problematic response to environmental crises. The authors argue that this approach often simplifies complex ecological relationships and reduces environmental protection to a series of economic transactions, which may not effectively address the underlying causes of environmental degradation. Chiapello and Engels [60] also highlight the ethical and practical issues involved in this commodification process, questioning whether market mechanisms can truly capture the value of environmental goods and services. They suggest that this market-based approach may exacerbate environmental problems by encouraging a focus on financial gains rather than genuine ecological sustainability. Such arguments highlight the need for a balanced and cautious approach to valuation and monetisation of natural capital and their effects on decision-making.

8.3   Strategies and frameworks

How can the world overcome the challenges finance for nature is currently facing? The precise answer may differ per jurisdiction and per sector. But on a global level, several strategies and frameworks have been crafted to tackle the challenges impacting nature and biodiversity finance. These efforts include summits, alliances focusing on framework development, and standards-setting organisations that create metrics and measures. In what follows, we provide a high-level overview of these strategies and frameworks.

Natural Capital Accounting

As mentioned earlier in the chapter, NCA is a framework that integrates natural resources and ecosystem services into national and corporate accounting systems. By valuing the contributions of nature to economies, NCA provides a clearer picture of the economic importance of biodiversity and ecosystems. It allows governments and businesses to make more informed decisions by highlighting the trade-offs between economic development and environmental sustainability. Globally, various initiatives, such as the UN's System of Environmental-Economic Accounting (SEEA), are working to standardise NCA practices. These efforts are crucial for integrating natural capital into mainstream economic decision-making, ensuring that the true costs of environmental degradation are recognised and addressed.

National biodiversity finance plans

Biodiversity finance plans are strategic frameworks designed to mobilise and allocate financial resources for the conservation and sustainable use of biodiversity. These plans involve a scan of the finance landscape for nature that, combined with an examination of the global catalogue of finance solutions and national consultations on emerging ideas, often results in solutions that had not previously been identified. They also include a baseline measurement of the national biodiversity finance gap and typically involve identifying financial requirements, priority setting, and sourcing funds from a mix of public, private, and philanthropic sources.

The plans also emphasise innovative financing mechanisms, such as biodiversity credits, green bonds, and payments for ecosystem services, to ensure sustainable and long-term funding. This tool is being used by more 130 countries, with early adopter Malaysia being able to introduce a system of Ecological Fiscal Transfers that channelled over $87 million in allocations to its states using ecological criteria [58]. With support from the Global Environment Facility (GEF), the United Nations Development Programme (UNDP) is supporting an additional 91 countries to develop national biodiversity finance plans, including addressing harmful subsidies and realignment of fiscal policies [61].

Taskforce on Nature-related Financial Disclosures

The TNFD is a global initiative that aims to develop a framework for organisations to report and act on nature-related risks and opportunities [62]. The TNFD framework provides guidance on how companies can disclose their impacts on nature, integrating biodiversity considerations into financial decision-making. By promoting transparency and accountability, the TNFD helps align financial flows with global biodiversity goals, supporting the transition to a nature-positive economy.

PERSPECTIVE 8.2   How the Taskforce on Nature-related Financial Disclosures helps shift global capital away from nature-negative towards nature-positive outcomes

By Tony Goldner, Executive Director, TNFD

Modelled on the work and global impact of the Taskforce on Climate-related Financial Disclosures (TCFD), the Taskforce on Nature-related Financial Disclosures was launched in 2021 with the support of the G20 and G7 and a network of global science and standard-setting partner organisations [63].

In simple terms, its goal is to help bring nature into the risk registry and onto the balance sheet of companies and financial institutions, based on a foundational realisation that the resilience of business and capital portfolios everywhere depends, to varying degrees, on the resilience of nature.

By shifting the mindset, behaviours, and ultimately actions of business and finance towards nature through the instrumentality of corporate reporting, the TNFD's ultimate goal is to help shift the flow of global capital away from nature-negative outcomes and towards nature-positive outcomes.

As a market-led initiative, the Taskforce comprises 40 private sector executives [64]. However, through its 'open innovation' design and development approach and extensive market engagement activities, it has gained the support and participation of stakeholders from around the world, including scientists, civil society leaders, conservation organisations, and Indigenous communities.

Initial market response to the release of the TNFD's corporate reporting recommendations, released in September 2023, was significant. In 2024, the first year following the publication of its disclosure recommendations and metrics and a suite of implementation guidance, the TNFD secured the voluntary adoption of 500 organisations across more than 40 jurisdictions [65]. They represented over $6 trillion in market capitalisation among publicly listed companies, over $16 trillion in assets under management among asset owners and asset managers, and 25% of the Global Systemically Important Banks (GSIBs).

Building on, and anchored by, its core disclosure recommendations, the TNFD in the fall of 2024 developed further implementation guidance, most notably on nature transition planning and financed nature impacts (the nature equivalent of financed emissions) [66]. Underpinning this implementation guidance is a measurement and metrics architecture that provides market participants for the first time with recommended metrics to support quantitative reporting across nature-related dependencies, impacts, risks and opportunities (DIROs). A critical part of that architecture for measurement is the metrics for assessing the 'state of nature' developed by the Nature Positive Initiative [67] of which the TNFD has been one of the founding members.

Within three years of its launch, and in close collaboration with key partners such as the Capitals Coalition, Science Based Targets Network (SBTN), Global Reporting Initiative (GRI) and the ISSB, the TNFD has played a key role in providing an integrated end-to-end assessment and reporting toolkit to enable corporate and financial institutions to take concrete, practical action aligned to the goals and targets of the Global Biodiversity Framework (GBF). Specifically, it contributes to Target 15 of the GBF, which calls on governments to introduce requirements by 2030 for assessment and disclosure of nature-related impacts, dependencies, and risks.

Biodiversity Credit Alliance (BCA)

This is a coalition that brings together diverse stakeholders, including a group of scientists, academics, conservation practitioners, and standard-setters, with direct links to Indigenous Peoples and Local Communities to define and develop relevant concepts that assist in the formation of a transparent and trusted biodiversity market [68].

The International Advisory Panel on Biodiversity Credits (IAPB)

This independent and global initiative aims to help unlock significant financial flows to nature through the development of high-integrity biodiversity credit markets. It provides expert guidance on the creation and implementation of biodiversity credit schemes and works closely with various organisations to establish best practices and ensure that biodiversity credits deliver real conservation benefits [69].

The taskforce on nature markets

This taskforce focused on creating guidelines and frameworks for nature-based markets, including biodiversity credits. It seeks to align these markets with global sustainability goals, ensuring that they contribute to biodiversity protection and restoration [70].

Nature investment standards programme

The Nature Investment Standards Programme is an initiative focused on developing standards that guide investments in Nature-based Solutions and biodiversity projects. This programme seeks to establish clear criteria for the financial sector to assess and manage biodiversity-related risks and opportunities [71].

IUCN's contribution to measuring Nature Positive

The IUCN is developing an approach to set and deliver verified, robust targets for species and ecosystems within a nature-positive framework. This approach involves setting targets for net-positive biodiversity outcomes, implementing biodiversity-friendly practices, and measuring progress against established benchmarks. IUCN's approach provides a roadmap for achieving biodiversity gains, fostering a proactive and positive contribution to global conservation efforts [72].

Plan Vivo's biodiversity standard

PV Nature is the Plan Vivo Foundation's (a registered UK charity that supports vulnerable rural communities and tackles climate change around the world, primarily through the development and application of its standards) biodiversity standard, aimed at generating the first high-integrity biodiversity certificates that deliver robust and credible outcomes for nature alongside social and climate benefits [73].

Verra's new biodiversity methodology

Verra, known for its role in carbon standards, has introduced a new biodiversity methodology that provides a rigorous framework for the quantification, monitoring, and verification of biodiversity outcomes. This methodology is designed to support the creation of biodiversity credits, enabling projects to generate measurable biodiversity gains that can be traded in voluntary markets. By setting clear and robust criteria, Verra's methodology ensures that biodiversity credits are based on scientifically sound and transparent practices [74].

Global Biodiversity Standard (GBS)

The GBS sets out a comprehensive framework for assessing and verifying biodiversity impacts and outcomes and is the world's only international certification that recognises and promotes the protection, restoration, and enhancement of biodiversity. It aims to provide a universal standard for measuring biodiversity performance across various sectors and geographies [75].

The Organization of Biodiversity Certificates (OBC)

The OBC is dedicated to creating standardised biodiversity certificates that quantify and verify positive biodiversity outcomes. It aims to provide a globally recognised framework for issuing biodiversity certificates, ensuring consistency and credibility in measuring and reporting biodiversity gains. These certificates can be used by companies and governments to demonstrate their contributions to biodiversity conservation, facilitating the integration of biodiversity into financial and corporate reporting [76].

8.4   Financial instruments for nature positivity

A second major aspect of making nature finance effective is to identify and use the right financial instruments to address the nature and biodiversity challenge. Broadly speaking four tools exist: credit markets, debt-for-nature swaps, payments for ecosystems, and traditional investing approaches, such as risk integration and meeting shifting investor demands. Among them, credit markets, and particularly carbon credit markets, are probably most developed, so we'll start our overview with them.

- *Credit markets*

A credit is a quantified amount of an ecosystem service, such as 1 tonne of carbon or a defined amount of biodiversity which can be sold in a market [77]. According to the nature markets taxonomy [77], credit markets are made up of two types of credits:

- *Nature-specific credits* reflect the value of ecosystem services. They may include voluntary biodiversity credits that represent a positive biodiversity outcome (or an activity that has been carried out and is likely to result in a positive biodiversity outcome) and that are not used to offset an equivalent negative impact on biodiversity elsewhere.
- *Nature-related carbon credits*, on the other hand, reflect the value of carbon sequestration or storage [78]. They are transferrable tokens representing the avoidance or removal of greenhouse gas emissions, measured in tonnes of carbon dioxide equivalent (tCO2e) [76].

While carbon credits are chiefly aimed at reducing or offsetting CO2 emissions, rather than protecting biodiversity, they are increasingly leveraging Nature-based Solutions – which harness the power of healthy ecosystems to remove and store carbon – to achieve this goal. Some projects are now valuing the positive biodiversity outcomes that can result from these nature-based carbon projects to create higher-value carbon credits [79].

A second way to distinguish credits is whether they are 'active' or 'pending'. It is a distinction that is used very often already for carbon credits, which are the most developed nature-related credits.

- 'Active' means the credits have been fully verified, issued, and are ready to be used, traded, or retired. These credits represent actual, quantified reductions or removals of carbon dioxide or other greenhouse gases that have already taken place.

- 'Pending' carbon credits, also known as 'forward' credits or 'future' credits, represent anticipated carbon reductions or removals that are expected to occur in the future. These credits have not yet been verified or issued because the carbon reduction activities have not been fully completed. Buyers of such credits are essentially investing in the project's future ability to generate carbon reductions.

Carbon credits can also either be human-made or nature-based and can be used as 'offsets' or 'insets'.

- 'Offsets' relate to credits purchased by an entity to compensate for its greenhouse gas emissions by funding projects that reduce or remove an equivalent amount of CO_2 or other greenhouse gases elsewhere.
- 'Insets', on the other hand, are credits generated from projects that reduce or remove emissions within a company's own supply chain or operations. In other words, insetting involves implementing carbon reduction projects that are directly related to the company's value chain. For example, a food company might work with its farmers to implement sustainable agricultural practices that increase soil carbon sequestration or reduce emissions from fertiliser use. The carbon credits generated from these activities would be considered insets because they are part of the company's supply chain.

- ***Carbon credits markets***

Carbon credits are traded in two types of markets: *Voluntary Carbon Markets* (VCMs) and *Compliance Markets*.

- VCMs are driven by organisations and individuals who choose to offset their emissions beyond what is legally required. These markets are often used by companies aiming to achieve corporate social responsibility goals or by individuals wanting to reduce their carbon footprint. Here buyers support projects that reduce emissions.
- Compliance Markets, on the other hand, are regulated by mandatory national, regional, or international carbon reduction regimes. They are usually aimed at energy-intensive emitters. In these markets, entities are legally obligated to reduce their emissions and can buy or trade carbon credits to meet their regulatory requirements. Such schemes include either a carbon tax or a cap-and-trade scheme, as in the EU markets, shifting economic incentives by making it more expensive to pollute. Seven of the 30 compliance markets globally allow some use of carbon offsets.

According to the London Stock Exchange Group (LSEG), the value of traded global markets for carbon dioxide (CO_2) permits reached a record 881 billion euros in 2023 [80]. The world's most valuable carbon market among them by far was the EU Emissions Trading System (EU ETS), worth around 770 billion euros, representing 87% of the global total. Most of the remaining value came from a limited group of other countries and regions, including the UK (4%), and the US (8%) [81], where California and the 11 Northeastern states gathered in the Regional Greenhouse Gas Initiative [82] have created regional compliance markets. (On a national level, however, the US does not have a carbon market.) Other (major) economies that have a carbon credit market include Australia, Canada, China, Indonesia, and Korea. Several more major economies, including Brazil, India, Mexico, Nigeria, Thailand, and Vietnam are either piloting the model or considering doing so, according to 2024 World Bank figures [83]. Yet despite these developments, only the EU, UK, and regional US markets have reached a significant carbon market value (more on that later).

By contrast, the total traded value in the VCM was only around $2.5 billion in 2023 [84], which is far below what experts believe is required to have a material impact on emissions. Notably, former Bank of England Governor Mark Carney indicated that a "$50–100 billion per annum market is an imperative to help reduce emissions" [85]. A recent report by the Taskforce on Scaling Voluntary Carbon Markets indicated that demand for voluntary carbon credits (VCC's) is set to rise by a factor of 15 by 2030, with the need for more than a 100-fold increase to achieve net zero by 2050 [86].

Beyond the obvious obstacle of reaching scale, VCMs face several other challenges, including concerns about the quality and integrity of carbon credits, lack of standardisation, and issues with transparency and verification. One significant challenge is ensuring that the carbon credits represent real, additional, and permanent emissions reductions. For example, some forest conservation projects have been criticised for issuing credits for avoided deforestation that may not have been at risk in the first place, leading to questions about the true environmental impact. This lack of robust oversight and standardised methodologies can undermine the credibility of the VCM, making it difficult for companies and individuals to trust that their purchases are genuinely contributing to climate mitigation.

To better understand just how carbon markets work in practice, let's have a closer look at the EU Emissions Trading System (EU ETS), the largest carbon market in the world.

- *The European carbon compliance market and how it operates*

Launched in 2005, the EU ETS has become a cornerstone of the EU's strategy to combat climate change. The market regulates around 11,000 installations across different sectors, from heat and power generation to energy intensive industry sectors [87]. It operates on a cap-and-trade principle, where a cap is set on the total amount of greenhouse gases that can be emitted by installations covered by the system. Within this cap, companies receive or purchase emission allowances, which they can trade with one another as needed. These allowances can be allocated for free or auctioned, depending on the sector and the specific rules in place. High-risk industries, such as those prone to carbon leakage (where companies might move production to countries with looser emission constraints), often receive a significant portion of their allowances for free to maintain competitiveness.

Each allowance permits the holder to emit one tonne of carbon dioxide or its equivalent in other greenhouse gases. Companies that reduce their emissions can sell their excess allowances to those that are struggling to stay within their limits. This creates a financial incentive for businesses to reduce their emissions, as they can profit from selling unused allowances.

Most allowances are auctioned by EU member states. Companies must purchase these allowances to cover their emissions. The auctioning process is centralised through appointed auction platforms, ensuring transparency and equal access for all participants. There are daily and monthly auctions which vary in volume. To partake in the auctions of buying and selling future contracts on European Union Allowances (EUAs), you must either be a direct market participant with a registry account or use a broker or bank to facilitate this on your behalf. Once the emission allowances have been put into circulation by EU companies, they can be traded by any party that has an EU registry account in the secondary market. Within the registry account, parties can buy and sell EUAs over the counter.

The cap on total emissions within the EU ETS is reduced annually, decreasing the number of allowances available. This scarcity drives up the price of allowances over time, encouraging companies to invest in cleaner technologies to reduce the need to purchase additional allowances. At the end of each compliance period (usually a year), companies must surrender enough allowances to cover their total emissions. If they fail to do so, they face hefty fines and are required to make up the shortfall in the next period. This ensures that the cap is respected and emissions reductions are achieved.

Through this system of distribution and trading, the EU ETS incentivises companies to lower their emissions while allowing flexibility in how they meet their regulatory obligations. The gradual tightening of the cap ensures that overall emissions in the EU continue to decrease, driving progress toward the EU's climate goals.

REDD+ and the African carbon markets initiative

REDD+

Carbon credit markets can be deployed with the exclusive ambition to reduce carbon emissions, but they can also play a dual role, creating nature-positive outcomes alongside carbon-negative ones. One prime example in this regard is the sovereign carbon markets programme known as REDD+. The REDD+ mechanism was established by the UNFCCC to protect forests as part of the

Paris Agreement. ('REDD' stands for 'Reducing emissions from deforestation and forest degradation in developing countries'. The '+' stands for additional forest-related activities that protect the climate, namely sustainable management of forests and the conservation and enhancement of forest carbon stocks) [88].

REDD+ is the world's largest single category of carbon credit projects, comprising 28% of all credits issued since 2016, while also producing sovereign credits [89]. Its aim is to incentivise developing nations to conserve their forests and reverse deforestation to count towards these countries' Nationally Determined Contributions (NDCs). While REDD+ primarily targets developing countries due to their significant tropical forests, which are crucial for carbon storage, the principles of sustainable land management and conservation are relevant globally.

The COP26 agreement on sovereign carbon credits, which facilitated the creation of a rulebook for global trading of sovereign credits and UNFCCC REDD+, offers crucial optimism for the voluntary market's transition to Paris Agreement compliance, aligning in terms of standards, scale, and sophistication. As such, sovereign carbon credits could stimulate the development and improvement of the voluntary market [90]. Specifically for low- and middle-income countries, experts have pointed out that developing countries are driving the market towards a jurisdictional REDD+ (sovereign or sub-sovereign) approach which expands the scope of REDD+ beyond individual projects or specific areas. This enables the voluntary purchase of a compliance grade asset with the potential to push more market participants to finance progress towards the Paris Agreement net-zero targets [91].

Africa Carbon Markets Initiative (ACMI)

Another example of a nascent but promising carbon market is the Africa Carbon Markets Initiative (ACMI). The ACMI was launched at COP 27 in 2022 to harness the potential of carbon markets in Africa, aiming to unlock billions of dollars in investment while supporting climate action and sustainable development across the continent. It seeks to boost the production and sale of high-quality African carbon credits by promoting best practices, providing capacity-building, and facilitating market access.

Achievements so far, which build on past momentum, include broadening its scope to integrate compliance carbon markets and mechanisms under Article 6 of the Paris Agreement, in addition to the initially targeted voluntary carbon markets. It has aggregated investment intentions totalling $1 billion for high-quality African carbon credits by 2030, with $250 million already pledged, helping to boost confidence in African carbon markets. It has also worked with development partners to create a market growth pathway that fosters high integrity and transparency [89].

- *How global carbon credit markets can expand going forward*

For carbon markets to be effective, that is, to drive down emissions over time, a prime requirement is that the price credits trade at is high enough (and their supply low enough). Only in that case will their buyers be incentivised to reduce their emissions rather than offset them by buying credits. But what is this effective price? In 2022, the World Bank estimated it to be $50 to $100 per tonne of CO_2 by 2030, to meet the temperature goals of the Paris Agreement [92]. Unfortunately, the only major scheme to meet this threshold pricing is the EU ETS, whose price at the time of writing hovered around $70 [93].

Furthermore, according to a recent Nasdaq study, structural reforms are needed to unlock global carbon markets. *Price transparency, market inefficiencies*, and *fragmentation* remain critical structural barriers to scale while carbon credit registry reform is the most important facilitator of growth [94]. Let us look at this in a bit more detail.

First, structural issues within the VCM have led to low confidence in the valuation of carbon assets, making it difficult for investors to accurately determine prices or benchmark credits on the demand side. This lack of *pricing transparency* hampers brokers' ability to trade and prevents investors from holding onto the assets, resulting in trading volumes remaining at an artificially low level. Secondly, this restriction on volumes limits supply in commercial banks, as financiers struggle to model risk accurately and deploy capital effectively, leading to market inefficiencies. Finally, inconsistencies across different credit types and a reliance on manual processes for transactions, including due diligence and pricing, not only hinder trading but also restrict access for foreign investors to local markets, resulting in continued market fragmentation.

To overcome these hurdles, *carbon registries* are seen as a key driver to scaling carbon markets. They would provide a standardised, transparent, and credible system for tracking, verifying, and trading carbon credits. They would essentially enhance market integrity, attract investment, facilitate trade, and support both compliance and voluntary carbon markets.

Other market-based developments aimed at overcoming current shortcomings, or at expanding the markets' reach, include a better functioning of *digital spot markets* and standardised *futures contracts* linked to specific project attributes or compliance needs.[2] As the use of futures contracts grows – allowing buyers to secure credits for future use – these exchanges may gain in popularity. Additionally, *crypto exchanges* are emerging as blockchain technology offers new opportunities for innovation and growth in the market. Several leading institutions are working to ensure that tokens marketed as carbon credits comply with the standards set by the original issuing bodies [95].

Biodiversity credit markets

Biodiversity credits represent a financial mechanism aimed at supporting and incentivising the conservation and restoration of natural habitats and species. These credits function similarly to carbon credits but are focused on biodiversity outcomes, allowing businesses and governments to offset their environmental impact by investing in projects that protect or enhance biodiversity. Target 19 of the GBF identifies biodiversity credits (not offsets) as a novel approach to channel private sector investment into biodiversity. To support this, organisations like the International Advisory Panel on Biodiversity Credits (IAPB) and the Biodiversity Credit Alliance are working with governments, NGOs, philanthropic entities, and private market players to develop the essential frameworks for implementing these schemes.

The biodiversity credits market is still at a very nascent stage, with a report by the World Economic Forum stating that with effective progress across multiple fronts, global demand for biodiversity credits could reach $2 billion in 2030, which is just 1% of the total finance required to meet the 2030 goal, and $69 billion in 2050. With less effective progress, global demand could reach $760 million in 2030 and up to $6 billion in 2050 [96].

Regulated national biodiversity offset schemes are currently mobilising jointly around $6 billion to $9 billion annually [97], while on the voluntary side there is currently very little trading and associated investment in biodiversity outcomes.

In terms of businesses, only half of Fortune 500 companies acknowledge biodiversity loss in their sustainability reporting, while just 5% have developed quantified biodiversity-related targets [79]. Unlocking scale in biodiversity credit markets involves dealing with demand and supply-side challenges, consolidating common principles, standards, and methods, and building a supportive policy environment.

The complexity of biodiversity credits lies in the difficulty of measurement and standardisation. Unlike carbon credits, which can be quantified in metric tonnes of CO_2 equivalent, biodiversity encompasses a wide range of ecological factors, such as species diversity, habitat quality, and ecosystem functions [79]. These factors are often context-specific and cannot be easily reduced to a single unit of measurement. This adds complexity in establishing standardised metrics and valuation methods for biodiversity credits, making it challenging to ensure consistency, comparability, and credibility across different projects and regions.

Five key design challenges have been identified to unlock high-integrity biodiversity credit markets. These include delivering credible, timely, and cost-effective measurement and monitoring of biodiversity status, improvements, and maintenance. Other challenges involve scaling sustained demand for credits with integrity, securing sufficient supply, and ensuring fair pricing and equitable distribution of benefits to project developers, governments, Indigenous peoples and local communities. Additionally, robust governance and transparent institutional frameworks are essential [98]. These challenges build on ten recommendations outlined in a report by GEF and the IIED.96 [99].

• *Overview of current biodiversity schemes and methodologies*

Biodiversity credit markets are expected to develop under private sector-led schemes via voluntary standards or under government-led schemes via specific policy or legislation, or a potential combination thereof [100].

Research by Pollination Group indicates that the private sector is currently leading the development of biodiversity credit schemes and initiatives. Twenty-six are currently active: eight are international; there are four in Australia; four in Colombia; two in New Zealand; one each in South Africa, France, Switzerland, Sweden, Scotland, Brazil; and another one in the African continent [101]. Out of these 26 initiatives, Pollination reported that the majority failed to establish requirements for obtaining free, prior, and informed consent (FPIC) and did not incorporate co-ownership, partnership, or benefit-sharing models with Indigenous Peoples and local communities, leading to the much-discussed equity concerns of credit schemes. The study "Investing in Africa: Investing in Nature" explores the biodiversity credits landscape across Africa, outlining three potential pathways for market evolution: a localized, community-driven development scenario; a global, market-oriented scenario; and a coordinated, policy-supported scenario [102].

- *Global action on biodiversity credits*

Several global initiatives are collectively working to create the infrastructure, standards, and market conditions necessary for biodiversity credits to become a viable tool for conservation finance globally.

These include the Biodiversity Credit Alliance, a coalition that brings together diverse stakeholders, including governments, NGOs, and private sector players, to develop frameworks and standards for biodiversity credits. The International Advisory Panel on Biodiversity Credits provides expert guidance on the creation and implementation of biodiversity credit schemes, working closely with various organisations to establish best practices and ensure that biodiversity credits deliver real conservation benefits. The Taskforce on Nature Markets is focused on creating guidelines and frameworks for nature-based markets, including biodiversity credits. It seeks to align these markets with global sustainability goals, ensuring that they contribute to biodiversity protection and restoration.

Country approaches so far include the EU's Green Deal and Biodiversity Strategy which is exploring the integration of biodiversity credits into its broader environmental and climate policies. This consists of developing regulatory frameworks that support the use of biodiversity credits as part of the EU's Green Deal and Biodiversity Strategy. Actions in emerging biodiversity markets include Australia, Brazil, and South Africa which are developing national frameworks for biodiversity credits, often linked to specific ecosystems or species. These frameworks aim to attract private investment in biodiversity conservation through market-based mechanisms.

- *Payment for ecosystem services (PES)*

PES is a market-based instrument that incentivises the conservation and sustainable management of natural ecosystems by compensating landowners or resource stewards for the ecological benefits their lands provide. In this context, ecosystem services – such as clean water, carbon sequestration, biodiversity, and soil fertility – are recognised for their critical contributions to both the environment and the economy.

Such programmes allow for the translation of the services that ecosystems provide for free into financial incentives for their conservation, targeted at the local actors who own or manage the natural resources [103]. For example, a government or private entity might pay farmers to maintain forest cover, which in turn supports biodiversity, water regulation, and carbon storage. The concept is such that whoever provides a service should be paid for doing so.

PES helps bridge the gap between environmental protection and economic growth, providing a tangible economic value to ecosystems that are often undervalued or ignored in traditional markets.

Payment mechanisms for PES schemes can be performance-based, where payments are tied to the actual delivery of ecosystem services, such as a recorded increase in biodiversity. Alternatively, they can be input-based, where payments are made for the implementation of specific land or resource management practices [104].

PES financing is usually mobilised as direct user-financing and third-party financing (i.e. where governments or organisations act on behalf of beneficiaries) [105].

Most PES programmes are generally limited in geographic scope and are funded by users of specific environmental services. Most of these programmes receive their financing from government sources, with additional funding coming from multilateral banks and international non-governmental organisations.

Key design features for successful PES implementation include ecosystem valuation – quantifying impacts and assigning economic value – along with supportive

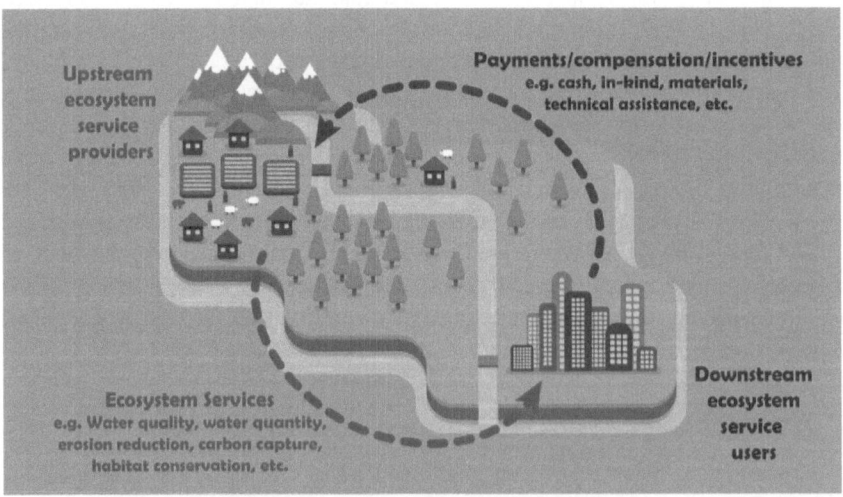

Figure 8.3 Payments for Ecosystem Services (PES)

Source: Bosco Lliso (Licensed under CC BY-SA 4.0)

legal and institutional frameworks, and effective stakeholder organisation [106]. Where resource tenure or land tenure/use rights are not defined, significant challenges abound. Designs should account for monitoring and reporting, additionality, leakage, and permanence, while also addressing equity and distributional concerns [104].

- *Debt-for-nature swaps*

A debt-for-nature swap is an agreement that reduces a developing country's debt stock or service in exchange for a commitment to protect nature from the debtor-government. It is a voluntary transaction whereby the donor(s) cancels the debt owned by a developing country's government. The savings from the reduced debt service are then invested in conservation projects [107].

The uniqueness of each country's debt position and structure enables this instrument to be customised to fulfil specific fiscal needs, bringing much-needed flexibility into the structuring.

Debt-for-nature swaps are estimated to have the capacity to provide $100 billion to go towards restoring nature and to help countries adapt to climate change [108]. The countries most vulnerable to climate change, despite having contributed the least to global warming, often face heavy debt burdens. Fifty-eight of the developing countries most vulnerable to climate change have almost $500 billion of debt servicing payments due in the next four years [109].

Debt-for-nature swaps offer a way to channel finance into nature and biodiversity without adding to already-existing debt burdens. This mechanism becomes a triple win since countries can reduce their external debts, nature and biodiversity outcomes can be achieved, while banks profit from participating in the transactions [110]. There are around 140 swaps in total, including in Belize, Gabon, Philippines, Costa Rica and Seychelles.

Credit enhancement, which enables capital raising for debt retirement at a lower cost, as well as political risk insurance, plays a major role in reducing any structural and transaction risks, providing confidence to investors and reducing the need for concessional capital.

Ecuador, for example, has since sealed one of the world's largest debt-for-nature swaps on record, a new 'blue bond' that will funnel at least $12 million a year into conservation of the Galapagos Islands. This project, done with the help of Credit Suisse, bought back around $1.6 billion of the country's debt at a near 60% discount due to the devaluation of existing Ecuador sovereign bonds. The transaction has an $85 million 'credit guarantee' from the Inter-American Development Bank and $656 million of political risk insurance from the U.S. International Development Finance Corp (DFC) [111].

- *Traditional finance tools and shifting investor preferences*

As the financial sector increasingly recognises the interconnectedness of economic stability and environmental health, nature-positive finance is emerging as a critical investment strategy. Investors who prioritise nature-positive considerations in their portfolios are leading the charge in this paradigm shift. Their insights reveal a growing commitment to aligning financial returns with the preservation and enhancement of natural ecosystems.

One key insight from these forward-thinking investors is the emphasis on *risk management*. Traditional investments often overlook environmental risks, which can lead to significant long-term financial losses. By integrating nature-positive criteria into their decision-making processes, investors not only mitigate environmental risks but also ensure the resilience of their portfolios. This approach recognises that stable natural ecosystems underpin economic activity, and that the degradation of these systems can pose substantial risks to financial stability. In recent years, substantial work has been done on the impact of nature loss on banking systems in various countries, prompting investors to place nature and biodiversity at the forefront of their risk discussions.

In addition to risk management, investors are increasingly aware of the potential for competitive or at least 'sufficient' returns from nature-positive investments, especially in the long run. Demand for sustainable funds, notably, has continued to rise into 2024 [112], continuing a trend that started several years ago – despite recent negative headlines and a slowing of flows into ESG-related funds in the US. The shift is most pronounced in Europe [113], where a new generation of investors arguably value not only financial returns but also the positive impact of their investments on the environment. All in all, by 2023, the funds labelled by the finance industry as 'sustainable' grew to a total of $3.4 trillion under management (compared to more than $40 trillion in traditional funds), and returned more to investors than did the traditional ones in every of the past five years, except 2022 [114].

Compared to this broader category of 'sustainable funds', much fewer data are available about 'nature-positive funds' It has not been easy to get global figures on Nature Positive funds, although according to large investors, consulting firms and private banks such as Federated Hermes, McKinsey, and Lombard Odier, "the demand for nature and biodiversity investments is growing" [115], and will in fact be "unstoppable" in the long run given that the need to become nature-positive and climate-resilient is present in almost every major industry.

Conclusion

As the global focus on sustainability intensifies, the financial landscape is rapidly evolving, with nature finance emerging as a critical component of sustainable development. Investors, increasingly disillusioned by traditional methods that fail to deliver measurable impact, are turning to direct investments and strategies that go beyond mere offsets and insets. These investments, which include renewable energy transitions, sustainable product development, and circular economy initiatives, represent a proactive approach to reducing carbon footprints and conserving nature and biodiversity. However, they also demand substantial upfront capital and a long-term commitment.

The shift towards nature finance introduces new complexities. Geopolitical dynamics, power imbalances between countries, and social equity concerns underscore the multifaceted nature of these investments. The increasing costs associated with transitioning to sustainable practices raise important questions about broader economic implications, including the potential for higher food and energy prices for consumers. Yet economic activity is sustained by the stability of natural ecosystems. To effectively navigate this landscape, there is a pressing need to reframe the conversation in terms that resonate with investors. The language of risk, familiar to financial institutions and investors, must be adapted to highlight the investible opportunities inherent in ecological health. By viewing the reduction of ecological risk as a pathway to creating investible assets, the case for nature finance becomes more compelling and accessible to the financial community.

Two critical themes emerge from these discussions. First, risks are escalating as natural assets degrade, threatening essential ecosystem services. Second, valuing natural capital is crucial to mitigating these risks. The link between nature degradation, biodiversity loss, and systemic risks has long been recognised, as highlighted in global risk reports over the years. Public finance currently shoulders the bulk of investments in Nature-based Solutions, yet this funding remains woefully inadequate relative to the immense needs and opportunities. Private finance, which is critically low, must be significantly scaled to make a meaningful impact.

Given this context, why is there still so little investment? Various reasons for this have been explored, varying from market imperfections and a lack of investible instruments to transparency issues and short-term profit motives. Additionally, the inherent complexity of nature and biodiversity raises important questions about how to measure them effectively, much like we do with carbon. As a result, valuing and monetising nature requires a holistic approach that integrates both use and non-use values of nature. The total economic value of nature must recognise that the intrinsic worth of natural ecosystems extends beyond their immediate utility, else the topic on monetisation will continue to be debated. A systems thinking approach that ensures human and social capital aspects are not overlooked in the pursuit of nature-positive investments is essential with the ultimate goal of nature finance being the achievement of ecological health.

Creating the right environment through regulation and market mechanisms is essential. Key enablers include establishing a taxonomy and ensuring transparency to harmonise approaches to NCA. It is also crucial to align capital and credit flows with the Global Biodiversity Framework targets and prioritise the phasing out of harmful subsidies. Without investible instruments, private finance will remain disengaged, and the full value of nature will continue to be overlooked.

As we advance in the field of nature finance, a comprehensive approach is required to address these multifaceted challenges. It is crucial for us to build an environment conducive to investment and align all financial flows with ecological health. Only then can we create the world that we not only want but the world that we need.

ADDITIONAL PERSPECTIVES

PERSPECTIVE 8.3   How nature rose to the top of the UNEP finance initiative agenda

By Jessica Smith & Romie Goedicke den Hertog,
UNEP FI Nature Leads

Throughout its nearly 35-year history, the United Nations Environment Programme Finance Initiative (UNEP FI) [116] has been a key player in uniting the UN with financial institutions worldwide to drive the sustainable finance agenda. To urgently address the triple planetary crisis, climate change, biodiversity loss, and pollution, UNEP FI has been at the forefront of integrating ecosystems and natural capital considerations into mainstream finance.

Early initiatives, including the Natural Capital Finance Alliance and the ENCORE tool (short for 'Exploring Natural Capital Opportunities, Risks and Exposure'; [117] developed with UNEP World Conservation Monitoring Centre [118] and Global Canopy, a data-driven not-for-profit that targets the market forces destroying nature) [119], laid the groundwork for this pivotal work.

In the early 2020s, a significant shift gave way for more action on nature by our membership through two influential publications. Indebted to Nature [120], a pioneering report by the Dutch Central Bank and Netherlands Environmental Assessment Agency, investigated the risks associated with biodiversity loss for investments by Dutch financial institutions, highlighting nature loss as a financial risk. And Financing Nature: Closing the Global Biodiversity Financing Gap [121], by the Paulson Institute, The Nature Conservancy, and Cornell Atkinson Center for Sustainability, made a compelling economic case for valuing nature and identified fundamental mechanisms to close the biodiversity financing gap.

These impactful publications catalyzed forward-thinking financial institutions to recognise nature as a substantial risk and understand their pivotal role in addressing the global biodiversity funding gap. Building on its extensive experience, UNEP FI worked alongside a supportive ecosystem of actors that proved instrumental in rapidly advancing action on nature among UNEP FI members. UNEP FI contributed to this evolving space, providing support for the Global Biodiversity Framework via a statement from the global financial sector. UNEP co-chaired the Taskforce on Nature-related Financial Disclosures, which has significantly advanced nature-related standards and frameworks. In addition, UNEP FI provided technical support, including piloting the draft beta framework with 50 members in 25 countries.

But there is no time to rest on our laurels yet. While our membership's awareness and early action on nature are growing, we still are not catching up to the curve in biodiversity decline. At the time of writing (November 2024), UNEP FI is translating global frameworks into action by launching first-of-its-kind publications such as our guidance for banks on nature target setting last year and the Nature Positive Insurance Working Group. For Nature Positive to become a reality, we must go deep and wide.

Ahead of the CBD COP16 in Cali, Colombia, UNEP FI focused its attention on translating the GBF's goals and targets into meaningful action for its members. This involved examining the effectiveness of private sector support for biodiversity, ensuring alignment with the GBF's objectives, and contributing to global conservation efforts. The goal was and is to shape nature considerations into a practical and robust model for the international financial sector, fostering incremental changes that can substantially contribute to this cause.

While nature is rising on the agenda of our membership, we hope in parallel to see the biodiversity curve bend upwards, helping us all prepare for a future in which we live in harmony with nature.

PERSPECTIVE 8.4 The role of the principles of responsible investment in becoming Nature Positive

By David Atkin, CEO, UNPRI

The UN supported 'Principles for Responsible Investment' (UNPRI) is the leading proponent of responsible investment [122]. Its six principles stem from a recognition of the need for institutional investors to act in the best long-term interests of their beneficiaries, and therefore to incorporate environmental, social and governance (ESG) issues into investment analysis and decision-making processes.

Over the past two decades, the PRI has supported its more than 5,000 signatories around the world in addressing nature-related risks, particularly linked to deforestation as it is a key driver of both biodiversity loss and climate change. Active ownership and collaborative initiatives have been particularly important in driving more sustainable corporate practices and reducing their impact on nature.

Despite the actions from individual investors and companies, we are not seeing the nature crisis tackled at the speed and scale required. Today we need transformative change to achieve a net zero and nature-positive future, with whole-of-government policy reform and contributions from the whole-of-society.

Investors have an instrumental role to play, including through progressing their responsible investment practices and taking part in collaborative initiatives. They also need to drive an enabling environment for responsible investment, both locally and globally, through policies and financial market practices.

Nature Positive is aligned with the goal for long-term prosperity, in line with responsible investment, and drives proactive, meaningful and ambitious contributions from investors. The PRI is working side-by-side with its signatories, representing over half the world's institutional capital, to play their part in the realisation of this mission.

PERSPECTIVE 8.5   The role of philanthropy in reversing biodiversity loss

By Andrew Steer, former President and CEO, Bezos Earth Fund and World Resources Institute

Climate and nature are two sides of the same coin. To address climate change, we must reverse biodiversity loss. And to address biodiversity loss, we must address climate change. No individual group – whether government, philanthropy, or private sector – can solve the problem on its own.

Philanthropy can play a helpful role. It can take risks, knowing that sometimes, things fail. It can act quickly. It can help catalyze finance and action from others.

The Bezos Earth Fund [123] is the result of a very generous donation ($10 billion) from our Chair, Jeff Bezos. Managing these funds is a big responsibility. It's vital that every dollar not only delivers project success but helps play a bigger role in bringing system-wide change.

We give particular focus to protecting and restoring our natural world. Nature has intrinsic value, and it is integral to fighting climate change. In our work to preserve and restore forests, marine areas, grasslands, and more, we center the Indigenous Peoples and local communities who are – and have long been – the stewards of these precious places.

In 2021 we pledged to invest $1 billion toward 30x30 – an initiative that aims to protect 30% of the planet by 2030. We worked with 10 other philanthropies to join us and created the 'Protecting Our Planet Challenge' [124], and together pledged $5 billion by 2030 – the largest pledge ever made for nature. This was crucial to secure the adoption of 30x30 as part of the United Nations Global Biodiversity Framework, which includes 196 countries.

To date, we've committed more than $500 million in key biodiversity areas that include Gabon and the Congo Basin, the Tropical Andes of Colombia, the Eastern Tropical Pacific, the Brazilian Amazon, the United States, Canada, and the Pacific Ocean.

Through these efforts, we're supporting Indigenous groups to secure rights to their ancestral lands and the protection of their territories against illegal logging and mining and helping protect some of our world's most precious ecosystems and species, including forest elephants, mountain gorillas, sea turtles, sharks, and more.

Notes

1 Dorothy Maseke, lead of FSD Africa's Natural Capital, Biodiversity Finance & TNFD Programme, as well as the Head of the African Natural Capital Alliance (ANCA) Secretariat – a multistakeholder initiative set up by the United Nations Economic Commission for Africa (UNECA) and FSD Africa.
2 See, for example, the Carbon Trade Exchange (https://ctxglobal.com/about/), which offers digital spot trading, and the ICE Carbon Futures Index Family, which trades in futures of the world's largest carbon markets, including the EU ETS (https://www.ice.com/fixed-income-data-services/index-solutions/commodity-indices/carbon-futures)

References

1 Levitt, J.N., 2005. From Walden to Wall Street: Frontiers of Conservation Finance. Washington, DC: Island Press.
2 National Park Service, n.d. Boston Common. Available at: https://www.nps.gov/places/boston-common-ma.htm
3 National Park Service, n.d. Yellowstone Act 1872. Available at: https://www.nps.gov/parkhistory/online_books/anps/anps_1c.htm
4 Troster, L., 2014. Beyond the Letter of the Law: The Jewish Perspective on Ethical Investing and Fossil Fuel Divestment. Available at: https://sojo.net/articles/disinvest-reinvest/beyond-letter-law-jewish-perspective-ethical-investing-and-fossil-fuel
5 Congregational Research Service, 2010. Islamic Finance: Overview and Policy Concerns. Available at: https://crsreports.congress.gov/product/pdf/RS/RS22931/6
6 Discipleship Ministries, n.d. The Use of Money by John Wesley, pp. 4–6. Available at: https://www.umcdiscipleship.org/articles/the-use-of-money-by-john-wesley
7 Oreskes, M., 1985. 'Protests at Columbia: Students and issues have changed since the '60s', The New York Times, 18 May. Available at: https://www.nytimes.com/1985/04/13/nyregion/protests-at-columbia-students-and-issues-have-changed-since-the-60-s.html
8 Lansing, P., 1981. 'The divestment of United States companies in South Africa and aparthied', Nebraska Law Review. Available at: https://digitalcommons.unl.edu/cgi/viewcontent.cgi?article=2025&context=nlr
9 Resor, J.P., n.d. Debt-for-Nature Swaps: A Decade of Experience and New Directions for the Future, Unasylva – No. 188 – Funding Sustainable Forestry. Available at: https://www.fao.org/4/w3247e/w3247e06.htm
10 Wachira, C., 2023. 'First "debt-for-nature" swap complete in Gabon', Global Finance Magazine. Available at: https://gfmag.com/sustainable-finance/first-debt-for-nature-swap-complete-in-gabon/#:~:text=This%20was%20Africa's%20first%2Dever,long%20conservation%20and%20refinancing%20project
11 World Economic Forum, 2023. Climate Finance: What Are Debt-for-Nature Swaps and How Can They Help Countries? Available at: https://www.weforum.org/agenda/2024/04/climate-finance-debt-nature-swap/#:~:text=The%20first%20debt%2Dfor%2Dnature,around%20140%20swaps%20in%20total
12 World Economic Forum, 2023. What Are Green Bonds and Why Is This Market Growing So Fast? Available at: https://www.weforum.org/agenda/2023/11/what-are-green-bonds-climate-change/#:~:text=The%20first%20green%20bonds%20were,have%20helped%20spur%20this%20expansion
13 Intergovernmental Science-Policy Platform on Biodiversity and Ecosystem Services, n.d. About: What is IPBES? Available at: https://www.ipbes.net/about
14 United Nations Environment Programme, 2023. State of Finance for Nature 2023. Available at: https://www.unep.org/resources/state-finance-nature-2023
15 NGFS, 2022. NGFS Acknowledges That Nature-Related Risks Could Have Significant Macroeconomic and Financial Implications. Available at: https://www.ngfs.net/

en/communique-de-presse/ngfs-acknowledges-nature-related-risks-could-have-significant-macroeconomic-and-financial

16 HM Treasury, 2021. The Economics of Biodiversity: The Dasgupta Review – Full Report. Available at: https://www.gov.uk/government/publications/final-report-the-economics-of-biodiversity-the-dasgupta-review

17 World Economic Forum, 2020. New Nature Economy Report II: The Future of Nature and Business. Available at: https://www3.weforum.org/docs/WEF_The_Future_Of_Nature_And_Business_2020.pdf

18 IUCN Red List, 2024. Available at: https://www.iucnredlist.org/

19 Convention on Biological Diversity, n.d. Target 19. Available at: https://www.cbd.int/gbf/targets/19

20 The Nature Conservancy, 2020. Closing the Nature Funding Gap: A Finance Plan for the Planet, 14 September. Available at: https://www.nature.org/en-us/what-we-do/our-insights/perspectives/closing-nature-funding-gap-global-biodiversity-finance/

21 The 10 Point Plan for Financing Biodiversity: A Call to Action. Available at: https://www.financebiodiversity.org/call-to-action.

22 Koplow, D. and Steenb, R., 2022. Protecting Nature by Reforming Environmentally Harmful Subsidies: The Role of Business. Available at: https://www.earthtrack.net/sites/default/files/documents/EHS_Reform_Background_Report_fin.pdf

23 OECD, 2021. Biodiversity, Natural Capital and the Economy. Available at: https://www.oecd.org/en/publications/2021/05/biodiversity-natural-capital-and-the-economy_940af1d4.html

24 Business for Nature, 2023. COP16 Business Statement. Available at: https://www.businessfornature.org/cop16-business-statement

25 Dixon-Declève, S., Gaffney, O., Randers, J., Rockström, J. and Stoknes, P.E., 2022. Earth-4All: The Book. Available at: https://www.clubofrome.org/publication/earth4all-book/

26 Stockholm Resilience Centre, n.d. Planetary Boundaries. Available at: https://www.stockholmresilience.org/research/planetary-boundaries.html#:~:text=It%20stated%20that%20society's%20activities,the%20boundaries%20into%20unprecedented%20territory

27 Finance for Biodiversity Foundation, n.d. Finance for Biodiversity. Available at: https://www.financeforbiodiversity.org/

28 European Commission, 2023. Sustainable Finance Package 2023. Available at: https://finance.ec.europa.eu/publications/sustainable-finance-package-2023_en

29 European Commission, n.d. EU Taxonomy for Sustainable Activities. Available at: https://finance.ec.europa.eu/sustainable-finance/tools-and-standards/eu-taxonomy-sustainable-activities_en

30 Ministry of Foreign Affairs Japan, 2016. G7 Ise-Shima Leaders' Declaration. Available at: https://www.mofa.go.jp/files/000160266.pdf

31 Harvey, F., 2023. 'G20 poured more than $1tn on fossil fuel subsidies despite Cop26 pledges – Report', The Guardian, 23 August. Available at: https://www.theguardian.com/environment/2023/aug/23/g20-poured-more-than-1tn-on-fossil-fuel-subsidies-despite-cop26-pledges-report#:~:text=G20%20leaders%20agreed%20to%20phase,agreed%20to%20accelerate%20these%20efforts

32 Sustainable Markets Initiative, n.d. Enabling Taskforce: Natural Capital Investment Alliance. Available at: https://www.sustainable-markets.org/taskforces/ncia/

33 Dasgupta, P., 2021. The Economics of Biodiversity: The Dasguta Review (Full Report). Available at: https://assets.publishing.service.gov.uk/media/602e92b2e90e07660f807b47/The_Economics_of_Biodiversity_The_Dasgupta_Review_Full_Report.pdf

34 HM Treasury, Department for Energy Security and Net Zero, and Department for Environment, Food & Rural Affairs, 2023. Mobilising Green Investment: 2023 Green Finance Strategy. Available at: https://assets.publishing.service.gov.uk/media/

643583fb877741001368d815/mobilising-green-investment-2023-green-finance-strategy.pdf

35 Ortiz, E.M. and Kellenberg, J., n.d. Program of Payments for Ecological Services in Costa Rica. Available at: https://www.cbd.int/financial/pes/costarica-pesprogram.pdf

36 Wei, F., Cui, S., Liu, N., Jiang Chang, J., Ping, X., Ma, T., Xu, J., Swaisgood, R.R., and Locke, H., 2021. 'Ecological civilization: China's effort to build a shared future for all life on Earth', National Science Review, 8(7), July 2021, pp. 2–12. Available at: doi.org/10.1093/nsr/nwaa279

37 National Financial Regulatory Administration, 2023. CBIRC Releases the Green Finance Guidelines for Banking and Insurance Sectors. Available at: https://www.cbirc.gov.cn/en/view/pages/ItemDetail.html?docId=1055048&itemId=980

38 United Nations, n.d. Natural Capital Accounting and Valuation of Ecosystem Services – China. Available at: https://seea.un.org/content/china-0

39 New Zealand Treasury, 2022. Wellbeing Budget 2022: A Secure Future. Available at: https://www.treasury.govt.nz/publications/wellbeing-budget/wellbeing-budget-2022-secure-future

40 Kahui, V., 2024. 'Can granting legal "personhood" to nature stem biodiversity loss?', Greenpeace Aotearoa, 26 April. Available at: https://www.greenpeace.org/aotearoa/story/granting-legal-personhood-nature-biodiversity-loss/

41 Environment and Climate Change Canada, 2024. Nature Smart Climate Solutions Fund. Available at: https://www.canada.ca/en/environment-climate-change/services/environmental-funding/programs/nature-smart-climate-solutions-fund.html

42 World Bank, 2023. Assessment and Options Analysis of Climate and Nature Financing Instruments and Opportunities. Available at: https://documents1.worldbank.org/curated/en/099060123121542587/pdf/P1812770472a3805008b6306b88d19eee9e.pdf

43 Nature Finance, 2022a. Global Nature Markets Landscaping Study. Available at: https://www.naturefinance.net/wp-content/uploads/2022/12/GlobalNatureMarkets-LandscapingStudy.pdf

44 Helbling, T., n.d. 'Externalities: Prices do not capture all costs', Finance and Development Magazine, International Monetary Fund. Available at: https://www.imf.org/en/Publications/fandd/issues/Series/Back-to-Basics/Externalities#:~:text=Social%20costs%20grow%20with%20the,lead%20to%20lower%20production%20levels

45 Vionnet, S. and Blower, L., 2021. 'Corporate natural capital accounting: From building blocks to a path for standardization', Transparent. Available at: https://capitalscoalition.org/wp-content/uploads/2021/04/Transparent-benchmarking-final.pdf

46 United Nations Statistics Division, 2023. Findings from the 2023 Global Assessment of Environmental-Economic Accounting. Available at https://seea.un.org/news/findings-2023-global-assessment-environmental-economic-accounting

47 NGFS, 2023. Nature-Related Financial Risks: A Conceptual Framework to Guide Action by Central Banks and Supervisors. Available at: https://www.ngfs.net/sites/default/files/medias/documents/ngfs_conceptual-framework-on-nature-related-risks.pdf

48 Institute for Economics and Peace, 2023. Ecological Threat Report 2023. Available at: https://www.economicsandpeace.org/wp-content/uploads/2023/12/ETR-2023-web.pdf

49 Taskforce on Nature-related Financial Disclosures (TFND), 2023. TNFD Publishes Scoping Study Exploring Global Nature-Related Public Data Facility. Available at: https://tnfd.global/tnfd-publishes-scoping-study-data-facility/

50 Taskforce on Nature-Related Financial Disclosures (TFND), 2023. TNFD Publishes Scoping Study Exploring Global Nature-Related Public Data Facility. Available at: https://tnfd.global/tnfd-publishes-scoping-study-data-facility/

51 Taskforce on Nature-Related Financial Disclosures (TFND), 2022. Discussion Paper: A Landscape Assessment of Nature-Related Data and Analytics Availability. Available

at: https://tnfd.global/wp-content/uploads/2022/03/220321-TNFD-Data-discussion-paper-FINAL.pdf

52 Convention on Biological Diversity, 2022. Supplementary Information on Effective and Feasible Pathways for Closing the Biodiversity Finance Gap: A Note by the Panel of Experts on Resource Mobilization. Available at: https://www.cbd.int/doc/c/d6df/7c c2/0fe75538dc7109ed9d866c9f/sbi-03-inf-47-en.pdf

53 Coalition for Private Investment in Conservation (CPIC), 2021. Conservation Finance 2021: An Unfolding Opportunity. Available at: https://www.cpicfinance.com/wp-content/uploads/2021/12/CPIC-Conservation-Finance-Report-2021-1.pdf

54 Bioy, H., 2023. Investing in Times of Climate Change 2023. Available at: https://www.morningstar.com/funds/investing-times-climate-change-2023

55 Mahmood, R. and Guo, S., 2023. 'Biodiversity funds: Welcome to the jungle', MSCI. Available at: https://www.msci.com/www/blog-posts/biodiversity-funds-br-welcome/04075535373

56 Cuckston, T., 2019. 'Seeking an ecologically defensible calculation of net loss/gain of biodiversity', Accounting, Auditing & Accountability Journal, 32, pp. 1358–1383. Available at: https://doi.org/1108/AAAJ-01-2018-3339

57 Thomson, I., 2021. 'Designing environmental impact-valuation assemblages for sustainable decision-making', In J. Bebbington, C. Larrinaga, B. O'Dwyer and I. Thomson (eds.), Routledge Handbook of Environmental Accounting. 1st edn. Routledge, pp. 236–250. Available at: https://doi.org/10.4324/9780367152369-20

58 BIOFIN, 2023. A New Era of Biodiversity Finance Plan Emerges. Available at: https://www.biofin.org/news-and-media/new-era-national-biodiversity-finance-plans-emerges

59 We Value Nature, n.d. Risks of Perverse Outcomes from Accelerating Natural Capital Thinking: A Reflection. Available at: https://wevaluenature.eu/sites/default/files/2022-04/We_Value_Nature_Natural_Capital_Thinking_Briefing_Paper.pdf

60 Chiapello, E. and Engels, A., 2021. 'The fabrication of environmental intangibles as a questionable response to environmental problems', Journal of Cultural Economy, 14(5), pp. 517–532. Available at: https://doi.org/10.1080/17530350.2021.1927149

61 Neto, M., 2024. Evaluation in Ecosystem Management and Biodiversity Conservation Projects and Programmes Help Advance a Just Green Transition. Available at: https://www.undp.org/speeches/evaluation-ecosystem-management-and-biodiversity-conservation-projects-and-programmes-help-advance-just-green-transition

62 Taskforce on Nature-Related Financial Disclosures, 2024. Homepage. Available at: https://tnfd.global/

63 Taskforce on Nature-Related Financial Disclosures, n.d. About Us: History. Available at: https://tnfd.global/about/#history

64 Taskforce on Nature-related Financial Disclosures, n.d. About Us: The Taskforce. Available at: https://tnfd.global/about/the-taskforce/

65 Taskforce on Nature-Related Financial Disclosures, 2024. Over 500 Organisations and $17.7 Trillion AUM Now Committed to TNFD-Aligned Risk Management and Corporate Reporting, 25 October. Available at: https://tnfd.global/over-500-organisations-and-17-7-trillion-aum-now-committed-to-tnfd-aligned-risk-management-and-corporate-reporting/

66 Taskforce on Nature-Related Financial Disclosures, 2024. TNFD Publishes Draft Guidance on Nature Transition Planning at COP16. Available at: https://tnfd.global/tnfd-transition-plans-paper-published/

67 Nature Positive Initiative, 2024. State of Nature Metrics. Available at: https://www.naturepositive.org/metrics/

68 Biodiversity Credit Alliance, 2024. Homepage. Available at: https://www.biodiversitycreditalliance.org/

69 International Advisory Panel on Biodiversity Credits, 2024. Homepage. Available https://www.iapbiocredits.org/

70 Nature Finance, 2024. Taskforce on Nature Markets. Available at: https://www.nature-finance.net/making-change/nature-markets/taskforce-on-nature-markets/

71 BSI Standards, 2024. About the Nature Investment Standards Hub. Available at: https://nature-investment.bsigroup.com/

72 International Union for Conservation of Nature, 2023. Measuring Nature-Positive: Setting and Implementing Verified, Robust Targets for Species and Ecosystems. Available at: https://iucn.org/sites/default/files/2023-11/iucn-nature-positive-contribution-v1.0.pdf

73 Plan Vivo Foundation, n.d. About PV Nature. Available at: https://www.planvivo.org/pv-nature

74 Verra, 2022. New Biodiversity Methodology. Available at: https://verra.org/new-biodiversity-methodology/#:~:text=Verra%20is%20developing%20a%20biodiversity,certification%20of%20nature%2Dpositive%20investments

75 The Global Biodiversity Standard, n.d. About the Standard. Available at: https://www.biodiversitystandard.org/

76 Organization for Biodiversity Certificates, 2024. About Us. Available at: https://www.obiocert.com/

77 HM Government, 2023. Nature Markets: A Framework for Scaling Up Private Investment in Nature Recovery and Sustainable Farming. Available at: https://www.gov.uk/government/publications/nature-markets

78 Nature Finance, 2022. Nature in an Era of Crises. https://www.naturefinance.net/wp-content/uploads/2022/09/TNMNatureInAnEraOfCrises.pdf

79 Rao, R., Choi, E.S. and Czebiniak R.P., 2024. Can 'Biodiversity Credits' Boost Conservation? https://www.wri.org/insights/biodiversity-credits-explained

80 George, V., 2024. 'LSEG: Global carbon market value reached record $949b in 2023', Carbon Herald, 5 February. https://carbonherald.com/lseg-global-carbon-market-value-reached-record-949b-in-2023/

81 George, V., 2024. 'LSEG: Global carbon market value reached record $949b in 2023', Carbon Herald, 5 February. https://carbonherald.com/lseg-global-carbon-market-value-reached-record-949b-in-2023/

82 Regional Greenhouse Gas Initiative, n.d. About. Available at: https://www.rggi.org/rggi-inc/contact

83 World Bank Group, 2024. State and Trends of Carbon Pricing Dashboard, Compliance Mechanisms. Available at: https://carbonpricingdashboard.worldbank.org/compliance/instrument-detail

84 Rimmer, A., 2024. Infrastructure, Integrity and Innovation: How to Bring the Voluntary Carbon market to scale. Available at: https://www.lseg.com/en/insights/infrastructure-integrity-innovation-how-to-bring-voluntary-carbon-market-to-scale

85 Hook, L. and Temple-West, P., 2020. 'Carney calls for "$100bn a year" global carbon offset market', Financial Times, 3 December. Available at: https://www.ft.com/content/8ed608b2-25c8-48d2-9653-c447adbd538f

86 Ferris, N., 2022. 'The mission to mend the voluntary carbon offset market', Energy Monitor, 19 August. Available at: https://www.energymonitor.ai/policy/carbon-markets/the-mission-to-mend-the-voluntary-carbon-offset-market/

87 Pandya, H. and Archer, T., 2023. 'Understanding the compliance and voluntary carbon trading markets', Deloitte, 4 July. Available at: https://www.deloitte.com/uk/en/services/risk-advisory/blogs/2023/understanding-the-compliance-and-voluntary-carbon-trading-markets.html

88 UNFCCC, n.d. What is REDD+? Available at: https://unfccc.int/topics/land-use/workstreams/redd/what-is-redd

89 African Carbon Markets Initiative, 2024. Status and Outlook Report 2024. Available at: https://africacarbonmarkets.org/wp-content/uploads/2024/07/ACMI_Status-and-Outlook-Report-2024.pdf

90 Deutsche Bank, 2022. UNFCCCREDD+ and the Power of Sovereign Carbon. Available at: https://www.deutschewealth.com//content/dam/deutschewealth/cio-perspectives/cio-special-assets/unfccc-redd-sovereign-carbon/CIO-Special-UNFCCC-REDD-power-of-sovereign-carbon.pdf

91 Nature Finance, 2023. From Links to Linkages: Integrating Renewable Natural Capital into Sovereign Debt Instruments. Available at: https://www.naturefinance.net/wp-content/uploads/2023/11/FromLinksToLinkages-3.pdf

92 Bloomberg, 2022. The Untapped Power of Carbon Markets in Five Charts, 16 September. Available at: https://about.bnef.com/blog/the-untapped-power-of-carbon-markets-in-five-charts/

93 Trading Economics, 2024. EU Carbon Permits, 5 Year Price Evolution. Available at: https://tradingeconomics.com/commodity/carbon

94 Nasdaq, 2024. Nasdaq Study Shows Structural Reform Needed to Unlock Global Carbon Markets. Available at: https://www.nasdaq.com/press-release/nasdaq-study-shows-structural-reform-needed-to-unlock-global-carbon-markets-2024-03-0

95 Ecosystem Marketplace (2024) Today's VCM, Explained in Three Figures. Available at: https://www.ecosystemmarketplace.com/articles/todays-vcm-explained-in-three-figures/

96 World Economic Forum, 2023. Biodiversity Credits: Demand Analysis and Market Outlook. Available at: https://www3.weforum.org/docs/WEF_2023_Biodiversity_Credits_Demand_Analysis_and_Market_Outlook.pdf

97 Cumin, V. and Bromley, H., 2023. Biodiversity Finance Factbook. Available at: https://assets.bbhub.io/professional/sites/24/REPORT_Biodiversity_Finance_Factbook_master_230321.pdf

98 Nature Finance, 2023. Harnessing Biodiversity Credits for People and Planet. Available at: https://www.naturefinance.net/wp-content/uploads/2023/06/HarnessingBiodiversityCreditsForPeopleAndPlanet.pdf

99 Global Environment Facility, 2023. Innovative Finance for Nature and People: Opportunities and Challenges for Biodiversity-Positive Carbon Credits and Nature Certificates. Available at: https://www.thegef.org/sites/default/files/documents/2023-03/GEF_IIED_Innovative_Finance_Nature_People_2023_03_1.pdf

100 Waterford, L., Wilder, M., Crowley, H., Frederighi, P., Denman, S. and Atouguia M., 2023. 'Biodiversity credit markets: The role of law, regulation and policy', Taskforce on Nature Markets. Available at: https://cdn.prod.website-files.com/623a362e6b1a3e2eb749839c/6452340b9bcbb3ef3f82e6b6_BiodiversityCreditMarkets.pdf

101 Pollination Group, 2023. Review Frameworks for Biodiversity Credit schemes. Available at: https://pollinationgroup.com/wp-content/uploads/2023/10/Review-Frameworks-for-biodiversity-credit-schemes-Pollination-October-2023.pdf

102 Nature Finance, 2024. Investing in Africa. Available at: https://africannaturalcapitalalliance.com/theancaplus/uploads/2024/10/101424_Africa-Landscapes_FINAL_10.pdf

103 Intergovernmental Science-Policy Platform on Biodiversity and Ecosystem Services, n.d. Policy Instrument: Payment for Ecosystem Services. Available at: https://www.ipbes.net/policy-support/tools-instruments/payment-ecosystem-services

104 Smith S, Rowcroft, P., Everard, M., Couldrick, L., Reed, M., Rogers, H., Quick, T., Eves, C. and White, C. (2013) Payments for Ecosystem Services: A Best Practice Guide. London: Defra.

105 OECD, 2013. 'Scaling-up finance mechanisms for biodiversity', Chapter 4: Payments for Ecosystem Services. Available at: https://www.oecd-ilibrary.org/docserver/9789264193833-6-en.pdf?expires=1724636141&id=id&accname=guest&checksum=862580F981C7D8BD842447BD74FB7BEA

106 Turpie, J., Marais, C. and Blignaut, J., 2008. 'The working for water programme: Evolution of a payments for ecosystem services mechanism that addresses both

poverty and ecosystem service delivery in South Africa', Ecological Economics, 65, pp. 788–798. Available at: https://doi.org/10.1016/j.ecolecon.2007.12.024

107 United Nations Development Programme, 2024. Debt for Nature Swamps. Available at: https://sdgfinance.undp.org/sdg-tools/debt-nature-swaps

108 IIED, 2024. Debt Swaps could Release $100 Billion for Climate Action. Available at: https://www.iied.org/debt-swaps-could-release-100-billion-for-climate-action

109 White, N., 2022. 'Debt-for-nature swaps gain traction among developing countries', Bloomberg, 7 November. Available at: https://www.bloomberg.com/news/articles/2022-11-07/debt-for-nature-swaps-offer-option-for-developing-countries

110 World Economic Forum, 2024. Climate Finance: What Are Debt-for-Nature Swaps and How Can They Help Countries? Available at: https://www.weforum.org/agenda/2024/04/climate-finance-debt-nature-swap/

111 Jones, M. and Campos, R., 2023. 'Ecuador seals record debt-for-nature swap with Galapagos bond', Reuters, 10 May. Available at: https://www.reuters.com/world/americas/ecuador-seals-record-debt-for-nature-swap-with-galapagos-bond-2023-05-09/

112 Morningstar, 2024. Global Sustainable Fund Flows: Q3 2024 in Review. Available at: https://www.morningstar.com/lp/global-esg-flows

113 ISS Insights, 2024. Sustainable Funds Continue to Outgrow Peers in 2023, 19 March. Available at: https://insights.issgovernance.com/posts/sustainable-funds-continue-to-outgrow-peers-in-2023/

114 Morgan Stanely, 2024. Sustainable Funds Outperformed Peers in 2023. Available at: https://www.morganstanley.com/ideas/sustainable-funds-performance-2023-full-year

115 McKinsey, 2024. Nature-Positive Investments: Good for the Planet and Long-Term Value, 18 April. Available at: https://www.mckinsey.com/industries/agriculture/how-we-help-clients/natural-capital-and-nature/voices/nature-positive-investments-good-for-the-planet-and-long-term-value

116 UN Environment Programme Finance Initiative, n.d. About Us. Available at: https://www.unepfi.org/about/

117 Exploring Natural Capital Opportunities, Risks and Exposure, n.d. Available at: https://encorenature.org/en

118 UN Environment Programme World Conservation Monitoring Centre (UNEP-WCMC), n.d. About Us. Available at: https://www.unep-wcmc.org/en/about

119 Global Canopy, n.d. What We Do. Available at: https://globalcanopy.org/what-we-do/

120 DNB, 2020. Indebted to Nature. Available at: https://www.dnb.nl/media/4c3fqawd/indebted-to-nature.pdf

121 Paulson Institute, 2020. Financing Nature: Closing the Global Biodiversity Financing Gap. Available at: https://www.paulsoninstitute.org/conservation/financing-nature-report/

122 Principles for Responsible Investment, n.d. About Us. Available at: https://www.unpri.org/about-us/about-the-pri

123 Bezos Earth Fund, n.d. Who We Are. Available at: https://www.bezosearthfund.org/who-we-are (Accessed: 2 December 2024).

124 Protecting Our Planet Challenge, n.d. The Largest-Ever Private Funding Commitment to Biodiversity Conservation. (Accessed: 2 December 2024).

9 Nature-Positive governance

Carlos Manuel Rodríguez and
Sonja Sabita Teelucksingh

Introduction

Arturo Echandi Jimenez, a third-generation coffee farmer, was born in 1907 in San José, Costa Rica. Arturo was raised in an agricultural-based economy, with all development policies aimed at increasing agricultural development, and in particular coffee production. The best agricultural lands in the highlands were well divided between coffee growers and subsistence agriculture. At that time, Costa Rica had very high forest cover.

The Mexican Revolution of 1910 had a significant impact on public policies concerning land tenure and agrarian reforms all over Latin America. As such, the Costa Rican government initiated several legal and institutional reforms that aimed to (i) give land to the landless, (ii) bring all available land under agricultural production, and (iii) increase agricultural productivity and human development. In 1942, Arturo, by then 35 years of age, acquired a 440-hectare property in the Orosi Valley of Costa Rica as a beneficiary of a government-run land tenure programme. He established a coffee farm, dedicating 120 acres to coffee, leaving the remainder of the land intact and covered in pristine old-growth cloud forests.

As part of Costa Rica's agrarian reforms introduced in 1962, new policies and incentives were developed to increase productivity and expand the agricultural frontier. One notable policy was a small 10-article law called 'The Tax on Unproductive Lands', intended to tax all farms and properties that were not under economic production. As the soils of Arturo's 300 hectares of pristine forest were not suitable for coffee production, Arturo decided to convert his forests into pastureland for cows. This was, for many years, a productive endeavour. However, in around 1975, Arturo noticed a big decline in milk production, and by 1985 he was forced to cease this part of his operation. By this time, coffee production was also at risk. Pasture activities had degraded the farm's soil, and the total deforestation of his land had compromised its water sources.

Arturo Echandi Jimenez was my grandfather, and this is his story – a story mirrored by many thousands of Costa Rican farmers over the last two generations. As a boy, I saw this play out on a larger scale before my very eyes. My family would take the four-hour drive to the Costa Rican coast on the weekends, and as a game to keep us entertained, my cousins and I would compete by counting the hundreds

DOI: 10.4324/9781003474043-11

of timber trucks and cattle trucks that passed us by heading in opposite directions. The national policies incentivised and directly generated widespread deforestation – Costa Rica lost 50% of its forest cover in just two decades. This led to a loss of biodiversity, loss of soils, and loss of fresh water. As in the case of my grandfather, such policies ultimately limited the very income expansion and economic development the country's leaders were aiming to stimulate.

In 1979, in recognition of the high levels of deforestation and concomitant productivity loss, Costa Rica's Ministry of Finance included in its budget a set of forest incentives. These were aimed at stimulating the planting of trees as a commercial activity. This programme was very successful until the combination of a fiscal crisis, together with the World Bank's structural adjustment plans changed all incentive policies in preparation for free trade, globalisation, and competitiveness. Given the success of the programme, Costa Rica sought to find a replacement mechanism for its financing that was resilient to political changes and fiscal crises. This led to the establishment in 1997 of Costa Rica's pioneering 'Payments for Environmental Services' (PES) programme which aimed to address the market failure and create a different set of incentives targeted towards the mutually beneficial objectives of income generation and nature conservation [1]. Together with the same cousins with whom I used to count timber and cattle trucks as a child, I worked on restoring the forests on my grandfather's land, and we placed the farm under the PES scheme for carbon offsets.

At the time of his death in 2004, my grandfather had a productive 440-hectare farm, with 120 hectares under coffee production, almost 300 hectares of restored forests, many springs and freshwater sources, and additional income for carbon offsets. Today, because of the PES programme, our farm receives $78 dollars per hectare of forest per year, for its carbon, water, and biodiversity services. This aggregates to a yearly amount that is almost equivalent to the profits of our coffee production. My grandfather's lands are now as diverse and dense as they were in the early 1960s, and its rich biodiversity has returned. These lands are a crucial part of a biological corridor that runs between two large national parks in Costa Rica, which contribute in multiple ways to Costa Rica's global environmental commitments.

This story of my grandfather, Arturo Echandi, was replicated across thousands of farming households in Costa Rica, and stands as a tangible example of how a nation's policies can simultaneously enhance family income, support environmental conservation, promote national development, and contribute to global environmental goals. Alternatively, it shows how the achievement of these multiple objectives can be constrained if the relevant – and harmonised – public policies are not in place.

Today, as I travel around many developing countries in my capacity as the CEO of the Global Environment Facility, I see my grandfather's story repeating itself – with national policies, public funding, direct incentives, subsidies, and institutions working against sustainable conservation and climate outcomes. In another layer of contradiction and complexity, these factors can also work against each other – public policies 'promote' unsustainable development of agriculture, fisheries, infrastructure, logging, and oil and mining, while at the same time central governments support nature conservation and sustainability, both domestically through other national policies, and internationally through commitments to global agreements. These misalignments and contradictions of public policies and governance continue to feed into the triple planetary crisis of climate change, biodiversity loss, and pollution, which inhibits individual livelihoods, economic development, national progress, and global environmental stability.

9.1   The evolution of the international environmental architecture

Since the nineteenth century, intellectual and political minds have been increasingly conscious of the state of the natural world and how this interacts with economic systems [2]. This realisation was initially characterised by disparate, stand-alone assessments and actions at the national level, such as the creation of national parks [3]. Over time, the need for action on a global scale became apparent.

Created in 1948, the International Union for Conservation of Nature (IUCN) was the world's first global environmental union, charged with monitoring the status of the natural world and what is needed to safeguard it [4]. The United Nations Conference on the Human Environment in Stockholm in 1972 was the first major international landmark towards a global approach to take stock of, address, and enhance human and natural systems [5]. The United Nations Environment Programme (UNEP) was created out of this landmark gathering in Stockholm in 1972 as a specialised UN Agency aimed at monitoring the state of the environment, informing policy-making with science, and coordinating the global response to environmental challenges [6].

Stockholm 1972 led to a significant and increasing awareness of environmental matters at both grassroots and global levels [5, 7]. Concerted, coordinated global action was beginning to take place in various forms – for example, the discovery of the hole over the ozone layer led to the 1987 Montreal Protocol on Substances that Deplete the Ozone Layer, a global agreement to phase out the production and consumption of ozone-depleting substances, and the development of the Multilateral Fund to facilitate its implementation [8].

In 1992, another major global environmental landmark took place with the United Nations Conference on Environment and Development in Rio [5]. With the Rio Conference came the establishment of three major international conventions known as the 'Rio Conventions' – the United Nations Framework Convention on Climate Change (UNFCCC), the United Nations Convention on Biological Diversity (CBD), and the United Nations Convention to Combat Desertification (UNCCD) [9]. A fledgling pilot programme recently created within the World Bank known as the Global Environment Facility (GEF) was given the mandate to act as the financial mechanism of these three conventions and to facilitate their implementation.[1]

Over time, there have been several additional Multilateral Environmental Agreements (MEAs) and initiatives that have been developed in response to growing global environmental challenges [10]. We have seen the creation of new international institutions [11], together with an expansion of the operations of existing institutions [12]. Many regional institutions such as regional development banks have environmental objectives within their portfolios. And at the national level, there has been an expansion of bilateral aid from developed to developing countries that reflect environmental goals and targets.

Both the Stockholm and Rio Conferences therefore mark significant milestones in global environmental law-making and governance, with their respective declarations bookmarking the evolution of international environmental law, institutions, and conventions [13]. These agreements set the framework for national actions to achieve sustainable economic development and international global environmental goals. They have formed the baseline for additional commitments and agreements both within [14] and parallel [15] to their existing frameworks, and they have established a blueprint in both form and process for several subsequent MEAs.[2]

9.2   The challenge of fragmentation at national and international levels

As I reflect on the evolution of global attention to the state of the environment that has taken place over the last 50 years, I see parallels with personal and professional experiences throughout my life – from counting the timber and cattle trucks on the highways of Costa Rica as a boy, to implementing a PES scheme with my cousins on my grandfather's farm, to becoming an environmental lawyer, to my role as Minister of Environment and Energy for Costa Rica several times, and now to my present position as the CEO of the Global Environment Facility (GEF).

I was a youth representative at the Rio Conference in 1992, witnessing the development of international agreements that I would eventually implement both nationally and globally, and the fundamental development of the institution that I now lead. As Minister of Environment and Energy, I was responsible for the delivery of Costa Rica's commitments to the Rio Conventions. And now as the CEO of the GEF, the core mandate of my work is to direct funds, policies, processes, and frameworks towards facilitating the delivery of all relevant countries to these commitments.

As a conservationist and as a Minister of Environment and Energy, I have witnessed the challenges of implementing international agreements from the bottom-up, at the country level. As the CEO of the GEF, I am now witnessing these challenges from the top-down, at the international level. In different positions over the course of my career, I have attended approximately 60% of the Conference of the Parties (COPs) of the three Rio Conventions (and now the Basel, Rotterdam and Stockholm, and Minamata Conventions as well). Increasing attention to environmental governance at the international level over the last 50 years has been possibly one of the most positive global developments of my lifetime, with a paradigm shift towards the perspective that nature is at the core of our human, social, and economic systems.

However, from my perspective, this evolution of global environmental architecture has taken place in response to specific political circumstances, rather than in a well-defined global institutional plan with clear roles, structures, obligations, and responsibilities for all multilateral environmental institutions. As a result, today we see a plethora of conventions, agreements, institutions, and aid mechanisms in the environmental space at the international level. Some of these are complementary to each other, others have overlapping mandates and objectives, and all of them need to have increasing levels of cooperation among themselves to achieve the required impact.

This fragmentation at the international level is also mirrored at the national level. I continue to see firsthand, in multiple domestic settings, institutional frameworks generating a lack of political coherence in development policies similar to those that caused my grandfather to deforest his land. Increasing attention to environmental governance and frameworks at the national level has been a positive development, but this has resulted in the creation of Ministries of Environment that operate separately to other ministries with whom common natural resources are shared.

The political reality, therefore, is that in many countries, several ministries simultaneously manage the renewable and non-renewable natural resources and landscapes without sufficient (or any) common planning and budgeting. This level of fragmentation limits an integrated management of common natural resources, which can result in contradictory domestic policies. This can be seen as one of the reasons why governments spend more financial resources on activities that destroy nature and contribute to climate change than they do in nature conservation. This fragmentation can also further challenge countries' ability to deliver on the international environmental agreements to which they have committed.

9.3   'Integration' as an evolving global paradigm

The 1992 Rio Conventions, by definition, established funding, targets, and mechanisms that, to a large extent, individually and separately addressed common environmental challenges. Despite a growing recognition of common objectives and the need for interlinkages, my 30-year professional experience with the global environmental agenda shows me that, for the most part, the challenges of climate change, loss of biodiversity, and land degradation have been addressed with a siloed approach. In my view, this was a significant limiting factor in terms of progress on the environmental agenda both nationally and internationally.

In 2014, the then-CEO of the GEF and my predecessor, Naoko Ishii, proposed an innovative paradigm – that funds towards climate and nature should be channeled through 'integrated approaches' that sought to tackle, in a holistic manner, the underlying and common drivers of environmental degradation. The GEF as an institution was the perfect testing ground for this approach. Despite the Rio Conventions at this point still operating as largely separate mechanisms that were insulated from one another, having the GEF as their common financial mechanism was the one indisputable link between them.

In its early years, the GEF was developing programming that was specific to the individual needs and objectives of each convention. Over time, due to several factors including a demand from countries for programming that spanned the different convention areas, there was a growing awareness that this largely insular approach would not result in the impact that was needed at the country level and at scale. This culminated in the introduction of a series of 'Integrated Approach Pilots' in 2014 [16].

Over the last 10 years, the GEF has been increasingly working, with greater effectiveness and efficiency, in the space of integrated programming, addressing the different needs, objectives, and goals of the multiple conventions we serve. Today, close to 50% of GEF resources are programmed through integrated approaches [17].

These initiatives at the GEF have directly stimulated a changing global paradigm in the international environmental architecture. Today, the concept of integration and the potential for synergies between and among different initiatives are beginning to be addressed through existing international MEAs (such as the Rio Conventions), more recently negotiated MEAs (such as the Minamata Convention and the Biodiversity Beyond National Jurisdiction Agreement), and new ones that are currently being negotiated (such as Plastic Pollution).

At the UNFCCC climate COP 26 in 2021, I experienced an identity crisis of sorts because of the emphasis that this COP placed on forests, a topic that was usually the purview of the Biodiversity COP. This climate COP was even informally branded the 'Nature COP', resulting in the Glasgow Leaders' Declaration on Forests and Land Use. In all my years of attending most of the COPs of the three Rio Conventions, I had never seen this kind of 'crossover' take place. Today, integration is rightly seen as the way to maximise limited resources that are available for climate and nature, and to maximise the impact of those resources.

While the concept of integration is recognized and is beginning to be applied to international agreements, this concept is not yet mirrored at the national scale by countries who ultimately are the ones to deliver on these international agreements. Costa Rica's trajectory in environmental governance is a good example of the creation of integrated institutional frameworks to address environmental challenges at the landscape level.

Before the 1980s, Costa Rica had multiple ministries managing the same natural resources, landscapes, and seascapes. In 1988, the management of all renewable and non-renewable resources were merged into the Ministry of Environment and Energy [18], which encompassed natural resources (mainly biodiversity), energy, mining, water, and the ocean – the ministry that I would end up leading three times over the course of my professional career.

Costa Rica at that time, and even to this day, is an anomaly in this regard. I remember well that some of my fellow environment ministers across the globe were sometimes skeptical of my loyalty to the environmental agenda (and vice versa), given that I was also responsible for mining. Even with this merged ministry, there were still conflicts of both an ideological and practical nature with my fellow Minister of Agriculture, who was the main source of rural policies that promoted land use change and forest degradation.

In my first term as minister, I realised that despite this merging, silos persisted. In 1994, there were three agencies in Costa Rica (the Forest Service, National Parks, and the Wildlife Service) managing protected areas and natural resources, with inter-agency conflicts and without any common planning. This was limiting our ambitions in the consolidation of protected areas and the reduction of deforestation. So, we examined the structure of the ministry and found opportunity for integration at the agency level. We therefore worked on merging these into one agency whilst simultaneously decentralising their operations and opening the planning processes to civil society.

Whilst multiple global agreements and institutions are beginning to work in an increasingly integrated manner towards global goals and targets, this paradigm is not yet universally accepted or reflected nationally, where central government agencies are concurrently managing the same environmental resources.

Where it exists, the evidence of this integrated approach to environmental conservation is clear – for example, Costa Rica's deforestation trends from 1980–2005 show a dramatic decline that can be directly linked to increased political coherence.

The persistence of fragmentation in the management of natural resources across ministries and sectors remains a significant barrier to the achievement of international goals and targets such as the SDGs. This lack of integration at both national and international institutional levels is largely what continues to constrain the world in achieving our multiple global commitments on climate and nature.

PERSPECTIVE 9.1: Operation Black Crab: a model of Nature-Positive governance

By Carlos Eduardo Correa, Former Minister of Environment and Sustainable Development, Colombia

On 16 November 2020, Hurricane Iota, a Category 5 storm, directly struck Colombian territory, pummeling the island of Providencia in the Archipelago of San Andrés. The storm left unprecedented devastation in its wake. At the time, I was serving as the Minister of Environment and Sustainable Development of Colombia. Although initial efforts focused almost exclusively on humanitarian needs, it became evident that the recovery of Providencia required a long-term vision integrating conservation and socioeconomic development.

The mangroves were dubbed "The Heroes of Iota," as this natural barrier helped mitigate the storm's impact. However, the destruction to the island was significant, wreaking havoc on the coastline and tropical dry forest, as well as the reefs of the Seaflower Biosphere Reserve (the third-largest coral barrier in the world). These events were catastrophic for an island of just 5,000 inhabitants who depend on nature and the tourism services it enables.

We deployed the National Environmental System's full scientific and technical capacity to restore the island. We knew we could not do it alone, and that local knowledge would be our best ally. Leveraging our Payments for Environmental Services system, we launched Operation Black Crab, named after a species on the island. This programme involved over 350 families in ecosystem restoration. The inhabitants of Providencia, with their deep knowledge of the territory, played a fundamental role in damage assessment, seed collection, mangrove restoration, and coral reef recovery.

Through this project, we managed to restore dozens of hectares of critical ecosystems, strengthen the island's resilience to extreme weather events, and create employment opportunities. Additionally, Operation Black Crab gave rise to the national programme 'A Million Corals for Colombia' – the largest reef restoration in the Americas – which cultivated the same number of coral fragments and restored 200 hectares of reef throughout Colombia (60% in the San Andrés Archipelago), with active participation from territorial environmental authorities, communities, schools and diving centres, artisanal fishing organisations, academia, and sector NGOs.

The experience in Providencia taught us lessons about the importance of nature-positive governance, especially during moments of crisis. As Operation Black Crab demonstrated, a multi-level governance structure will allow us to mobilise resources, share knowledge, and coordinate actions more efficiently, with all stakeholders working towards a common goal. In particular, we learned that involving local communities in decision-making and providing them with adequate incentives are key to achieving positive and sustainable environmental and social outcomes that are mutually reinforcing.

From my local, national, and international experience, I have committed to rethinking our public policies to reverse biodiversity loss. Nature-positive governance requires us to think innovatively and adopt policies and regulatory frameworks that redefine our relationship with nature, as we did with Operation Black Crab. Over the years, I have come to learn that one of the greatest obstacles to this is fragmentation in environmental governance at all levels, reflected in numerous international agreements and the lack of a common language. As this chapter discusses, this transition requires a strategy of policy coherence within and between local and international levels of governance. Furthermore, as my political experiences have shown me, this must have rigorous stakeholder engagement that includes governments, scientists, urban planners, Indigenous Peoples, local communities, academia, the private and financial sectors, multilateral organisations, and NGOs.

It is clear that nature-positive governance is the fundamental pillar for building a sustainable future. Only through a joint, coherent, and inclusive effort can we face current challenges and build a world where nature and humanity thrive in harmony. The time for reflection is over. We must now move to action.

9.4   Guiding principles for the achievement of Nature-Positive governance

Nature-positive governance is the ideal institutional setting that is needed for the world to move from nature negative to Nature Positive. Nature-positive governance can be defined as policy coherence at three levels of scale: domestic, international, and systemic (see Figure 9.1).

Domestic policy coherence refers to the policy coherence *within* a country's policy and institutional frameworks. Countries must develop mutually supportive regulatory and policy environments that both discourage/eliminate harmful practices and encourage large-scale finance for climate and nature [19]. However, they must do so while simultaneously improving planning and harmonisation across the executive and legislative branches for the effective implementation of such policies. Improved domestic policy coherence can greatly enhance domestic resource flows towards the investments required to achieve critical environmental goals, further catalyse the impact of international funding flows, and increase both national and global environmental benefits.

The need for domestic policy coherence is being progressively mainstreamed in global dialogues as a critical mechanism which, if left unattended, can hamper the world's ability to reverse the current environmental trends and to reach its crucial nature-positive targets [20, 21]. While mechanisms such as Payments for Environmental Services address the market failure of environmental stewardship, domestic policy coherence addresses its institutional failure.

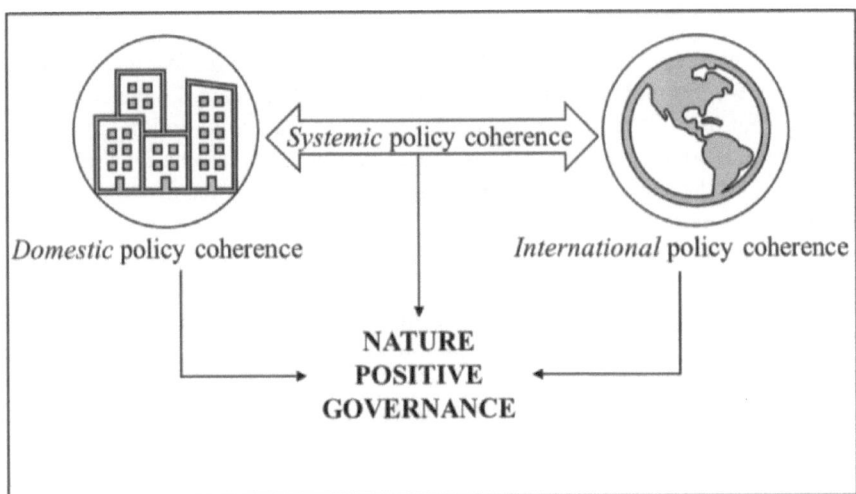

Figure 9.1 The three elements of nature-positive governance.

Source: Carlos Manuel Rodríguez and Sonja Sabita Teelucksingh, Global Environment Facility. Used with permission of the authors

International policy coherence refers to the policy coherence *within and amongst* international environmental agreements, the institutions that implement them, and the funding from multiple sources that serve their objectives. The current global financial and institutional architecture surrounding nature conservation is a complex landscape. It is growing even more complex with forthcoming new agreements and calls for more institutions to service those agreements. Within existing international institutions, another level of policy incoherence can be found where some activities may be simultaneously protecting nature whilst other activities may be unsustainably using these resources. Such activities need to be mapped and addressed at the institutional governance level.

Official Development Assistance (ODA) to nature financing is often taken to mean multilateral aid channeled through existing international financial architecture, such as the GEF. An often-overlooked and significant variable in the financing equation is that of bilateral aid. Early internal assessments by the GEF point to the fact that, for many donor countries, most of the funding relevant to nature is being channeled bilaterally. The key question is whether this aid is explicitly targeted and operationalised as a function of the recipient country's commitments to global goals and targets. Apart from the inefficiencies that may result from this fragmented landscape, making the most out of this multitude of existing and expanding financial opportunities and their processes can be a daunting task for recipient countries. The likelihood that available international resources are used to maximum impact at the country level is therefore small.

Systemic policy coherence refers to the policy coherence *between* a country's institutional and policy frameworks and the international environmental agreements to which it commits. Progress toward international goals and targets are the aggregation of progress at the respective national levels. Domestic policies that are inconsistent or contradictory with global environmental goals can not only inhibit a country's ability to deliver on these targets but can even work against the very global environmental benefits that need to be generated.

To achieve nature-positive governance, we propose three guiding principles: (i) the adoption of an institutional landscape/seascape approach, (ii) the alignment of private and public investments with Multilateral Environment Agreements, and (iii) increased and targeted mobilisation of financial resources towards climate and nature.

9.4.1   The adoption of an institutional landscape approach

A common paradigm in conservation literature is that of a landscape approach. This concept is defined as

> a multifaceted integrated strategy that brings together multiple stakeholders from multiple sectors to provide solutions at multiple scales. It can be broadly defined as a framework to address the increasingly widespread and complex environmental, economic, social, and political challenges that typically transcend traditional management boundaries [22].

A landscape approach therefore seeks to replace a sectoral approach with an integrated one, to achieve competing objectives in the context of competing sectors [23] It requires recognition of the interconnectedness between people and nature in places where the pursuit of productive land uses – such as agriculture, livestock, and mining – can compromise environmental and biodiversity goals.

We can extend the concept of a landscape approach to that of an *institutional landscape approach*. There is a clear parallel between the challenges of fragmentation that the landscape approach tries to address, and the challenges of fragmentation of international and national institutions in the context of climate and nature. Nature-positive governance requires an operational structure that is fundamentally rooted in integration instead of fragmentation.

At the global level, an increased awareness of the synergies of objectives across environmental conventions and agreements is slowly leading to operational changes in how these agreements are operationalised at the country level. Whether these changes are happening quickly enough to achieve the 2030 Global Goals and targets is in fact debatable. The approaches of institutions like the Global Environment Facility are therefore critical to creating increased momentum towards integration at all levels of spatial scale.

There are currently many international conversations taking place on the topic of effectiveness, efficiencies, and impacts of current international environmental institutions and the potential for further synergies between them.[3] At the same time, experience and evidence tell us that the creation of additional financial mechanisms will only serve to further fragment an already-fractured financing landscape that is simultaneously trying to come closer together. A fundamental parameter of impact is that of integration.

Similarly, at the national level, the fragmentation of roles and responsibilities in the governance of natural resources can compromise the effective management of natural resources. Improved planning, coordination, financing, and technologies are a step in the right direction. However, the goal of nature-positive governance requires stronger action, with the implementation of a systems-based institutional approach that coordinates across both the executive and legislative branches of government.

The adoption of an institutional landscape approach can therefore align policies, incentives, and programmes which will lead to the better management of nature. This will in turn facilitate the achievement of national commitments to international processes. How do we achieve this?

Firstly, there needs to be an assessment of existing international environmental agreements and institutions in the context of complementarities, duplications, and synergies. At a time when the efficiency and effectiveness of the limited global financing to nature conservation is paramount for maximum impact, a global stocktake and frank conversations are needed to determine areas of overlap, potential streamlining of objectives and/or mandates, and opportunities for synergies. Secondly, there needs to be an assessment at the country level of the political landscape surrounding environmental resources and the level of coherence among the various public actors. Thirdly, pathways need to be identified, at both methodological and operational levels, towards the adoption of the approach. Finally, the approach needs to be piloted, tested, and revised/adapted as needed. At its core is the generation of the relevant political dialogues by the executive branch of governments that can assess progress towards international commitments, budgets of the executive branch agencies that manage natural resources, and pathways to more integration across them.

These are, admittedly, complex, and context-specific endeavours. But we will not achieve global sustainability targets with the same institutional frameworks that generated the challenges that we need to solve today. At the Global Environment Facility, we are expanding our methodological and operational thinking in the direction of an institutional landscape approach, and we are piloting several innovative internal initiatives, as well as various programming efforts, through extensive engagements with selected countries.

9.4.2 The alignment of private and public investments with Multilateral Environment Agreements

Progress at the national scale is what leads to progress towards global goals and targets. Any disconnect between enabling environments at the country level, and its international commitments can lead to challenges in the sustained delivery of promised outcomes.

When a country signs an international agreement, which is usually done by the appropriate ministers, this signals the country's non-binding expression of intent to comply. The next step is to get approval within a country's own internal, national procedures. Once this is done, the country will ratify the treaty – that is, it will notify the other parties to the international agreement that they consent to be bound by the treaty. Once ratification is done, the treaty is now officially binding to the country. To take effect as international law, different treaties require different numbers of countries to ratify them. To date, the Montreal Protocol on Substances that Deplete the Ozone Layer is the first United Nations treaty to achieve universal ratification [24].

Once ratified, the implementation or achievement of a national commitment to an international agreement requires a significant level of coherence and coordination at the country level. This is usually done by the relevant ministry – therefore, in the case of environmental agreements, this would be done by the ministry responsible for the environment. The environment is placed in, and within, many different ministries in countries across the world. In some cases, it is a stand-alone ministry; in other cases, it is amalgamated with other portfolios (for example, Costa Rica's Ministry of Environment and Energy).

Across the spectrum of national governments in the world, there is an almost infinite list of combinations and possibilities of where the environmental portfolio is located within the national architecture of public administration. This seemingly simple fact directly affects how a country approaches its climate and nature agenda at the national level and in turn at the international level in the context of its global commitments. Furthermore, as previously discussed, in many cases it is likely that this portfolio does not fully oversee the economic activities that interact directly with, and impact, nature.

Relatedly, a country's commitment to an international treaty needs to be mirrored in national planning and institutional settings. While a binding treaty, by definition, transcends the national political cycles and inevitable turnover of administrations, the approach to implementation may differ across them. This disconnect may cause a lagged, fragmented, and/or unsustainable delivery on local action towards global commitments. An added element of internal complexity is that, depending on the local set-up, there may be different agencies, entities, or ministries responsible for different MEA commitments.

National governments impact climate and nature through multiple channels. While the public administration structures and the direction of spending of public funds is a clear driver of environmental change, governments are also responsible for the direction, establishment, and implementation of established domestic policy environments that can stimulate economic activity by other major actors, such as the private sector. Therefore, there needs to be a holistic planning approach at the country level with respect to MEA commitments, with both public and private funding flows aligned in the direction of those commitments.

Moving towards a greater alignment of private and public investments with MEAs by individual countries can be achieved through several initiatives. Firstly, there needs to be a holistic assessment of all commitments made by a country to the global environmental agreements. Secondly, these commitments need to be mapped to the relevant ministries and agencies to determine the local institutional landscape that is responsible for these commitments. Thirdly, a baseline of data and information should be established on the public administrative architecture relevant to the use of natural resources, existing funding flows from these entities, and existing policy frameworks that guide both public and private spending on activities that impact climate and nature. This baseline would highlight any contradictions and yield potential opportunities for further alignments between the MEAs and national implementation plans. Finally, this approach would need to be piloted and tested across a range of countries, inclusive of the relevant domestic political dialogues, to establish a methodological and operational framework that could be transferred and applied at scale.

PERSPECTIVE 9.2: Moving towards domestic policy coherence: designing Nature-Sensitive budgets

By Onno van den Heuvel, Senior Technical Advisor, Nature Finance, United Nations Development Programme (UNDP) Nature Hub

Biodiversity, many people would say, is the prerogative of conservation NGOs like the World Wildlife Fund, Ministries of Environment, and David Attenborough. But such thinking is based on a misconception. Biodiversity does not only stand for wildlife and a few beautiful national parks that we may want to preserve. It is the variety of all life and ecosystems of our planet that underpin our economies and societies. Our major ecosystems are closely intertwined with our economies. We need plants to sequester carbon and release oxygen, pollinators to grow our agricultural crops and sustainable fish stocks to enable a new generation of fishermen to maintain a livelihood. Unfortunately, most economic models to date are based on the assumption that our economies and our global ecosystems are not much connected.

When asking who is responsible for biodiversity in governments, most officials would point straight at the Ministry of Environment. Yet, as highlighted in this chapter, the responsibility for biodiversity conservation and its sustainable use must be among the mandates of multiple government organisations as well as international ones. Biodiversity after all is not an economic sector, but an overarching development priority. If we on the one hand create a series of national parks but on the other cut down rainforests to expand agriculture – the definition of domestic policy incoherence – we end up in a zero-sum game.

But how do we define the biodiversity contribution of each ministry? A very useful tool in doing so is a 'Biodiversity Expenditure Review'. Introduced in 2014 and used by more than 40 countries to date (for example, Colombia, Costa Rica, Ireland, Kazakhstan, Namibia, the Seychelles, India, and the Philippines), this review assesses public budgets for specific biodiversity objectives. Governments participating in this exercise found remarkable results. For example, it was often the Ministry of Agriculture which was found to be the main spender on biodiversity, with most countries having major public schemes such as organic farming or regenerative agriculture, which have an objective to sustain the underlying ecosystems. Such a review then enables the further promotion of the expenditure categories identified as Nature Positive, and a move towards domestic policy coherence.

Countries can in fact go one step further. Rather than a one-time exercise, the principles of the biodiversity expenditure review can be integrated into expenditure monitoring systems through a system of budget tagging. This introduces an official tag into the public budgeting system to henceforth identify certain programmes and their related expenditures as Nature Positive. A government can then periodically review how much is spent on nature-positive investments across the entire fiscal system. Mexico, Malawi, France, and Indonesia are among the frontrunners of testing this innovative practice.

Shouldn't every country do this? With support from the Global Environment Facility, 91 more countries will in fact soon commence a biodiversity expenditure review. They can then consider taking the next step of integrating such reviews into their expenditure monitoring systems. The time should be past when public budgets are designed as nature blind. The tools and experiences such as biodiversity expenditure reviews and budget tagging are now available to enable nature-sensitive budgeting. Actions such as these can move us closer to domestic policy coherence and therefore to nature-positive governance. Future generations will greatly benefit.

9.4.3   *Increased and targeted mobilisation of financial resources*
towards nature

Current international, political, and scientific conversations are converging around the conclusion that the world's environmental and sustainable development goals to 2030 can only be realised if the funding gap to climate and nature is narrowed. We can define the 'nature financing gap' as the difference between current financing for the environment, and what is needed to achieve critical Global Goals by 2030 and beyond.

The measurement of this integrated gap is a complex exercise, and one that is replete with assumptions. Notwithstanding a lack of unified concrete numbers, it is undeniable that (i) this gap exists, (ii) it is large, and (iii) it must be narrowed. Closing this financing gap requires resource mobilisation to climate and nature through a two-pronged approach: (i) increasing financial flows whilst (ii) simultaneously reducing financial needs.

With the financing gap continuing to widen, it is now recognised that ODA funds, whether through multilateral or bilateral channels, are necessary but insufficient. There is the need for new sources of financing from other entities such as the private sector and philanthropy. The need to mobilise resources from these non-traditional sources, and options for doing so, are now commonly discussed, and explicitly reflected, in international environmental negotiations.[4]

A move to nature-positive governance necessitates resource mobilisation at a higher scale. More than this, these resources need to be targeted towards the institutions and initiatives that can deliver sustained impact through holistic and integrated approaches at the international and national levels.

Alongside increased and targeted funding from all sources, positive impacts on the gap can, in parallel, be achieved through domestic policy coherence. Through these domestic enabling environments, countries can help to further catalyse the impact of the nature funding flows; alternatively, misaligned domestic policies can serve to lessen the impact of the very funds to the environment that are being increasingly required from ODA and other sources.

In recognition of the importance of this growing international agenda, the Global Environment Facility now has policy coherence explicitly identified as a foundational element to its programming, especially in our growing suite of integrated programmes. We currently have active engagement with the private sector [25], particularly through these integrated programmes and through the Global Biodiversity Framework Fund [26]. We also have increasing collaborations with philanthropic organisations, for example through our ongoing work with regional Conservation Trust Funds. As the financing mechanism for multiple environmental conventions, the GEF is well-placed to test and operationalise this concept and its impact.

Conclusion

My life as a conservationist has been fundamentally shaped by my grandfather's story. What it has taught me is that nature-positive governance, as the aggregate of national, international, and systemic policy coherence, is the missing link that has inhibited, and will continue to inhibit, our progress towards sustainable development.

Nature-positive governance as a political objective can be achieved through a paradigm shift that will generate more policy coherence, and integration, at all levels of scale. Conversely, a lack of policy coherence, and further fragmentation, will continue to challenge us in our achievement of the 2030 goals and targets and beyond. The creation of new laws, policies, agencies, and institutions to protect nature and reverse climate change will continue having limited impact until the governance and policies of international institutions and agreements, national public agencies, and the private sector are all aligned with global sustainability goals. We must explore tangible and bold steps to stimulate this paradigm shift.

In this venture, the role of universal, harmonised, and open-access data and information is key. We need full accounting for nature and to appropriately value ecosystem services in national and global economic decisions. Academia has a significant role to play, where the scientific and national/international policy communities need to work closely together to shape the relevant policy frameworks.

Whilst credible advancements on nature-positive solutions continue to take place across science, business, finance, and society, sustained and accelerated progress can only be achieved with nature-positive governance. This is the underlying, necessary condition for the needed systems transformation that will help us to achieve a nature-positive planet.

ADDITIONAL PERSPECTIVES

PERSPECTIVE 9.3   **For Nature Positive to take hold, the public sector must catalyze action**

By Brian O'Donnell, Director, Campaign for Nature

Much of the attention for advancing the nature-positive agenda has focused on the private sector, which is logical given that the private sector makes up 60% of the global economy. But we can't rely on a few shining examples of 'best in class' companies that are reducing their environmental damage. We need the entire class to engage in nature-positive business practices. Ultimately, this will require governments to play a leading role, providing the architecture that creates a fertile environment for investment and transformational change including increased public finance, incentives, and enhanced regulation.

The good news is there is a wealth of compelling examples of government actions catalysing better business practices, increased environmental investments, and recovery of lands, waters, and wildlife. In 1991, Michael Porter, a Harvard University economist, challenged conventional thinking stating, "Strict environmental regulations do not inevitably hinder competitive advantage against rivals; indeed, they often enhance it" [27]. In the decades since, the 'Porter Hypothesis' has been studied extensively, and much of the empirical evidence suggests that he was right.

Consider the Clean Water Act and other water quality regulations in the United States and China's 'War on Pollution' [28]. These efforts were not only very successful in significantly reducing pollution and helping to recover fish populations [29], but they also spurred economic innovation and enhanced productivity [30, 31].

Government policies have protected and restored nature throughout the world. In Baja California Sur, Mexico, in the 1980s, local scientists and fishermen noted the profound impacts of overfishing and pollution in the Cabo Pulmo region [32]. They successfully petitioned the Mexican government to establish a Marine Protected Area for the area in 1995. The resulting recovery of the marine biodiversity in the area was profound, which also brought about an economic recovery for fishermen and a robust tourism economy.

Governments have demonstrated numerous successful ways to enhance private-sector funding for nature conservation. In India, the Companies Act of 2013 encouraged companies to allocate at least 2% of their net profit to corporate social responsibility (CSR) which includes environmental initiatives [33]. (In the years following the implementation of the Act, the legislation has led to large increases in philanthropy, bringing CSR "from the fringes to the boardroom", as one observer put it, but it also led to an increase in companies avoiding and cheating the rules, The Guardian found) [34].

In the leadup to (biodiversity) COP15, a group of primarily US-based foundations announced the largest-ever philanthropic commitment to nature, $5 billion, to support protecting at least 30% of the world's lands and ocean by 2030 [35]. While this huge nature finance pledge was celebrated as a private sector initiative, US tax laws incentivised philanthropy through tax savings, which led to the creation of the foundations in the first place.

Even with these important private investments in nature, it's critical to remember that over 80% of biodiversity finance currently comes directly from governments [36]. Governments will have to secure additional resources from industries that are driving biodiversity loss, including the fossil fuels, mining, industrial agriculture, and fishing sectors. The US Government's Land and Water Conservation Fund is a key example of taxing an industry (offshore oil and gas) and allocating resources to conservation initiatives. Ultimately, as this chapter discusses, all public and private investments to nature conservation at the country level must be aligned with the Multilateral Environmental Agreements (MEAs) [37] to which the country has committed.

The opportunity for government action to catalyze conservation, restoration and enhanced nature funding is currently only scratching the surface of what is possible and what is required to halt and reverse biodiversity loss. Furthermore, governments can further stimulate private sector funding to nature through the establishment of appropriate domestic policy environments. The Global Biodiversity Framework set forth many important goals and targets to help us get to a nature-positive economy. Now it is up to governments to develop and implement the policies to make it possible.

PERSPECTIVE 9.4   Translating global nature commitments into transformative national action

By Kirsten Schuijt, Director General, WWF International

The Kunming-Montreal Global Biodiversity Framework (GBF) is a game-changing agreement with the potential to set the world on a nature-positive pathway. Its 23 targets to halt and reverse biodiversity loss by 2030, if all achieved, will deliver better conservation results on the ground and address the drivers of biodiversity loss, including those related to unsustainable global consumption and production. These are essential ingredients to deliver the GBF Nature Positive mission. Those global nature commitments will remain words on paper unless we all work together to translate them into transformative action on the ground.

For this to happen, collective efforts from all walks of life are essential. As a starting point, we will need ambitious updated GBF-aligned National Biodiversity Strategies and Action Plans (NBSAPs) that are immediately implemented.

At COP15, CBD parties agreed to revise their NBSAPs (or national targets) to reflect the mission, ambition, and targets of the GBF. Parties also committed to implement this process through a whole of government and whole of society approach, involving all sectors, rights-holders and key stakeholders. This offers an important opportunity to strengthen nature-positive governance at the national level. Approval of the revised NBSAPs at the highest national political level would further ensure that it will be fully implemented.

Strong and broad ownership over the revised NBSAPs, achieved through an inclusive revision process, will lay the foundations for the implementation of the immediate and transformative actions that are required to achieve the nature-positive mission of the GBF. This approach should not be a one-off undertaking but sustained throughout the implementation and monitoring process. NBSAPS are a starting point for action rather than an end in themselves. Continued participation and inclusivity are essential, in the elaboration of detailed and specific action plans, and in implementation and monitoring – the GBF monitoring framework is an important tool to enhance governance through greater transparency and accountability, and also to prepare the ground to increase national actions after the 2026 Global Review of collective progress – if it turns out then that progress is inadequate to achieve the nature-positive GBF mission by 2030.

Inclusive, nature-positive governance is not only important to address the drivers of biodiversity loss, but it should also enhance synergies. While many countries are in the process of enhancing their Nationally Determined Contributions (NDCs), and at the same time revising their NBSAPs, now is an opportune moment to seek greater synergies and address trade-offs between climate and nature actions.

Strong nature-positive governance needs to be enabled by adequate resources that will help to deliver appropriate actions in the right places by the appropriate stakeholders. This includes adequate, predictable, and easily accessible resources to support developing countries in delivering their NBSAPs, as well as Indigenous Peoples and Local Communities. National biodiversity finance plans therefore form an important part of the NBSAP package to create greater clarity on needs, resources and gaps. Greater transparency and agreement on accounting methods for financial commitments is also important, in particular with regards to the financing of synergistic actions.

Unfortunately, many countries have not yet delivered their updated NBSAPs despite having committed to do so by CBD COP16 in October 2024. While the process to update the NBSAPs offers opportunities for more effective, comprehensive and inclusive governance and actions, the urgency to conclude these revision processes is evident.

At the time of writing (Autumn 2024), we have less than six years left to deliver the GBF 2030 targets. We need to work together to support the actors and actions, starting with the inclusive development and implementation of the NBSAPs we need, and ensure that enough resources reach developing countries and Indigenous Peoples and Local Communities the soonest.

Notes

1 Over the last thirty years, the GEF's mandate has broadened to include issues on forest protection and preservation, chemicals and waste, international waters, and biodiversity in the high seas. The GEF is now a 'Family of Funds' that serves the needs of a widening range of international agreements. The three Rio Conventions remain at the core of the GEF, which continues to operate under the authority of the respective Conference of the Parties (COPs) and draws guidance from the process.
2 Such as the Basel Convention on the Control of Transboundary Movements of Hazardous Wastes and their Disposal (1989), the Rotterdam Convention which facilitates informed decision-making by countries with regard to trade in hazardous chemicals (1998), the Stockholm Convention on Persistent Organic Pollutants (2001), the recent agreement on Biodiversity Beyond National Jurisdiction (2023), and the ongoing negotiations on Plastic Pollution
3 This is now a topic of most of the Conference of the Parties (COPs) of all major Conventions and Agreements, particularly in the Convention on Biological Diversity where explicit discussions on synergies have been tabled. Interestingly, this is also a topic that is found in ongoing negotiations of potential new agreements, such as those of the Intergovernmental Negotiating Committee on Plastic Pollution currently taking place. Various works on synergies and optimisations of existing funds are being conducted through the G20, currently under the presidency of Brazil. And some existing international institutions, including the Global Environment Facility, are contributing to such works as well as internally examining the potential for cross-linkages with other existing institutions working in the nature and climate space.
4 For example, the Kunming-Montreal Global Biodiversity Framework Fund of the Convention on Biological Diversity of 2022 (which is in part being operationalised through the Global Environment Facility's family of funds) explicitly has resource mobilisation as one of its 23 targets. This is also a topic that is featured in the 2023 Agreement on Biodiversity Beyond National Jurisdiction (2023) and is a part of the ongoing discussions of the Intergovernmental Negotiating Committee on Plastic Pollution.

References

1 Pagiola, S., 2008. 'Payments for environmental services in Costa Rica', Ecological Economics, 65, pp. 712–724.
2 Caldwell, L.K. and Weiland, P.S., 1996. International Environmental Policy: From the Twentieth to the Twenty-First Century. Duke University Press.
3 Gissibl, B., Höhler, S. and Kupper, P., 2012. 'Introduction: Towards a global history of national parks', In B. Gissibl, S. Höhler and P. Kupper (eds.), Civilizing Nature: National Parks in Global Historical Perspective. New York, Oxford: Berghahn Books, pp. 1–28. Available at: https://doi.org/10.1515/9780857455277-004
4 International Union for Conservation of Nature, 2024. International Union for Conservation of Nature. Available at: www.iucn.org/about-iucn
5 Handl, G., 2012. Declaration of the United Nations Conference on the Human Environment (Stockholm Declaration), 1972 and The Rio Declaration on Environment and Development, 1992. United Nations Audiovisual Library of International Law.
6 United Nations Environment Programme, n.d. UNEP: 50 Years of Environmental Milestones. Available at: www.unep.org/environmental-moments-unep50-timeline
7 Panjabi, R.K.L., 1993. 'From Stockholm to Rio: A comparison of the declaratory principles of international environmental law', Denver Journal of International Law and Policy, 21, pp. 215.
8 Multilateral Fund for the Implementation of the Montreal Protocol, n.d. Available at: www.multilateralfund.org/
9 United Nations Convention to Combat Desertification, n.d. Rio Conventions. Available at: www.unccd.int/convention/partners/rio-conventions#:~:text=The%20three%20 global%20agreements%20–%20which,joint%20approach%20to%20restore%20our
10 For example, the Stockholm Convention on Persistent Organic Pollutants, the Minamata Convention on Mercury, and the Biodiversity Beyond National Jurisdiction Agreement.
11 For example, the Climate Investment Funds established in 2008, and the Green Climate Fund established in 2010.
12 For example, in addition to UNEP, several UN organizations such as UNDP now have robust environmental portfolios.
13 Sand, H.P., 2007. 'The evolution of international environmental law', In D. Bodansky, J. Brunnee and E. Hey (eds.), The Oxford Handbook of International Environmental Law. Oxford University Press, pp. 30–43.
14 Such as the Paris Agreement (2015) and the Kunming-Montreal Global Biodiversity Framework (2022).
15 Such as the Millenium Ecosystem Assessment (2000) and the Sustainable Development Goals (2015).
16 Global Environment Facility, 2014. GEF/A.5/07/Rev.01* Report of the Sixth Replenishment of the GEF Trust Fund. Global Environment Facility.
17 Global Environment Facility, 2022. GEF/C.62/03 Summary of Negotiations of the Eight Replenishment of the GEF Trust Fund. Global Environment Facility.
18 Devex, Ministry of Environment and Energy (Costa Rica). Available at: www.devex. com/organizations/ministry-of-environment-and-energy-ministerio-de-ambiente-y-energia-minae-costa-rica-126001#
19 Global Environment Facility, 2023. GEF/C.65/04 Enhancing Policy Coherence through GEF Operations. Global Environment Facility.
20 United Nations Department of Economic and Social Affairs, n.d. Sustainable Development Goals. Available at: https://sdgs.un.org/goals
21 Convention on Biological Diversity, 2022. Decision Adopted by the Conference of the Parties to the Convention on Biological Diversity: 15/4 Kunming-Montreal Global Biodiversity Framework. Available at: www.cbd.int/doc/decisions/cop-15/cop-15-dec-04-en.pdf

22 Reed, J., Deakin, L. and Sunderland, T., 2015. 'What are "integrated landscape approaches" and how effectively have they been implemented in the tropics: A systematic map protocol', Environmental Evidence, 4, pp. 1–7.

23 Sayer, J., Sunderland, T., Ghazoul, J., Pfund, J.L., Sheil, D., Meijaard, E., Venter, M., Boedhihartono, A.K., Day, M., Garcia, C. and Van Oosten, C., 2013. 'Ten principles for a landscape approach to reconciling agriculture, conservation, and other competing land uses', Proceedings of the National Academy of Sciences, 110(21), pp. 8349–8356.

24 United Nations Environment Programme, n.d. About Montreal Protocol. Available at: www.unep.org/ozonaction/who-we-are/about-montreal-protocol (Accessed: 2 August 2024).

25 Global Environment Facility, 2020. GEF/C.58/05 GEF's Private Sector Engagement Strategy. Global Environment Facility.

26 Global Environment Facility, 2023. GEF/C.64/06/Rev.02 Programming Directions for the Global Biodiversity Framework Fund. Global Environment Facility.

27 Porter, M.E., 1991. 'America's green strategy', Scientific American, 264, p. 168. Available at: http://dx.doi.org/10.1038/scientificamerican0491-168

28 Greenstone, M., He, G., Li, S. and Zou, E., 2021. 'China's war on pollution: Evidence from the first five years', NBER Working Papers. Accessible at: www.nber.org/system/files/working_papers/w28467/w28467.pdf

29 U.S. Government Accountability Office, 2022. 50 Years after the Clean Water Act— Gauging Progress, 17 October. Available at: www.gao.gov/blog/50-years-after-clean-water-act-gauging-progress

30 Li, H. and Rus, H., 2022. Water Governance and Water Innovation Adaptive Responses to Regulatory Change and Extreme Weather Events. Available at: https://papers.ssrn.com/sol3/papers.cfm?abstract_id=4002240

31 Zhuge, L., et al., 2020. 'Regulation and innovation: Examining outcomes in Chinese pollution control policy areas', Economic Modelling, 89, pp. 19–31. Available at: www.sciencedirect.com/science/article/abs/pii/S0264999318301044

32 Global Conservation, undated. Baja Sur Marine Protection. Mexico. Available at: https://globalconservation.org/projects/baja-sur-mexico

33 Ministry of Corporate Affairs, Help & FAQs, Companies Act. Available at: www.csr.gov.in/content/csr/global/master/home/helpandfaqs.html

34 Balch, O., 2016. 'Indian law requires companies to give 2% of profits to charity: Is it working?', The Guardian, 5 April. Available at: www.theguardian.com/sustainable-business/2016/apr/05/india-csr-law-requires-companies-profits-to-charity-is-it-working

35 WWF, 2021. WWF Hails Largest-Ever Private Funding Commitment for Biodiversity as a "Momentous Move for a 'Nature-Positive' World", 22 September. Available at: https://wwf.panda.org/wwf_news/?3762941/WWF-hails-largest-ever-private-funding-commitment-for-biodiversity-as-a-momentous-move-for-a-nature-positive-world

36 Paulson Institute, 2020. Financing Nature: Closing the Global Biodiversity Financing Gap. Available at: www.paulsoninstitute.org/conservation/financing-nature-report/

37 UN Food and Agricultural Organisation (FAO), undated. Building Capacity Related to Multilateral Environmental Agreements in African, Caribbean and Pacific Countries. Available at: www.fao.org/in-action/building-capacity-environmental-agreements/overview/what-are-meas/en

Glossary

Aichi Biodiversity Targets: A set of 20 global objectives established in 2010 under the Convention on Biological Diversity (CBD) during the 10th Conference of the Parties (COP10) in Aichi, Japan. The targets aimed to halt biodiversity loss by 2020, covering issues such as habitat protection, species conservation, and the sustainable use of natural resources.

Anthropocene: An unofficial epoch that describes the time during which humans have had a substantial impact on our planet's ecosystems and geology. It suggests that human actions, such as industrialisation, deforestation, and pollution have shaped the planet's environment, marking a departure from the Holocene, the climate-stable period in which human development flourished, and which began around 11,700 years ago (see subsequent info).

biodiversity: Biodiversity is the variety of life on Earth, including all species, ecosystems, and the ecological processes that sustain them. It supports ecosystem health, resilience, and services like clean air, water, food, and climate regulation. Conserving biodiversity helps maintain the balance of natural systems critical for the planet and future generations, making its conservation vital for planetary and human health.

carbon neutrality: Refers to the end-state in which a net-zero carbon footprint is achieved. The amount of carbon dioxide (CO_2) emissions produced by an individual, organisation, or activity is balanced by an equivalent amount of CO_2 being removed from the atmosphere or offset through activities such as planting trees or investing in renewable energy projects.

climate adaptation: Adjusting systems and practices to minimise the impacts of climate change by adapting infrastructure, policies, and behaviors to new conditions like rising temperatures and extreme weather. Strategies include resilient infrastructure, better water management, sustainable agriculture, and ecosystem protection to reduce vulnerability and help communities and economies thrive.

climate mitigation: Efforts to reduce or prevent Greenhouse Gas Emissions (GHGs), slowing or reversing climate change. Strategies include renewable energy, energy efficiency, reforestation, carbon capture, and sustainable agriculture. The goal is to limit global warming and its impacts on ecosystems, economies, and communities.

Corporate Social Responsibility (CSR): CSR initiatives and business models aim to contribute positively to society beyond profit-making, addressing issues such as environmental sustainability, fair labor practices, philanthropy, and community engagement. It involves businesses taking responsibility for their impact on the environment, society, and economy, aligning their strategies with broader societal goals.

Earth Overshoot Day: The date when humanity's demand for ecological resources and services exceeds what the Earth can regenerate in that year. It signifies the point at which resource consumption, such as land, water, and carbon emissions, surpasses the planet's annual regenerative capacity. After this day, humanity begins to deplete natural resources or draw down ecological reserves (e.g., forests, oceans). The date has shifted earlier each year – in 2024, Earth Overshoot Day fell on 1 August, whereas in the 1970s, it took place only towards the end of December.

Earth systems: Refer to the interconnected components of the Earth, including the atmosphere, biosphere, hydrosphere, and lithosphere, that interact to sustain life and regulate the planet's environment. These systems work together to maintain balance, with processes such as the water cycle, carbon cycle, and nutrient cycling driving ecological and climatic conditions. Human activities, such as deforestation and fossil fuel burning, can disrupt these natural processes, leading to environmental changes.

Environmental, Social, and Governance (ESG): Three key factors for evaluating sustainability and societal impact: environmental criteria assess a company's carbon footprint and waste; social factors examine relationships with employees, suppliers, and communities; governance looks at leadership, transparency, and accountability. ESG considerations guide investors and companies toward more responsible and sustainable practices.

Global Commons Alliance: A network of organisations working together to protect and restore the *global commons* – shared natural resources and ecosystems that are vital to all life on Earth, such as the atmosphere, oceans, forests, and biodiversity. The alliance aims to tackle global environmental challenges by promoting collective action, science-based targets, and sustainable solutions. It supports the creation of frameworks for nature-positive economic systems, aiming to ensure the health of the planet's critical systems for future generations.

Global Environment Facility (GEF): An international partnership providing financial support for projects tackling global environmental challenges like climate change, biodiversity, and land degradation. Established in 1991, the GEF funds projects in developing countries and collaborates with governments, NGOs, and stakeholders. It serves as a financial mechanism for agreements like the UNFCCC and CBD.

Holocene: The current geological epoch, which began approximately 11,700 years ago after the last Ice Age. It marks the period of stable climate conditions that allowed human civilisation to develop.

Kunming-Montreal Global Biodiversity Framework (GBF): A global agreement adopted in December 2022 at the 15th Conference of the Parties to the Convention on Biological Diversity (COP15). The GBF sets targets for protecting ecosystems, halting species extinction, and addressing the drivers of biodiversity decline, including habitat destruction, pollution, and climate change. The framework includes 23 action targets for 2030, such as increasing protected areas, mobilising finance for biodiversity, and integrating biodiversity into policies and business practices.

The Intergovernmental Panel on Climate Change (IPCC): An international scientific body established in 1988 which provides comprehensive assessments of climate change science, its impacts, and potential mitigation strategies. The IPCC produces reports based on the latest research, offering policymakers critical information to guide global climate action. Its reports are widely used to inform climate policy and negotiations, including the Paris Agreement.

Intergovernmental Science-Policy Platform on Biodiversity and Ecosystem Services (IPBES): An independent, intergovernmental body established in 2012 to provide policymakers with reliable scientific assessments on biodiversity and ecosystem services. IPBES produces reports on the state of biodiversity, the benefits ecosystems provide to people, and the consequences of biodiversity loss. IPBES supports global efforts to halt biodiversity decline, offering tools and knowledge for informed decision-making in areas such as conservation, sustainable use, and policy integration.

Montreal Protocol: An international treaty, signed in 1987, aimed at protecting the ozone layer by phasing out substances that deplete it, such as chlorofluorocarbons (CFCs). It is widely regarded as one of the most successful environmental agreements, with significant reductions in ozone-depleting chemicals globally.

Multilateral Environment Agreements (MEAs): Legally binding international treaties aimed at addressing global environmental issues, such as climate change, biodiversity loss, and pollution. MEAs involve multiple countries and seek collective action to protect the environment. Examples include the Paris Agreement on climate change and the Convention on Biological Diversity.

National Biodiversity Strategies and Action Plans (NBSAPs): Country-specific plans developed by governments to address biodiversity loss. They outline strategies, actions, and policies for conserving biodiversity and sustainable use of natural resources, aligning with international frameworks like the Convention on Biological Diversity (CBD).

Nationally Determined Contributions (NDCs): Commitments made by countries under the Paris Agreement to reduce greenhouse gas emissions and take action on climate change. Each country sets its own targets based on national circumstances, aiming to limit global warming to below 2°C and pursue efforts to limit it to 1.5°C.

natural capital: The world's stocks of natural assets, including all living things, as well as geology, soil, air, and water. It encompasses the ecosystem services these assets provide, such as pollination, climate regulation, and water purification, that are essential for human well-being and economic activities.

nature-based solutions (NbS): Strategies that use natural systems, processes, and ecosystems to address environmental challenges, such as climate change, disaster risk, and biodiversity loss. Examples include restoring wetlands for flood control or reforesting areas for carbon sequestration.

net-zero emissions: A balance between the amount of greenhouse gases emitted into the atmosphere and the amount removed or offset. This can be achieved through reducing emissions and using carbon capture, reforestation, or other strategies to absorb or offset the remaining emissions, aiming for no net increase in atmospheric greenhouse gases.

Other Effective Conservation Measures: Areas or practices outside formally designated protected areas that contribute significantly to biodiversity conservation. These include sustainable land-use practices, community-managed areas, or conservation initiatives that maintain or enhance ecosystems and biodiversity.

the Paris Agreement: The international treaty adopted on 12 December 2015 under the United Nations Framework Convention on Climate Change (UNFCCC), aiming to limit global warming to well below 2°C, with a goal of 1.5°C, compared to pre-industrial levels; 196 countries committed to reducing greenhouse gas emissions, enhancing climate resilience, and mobilising financial support for climate action.

Rio Conventions: Three key international agreements adopted at the Earth Summit in Rio de Janeiro in 1992: the Convention on Biological Diversity (CBD), the United Nations Framework Convention on Climate Change (UNFCCC), and the United Nations Convention to Combat Desertification (UNCCD).

Science Based Targets Network (SBTN): An initiative that helps companies set science-based targets for nature, aiming to align business practices with global sustainability goals. It provides frameworks and guidance for companies to reduce their environmental impact on biodiversity, ecosystems, and natural resources.

sixth mass extinction: The ongoing human-driven loss of biodiversity, characterised by the accelerated extinction of species on a global scale. It is caused by habitat destruction, unsustainable food production, climate change, pollution, overexploitation of land and water, and invasive species, marking the first mass extinction event to be primarily driven by human activity.

the Sustainable Development Goals (SDGs): A set of 17 global objectives (often referred to as the Global Goals) agreed upon by the United Nations in 2015, aiming to address a wide range of interconnected challenges including poverty, inequality, climate change, and environmental degradation, with the goal of creating a more sustainable and equitable world by 2030.

Task Force on Climate-related Financial Disclosures (TCFD): An organisation that provides recommendations for companies to disclose climate-related risks and opportunities in their financial reporting. Its goal is to help investors make informed decisions by improving transparency on climate change impacts on business operations and strategies.

Task Force on Nature-related Disclosures (TNFD): An initiative that develops a framework for companies to disclose their impacts and dependencies on nature. It aims to provide businesses with standardised, science-based guidance to assess, manage, and report on nature-related risks and opportunities.

The Economics of Ecosystems and Biodiversity (TEEB): A global initiative focused on raising awareness of the economic value of biodiversity and ecosystems. It highlights the importance of protecting nature for human well-being and promotes the integration of ecological considerations into economic decision-making.

tipping points: Critical thresholds in Earth systems where small changes can lead to significant, irreversible shifts in environmental conditions, such as the loss of ice sheets or frequent droughts in the Amazon rainforest.

UN Biodiversity Beyond National Jurisdiction (BBNJ): A landmark international agreement which aims to protect biodiversity in areas beyond national jurisdiction, covering the high seas (the agreement is often referred to as the 'High Seas Treaty'), through measures like marine protected areas and biodiversity governance.

UN Climate Change Conference of the Parties (COP): An annual summit where countries negotiate international agreements and actions to address climate change, including the Paris Agreement. The first was held in Berlin, Germany, in 1995, the latest in Baku, Azerbaijan.

UN Convention on Biological Diversity (CBD): An international treaty aimed at conserving biodiversity and promoting the sustainable use of natural resources. The treaty's implementation is guided and reviewed through its Conference of the Parties (COP), a biennial meeting where member states set global biodiversity targets and strategies. Major outcomes from the CBD COP include frameworks like the Kunming-Montreal Global Biodiversity Framework.

UN Convention on the Law of the Sea (UNCLOS): A framework governing the rights and responsibilities of nations in their use of the world's oceans, including marine biodiversity protection.

UN Environment Programme (UNEP): Established in 1972, UNEP serves as the primary UN body responsible for coordinating environmental activities and addressing challenges such as climate change, biodiversity loss, pollution, and sustainable development.

Index